D0936930

Wireless Access Networks

Fixed Wireless Access and WLL Networks — Design and Operation

Wireless Access Networks

Fixed Wireless Access and WLL Networks — Design and Operation

Martin P. Clark
Telecommunications Consultant,
Eppstein, Germany

JOHN WILEY & SONS, LTD
Chichester • New York • Weinheim • Brisbane • Singapore • Toronto

Baffins Lane, Chichester,
West Sussex PO19 1UD, UK

National 01243 779777
International (+44) 1243 779777
e-mail (for orders and customer service enquiries): cs-books@wiley.co.uk

Visit our Home Page on: http://www.wiley.co.uk or http://www.wiley.com

Other Wiley Editorial Offices

John Wiley & Sons, Inc., 605 Third Avenue,
New York, NY 10158-0012, USA

WILEY-VCH Verlag GmbH
Pappelallee 3, D-69469 Weinheim, Germany

Jacaranda Wiley, Ltd, 33 Park Road, Milton,
Queensland 4064, Australia

John Wiley & Sons (Asia) Pte Ltd, 2 Clementi Loop #02-01,
Jin Xing Distripark, Singapore 129809

John Wiley & Sons (Canada) Ltd, 22 Worcester Road,
Rexdale, Ontario M9W 1L1, Canada

Library of Congress Cataloging-in-Publication Data

Clark, Martin P.
Wireless access networks : fixed wireless access and WLL networks — design and operation / Martin P. Clark.
p.cm.
Includes bibliographical references and index.
ISBN 0-471-49298-1 (alk. paper)
1. Wireless communication systems. 2. Broadband communication systems. 3. Local loop (Telephony) I. Title.

TK5103.2. C63 2000
621.382′1 — dc21 00-033016

British Library Cataloguing in Publication Data

A catalogue record for this book is available from the British Library

ISBN 0 471 49298 1

Typeset in 10/12pt Times by Dobbie Typesetting Limited, Devon
Printed and bound in Great Britain by Bookcraft (Bath) Ltd
This book is printed on acid-free paper responsibly manufactured from sustaintable forestry, in which at least two trees are planted for each one used for paper production.

Contents

Acknowledgements

When reviewing the list of sources, reference bodies and helpers who have contributed to this work, I come to think of myself more as a coordinator than a sole author, and my especial thanks are due to the many who have encouraged me, contributed material or reviewed sections of text. In particular, I would like to thank my friend and colleague, Dietmar Kroll, for his attention to radio planning details.

No book on telecommunications as it is today could fail to recognise the contribution to world standards made by the International Telecommunications Union and its subordinate entities ITU-R and ITU-T. You will find references littered throughout the text. Particular copyright extracts (chosen by the author, but reproduced with the prior authorisation of the ITU) are labelled accordingly. The full texts of all ITU copyright material may be obtained from the ITU Sales and Marketing Service, Place des Nations, CH-1211 Geneva 20, Switzerland, Telephone: +41 22 730 6141 or Internet: Sales@itu.int.

I am indebted for photographs to Netro Corporation. To Netro I am also indebted an education about broadband wireless. Special thanks to the co-founders Gideon Ben-Efraim and Eli Pasternak.

Martin Clark
Frankfurt, Germany
December 1999

Part I

Fundamentals of Fixed Wireless

1

The Case for Fixed Wireless Networks

Radio is playing an important role in the restructuring of the public telecommunications industry — at a time of deregulation, structural change and rapid growth in demand. It is an effective means of communication between remote or mobile locations and over difficult terrain, where cable laying and maintenance is not possible or is prohibitively expensive. It is also an efficient means of broadcasting the same information to multiple receivers. The first part of the renaissance of radio in public telecommunications networks was its use in mobile telephone networks, but the next phase is already upon us, the use of radio as an 'access means' in 'fixed wireless networks'. In this chapter, we discuss when and why fixed wireless networks are attractive. We introduce the basic applications of fixed wireless networks, and consider the alternative technologies. We also discuss the economic, coverage and service advantages.

1.1 Radio

By using a very strong electrical signal as a transmitting source, an electromagnetic wave can be made to spread far and wide through thin air. That is the principle of radio, discovered by Heinrich Hertz in 1887. The radio waves are produced by transmitters, which consist of a radio wave source connected to some form of antenna. Examples of radio systems which are used in public telecommunications or broadcast networks include:

- low frequency radio and mast antenna used in public radio *broadcast* applications (LW (longwave), MW (medium wave), SW (shortwave));
- HF (High Frequency), VHF (Very HF), or UHF (Ultra-HF) radio system and mast antenna used for modern public radio *broadcast*, particularly for local radio stations and television broadcasting;
- troposheric scatter radio systems used in early 'over-the horizon' telecommunications radio systems, e.g. from ship-to-shore;
- microwave radio links, providing for *point-to-point* (*PTP*) connections between pairs of endpoints, using highly directional dish-type antennas;
- satellite radio systems, providing for both *point-to-point* (*PTP*) as well as *point-to-multipoint* (*PMP*) and *broadcast* applications over long distances and very large geographical areas;

- cellular radio systems, used to provide mobile telephone service from portable devices, car telephones and handheld mobile telephones;
- *fixed wireless access* networks.

Fixed wireless access systems, the main subject of this book, provide for the 'last mile' connection of the customer site or other remote location to the public telecommunications network. Fixed wireless access provides for a real alternative to 'wireline' *local loop* connections, and thus the common term *Wireless Local Loop* (*WLL*).

A subscriber radio unit, or *Terminal Station* (*TS*), is installed permanently at a fixed location (e.g. in a customer's premises) and directed at the base station (or *Central Station* (*CS*) a site connected to the 'backbone' network of the public telecommunications network provider).

Unlike a mobile telephone network, there is usually no scope for motion of a subscriber radio unit of a fixed wireless access system during operation — hence *fixed* wireless access. In effect, the fixed wireless access system provides for a direct alternative to or replacement for a fixed wireline connection. In practice, however, fixed wireless access systems are not installed to replace wireline systems, and there is not usually a viable alternative using wirelines.

1.2 The Use and Advantages of Radio for Fixed Network Access

In the modern 'wired-society', it may seem a retrospective step that the public telecommunications world should be considering fixed wireless access. Why use radio, when fibre can be used as a solution for all types of telecommunications signal transport, and can provide for previously undreamt of capacity? There are various reasons, which we present in this chapter. We first describe the particular attributes of networks and services for which radio is ideally suited. We then consider the technological alternatives to contrast the benefits.

The main advantages of radio for fixed wireless access are:

- the rapid manner in which access networks can be newly provided or existing networks can be complemented with new capacity;
- the speed and ease of adding further customer lines for existing services;
- the ability to offer new high bitrate access services;
- the ease of realisation and the low investment needed to add large coverage areas to an access network;
- the speed with which service can be installed for new customers;
- the ease with which service can be provided in remote places, over difficult terrain or in places where other building restrictions may prevent easy wireline installation (e.g. in 'historic' town centres);
- the ease of redeployment of subscriber radio hardware and consequent security of the investment associated with a radio access network;
- the inherent 'shared medium' nature of radio — providing an ideal medium for 'concentration' of connections as well as for 'broadcast' and 'asymmetric' transmission applications (i.e. those with different volumes of information to be transmitted in 'transmit' and 'receive' directions);

- the relative ease and efficiency with which redundant or 'back-up' connections can be realised with radio.

1.3 The Time to Consider Installing a Fixed Wireless Access Network

A fixed wireless access network is, of course, not suited to all occasions. It is, for example, always cheaper for a network operator to use his existing network (if he has one) rather than invest in new access network technology. Thus, an ex-monopoly operator is best advised to make the best of existing copper lineplant. A cable TV operator, meanwhile, should explore the potential of coaxial 'cable modem' technology; and a new 'city network' operator should install fibre for high capacity metropolitan networks. Nonetheless, even for these 'established' operators, there comes a time when the network capacity in a given region is not sufficient to meet demand. At this point in time, major investment is necessary to provide for new capacity. This is when even the 'established' operators are forced to pose themselves the same question that their newly licensed competition are asking ''What is the most economic and effective way to provide a new access network infrastructure in a given area, for a given service?'' The answer may be fixed wireless technology.

1.4 The Coverage Offered by Wireless

A large coverage area in a radio access network can be established by the installation of a single base station (or *central station*), and thus with limited investment. The subsequent investment by the network operator in subscriber radio units and their installation is a 'variable cost' investment which can be programmed to run in step with the operator's growth in actual customer connections and business revenue. The same is not true of an underground cable network. A new cable network requires a higher initial investment in conduits and in 'junction' and 'feeder' cables along the various business and residential streets, in order that individual new customer connections can be relatively quickly realised. Most of the investment in cable infrastructure is thus an upfront risk investment. As long as cables and/or conduits have spare capacity, new customer lines can be realised relatively cheaply. However, once the conduits are full a major further investment in 'overlay' capacity is necessary. This is the time to consider the alternatives.

1.5 Radio as a Means of Providing High Bit Rate Access Lines

Long established access networks are not always suitable for providing the high speed connection lines expected by today's Internet users. Many copper cable networks were designed and installed as the access network infrastructure for the post-war telephone networks. When they were installed, they were optimised for telephone technology and not conceived for the high bit rates of modern data communications. Thus even for 'established' operators, 'overlay' access network technology may be essential if they wish to offer their customers modern highspeed data, Internet and multimedia communications network access.

1.6 Wireless Technology Allows for a 'Pay-as-you-build' Network Structure

Once the radio base station is installed in a given area, a fixed wireless link to an individual customer's premises can be provided simply by the installation of a subscriber radio unit (a *terminal station*) at the customer site. There is no major building work necessary, no third parties to have to consult with (e.g. for digging up the road), and no detailed survey work about availability of lineplant within a given street to conduct. Provided there is capacity available at the base station, the equipment installation (subscriber radio unit) can be conducted typically within half a day.

Should the customer subsequently decide he needs a different type of access line, or even if he decides to defect to a competing telecom service provider, it is relatively easy for the operator to recover the subscriber radio equipment and redeploy it for another customer location. In contrast, a conduit and cable would simply remain idle — a written-off investment.

Fixed wireless access is particularly suited for use in historic towns, over difficult terrain and in remote rural areas where the cost of installation and the effort of meeting the planning restraints associated with establishing wireline infrastructure can be prohibitive.

1.7 Wireless as a Means of Access Line 'Back-up'

Radio is an ideal medium for 'back-up' and redundant networks, providing for the availability of physically diverse communications links from given customer's premises to the backbone network of his public telecommunications network provider. The radio path can either be a back-up path for a wireline connection (e.g. fibre or copper) or two separate radio paths could provide back-up for one another. This gives greater flexibility than in a typical wireline access infrastructure, where the lines usually all converge on a single switching location, even if a 'ring' topology is employed within the local network. We consider this in Figure 1.1.

Figure 1.1 illustrates a customer site (represented by the customer's PC) connected to a switch site (A) by means of a modern fibre network, using a 'ring' topology. The ring topology has gained favour in recent years, because of its inherent immunity to line failure. When one side of the ring is affected by an outage or line cut, then an alternative path is always available via the second side of the ring. The problem is that rarely does this network redundancy (double provision) extend to the switching locations. Thus, within a given city or area, all the access lines converge on a single switching site, and perhaps on a single telephone exchange. The switching site acts as a 'hub' location for the local area access network cabling. It is simply not practicable or economic to try to use multiple switching locations within the area, with each customer 'homed' by pre-laid wireline infrastructure on several switch sites.

In contrast to wireline infrastructures, it is much easier with radio systems to provide secondary (back-up) connections from customer locations to remote switching sites (site B

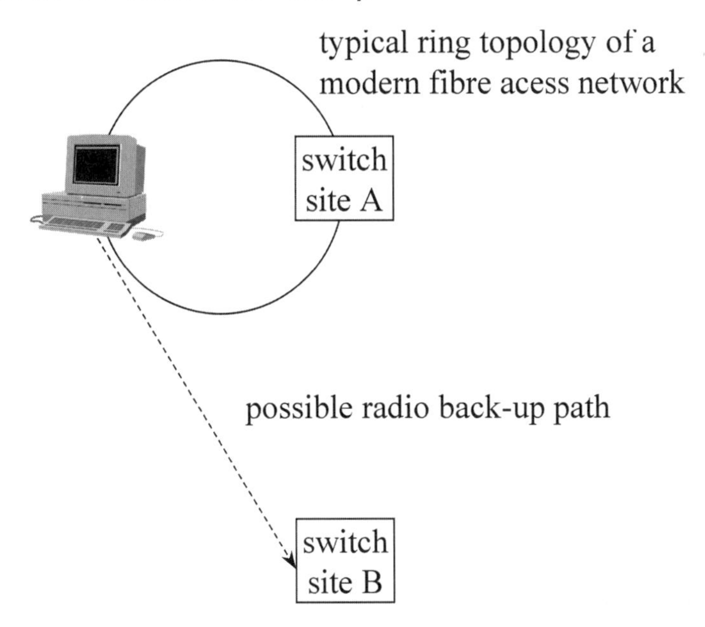

Figure 1.1 Back-up path alternatives in a local access network

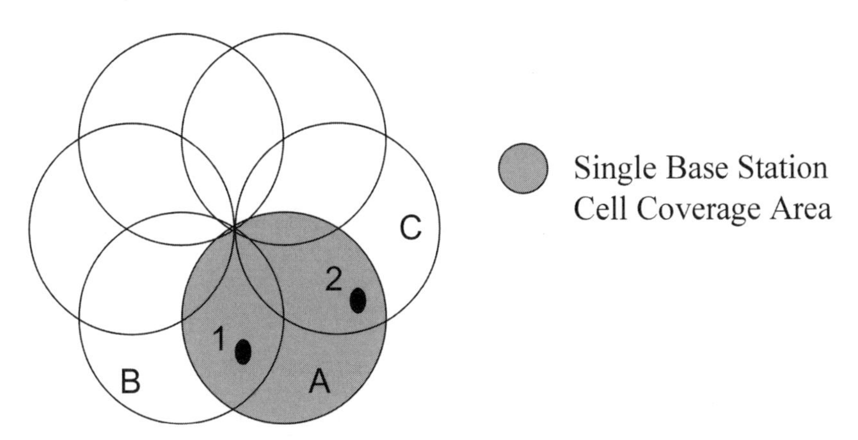

Figure 1.2 Overlapping radio cells create an access network offering redundant alternative access connections

of Figure 1.1). This could be by means of a relatively long *point-to-point* link (e.g. microwave link of up to about 20 km as in Figure 1.1). Alternatively, by overlapping the 'cell' areas of adjacent point-to-multipoint radio base stations (Figure 1.2), a robust radio network design can be achieved. Thus, a customer at point (1) can be hooked up to either base station A or base station B. Similarly, the customer at point (2) can be hooked up to either base station A or base station C.

The overlapping cells of Figure 1.2 also ensure that full coverage is retained, even in the case of a hardware outage (e.g. for maintenance or due to equipment or power failure). Such overlapping cell topology will also be relevant in fixed wireless access networks since it provides not only for 'back-up' customer connections as in Figure 1.2, but also eases some of the radio path and frequency planning as well as capacity considerations, as we shall discuss in Chapters 8 and 9.

1.8 Radio as an Ideal Mechanism for Access Port Concentration or Multiplexing

The most economic public telecommunication access networks share two important characteristics:

- the geographical coverage area of each hub or switching site is maximised (the greater is the coverage area, the less hub switching sites are required);
- the number of switching ports provided at the switch and the 'backhaul' capacity is optimised to the actual traffic demand and not hugely overprovided.

As we have already discussed, fixed wireless access has the potential to provide a central hub site (e.g. a switch location) with a large radius access network 'cell' with relatively low capital outlay. As we shall now discover, it also acts as a natural 'concentration' medium, allowing the switch port hardware and backhaul capacity to be optimised rather than massively overprovided. Let us consider the wireline access network illustrated in Figure 1.3.

Figure 1.3 illustrates a copper-cable based wireline access network. Most established public telephone networks have such an access network infrastructure. In this type of access network, each individual customer needs a separate cable connection from his

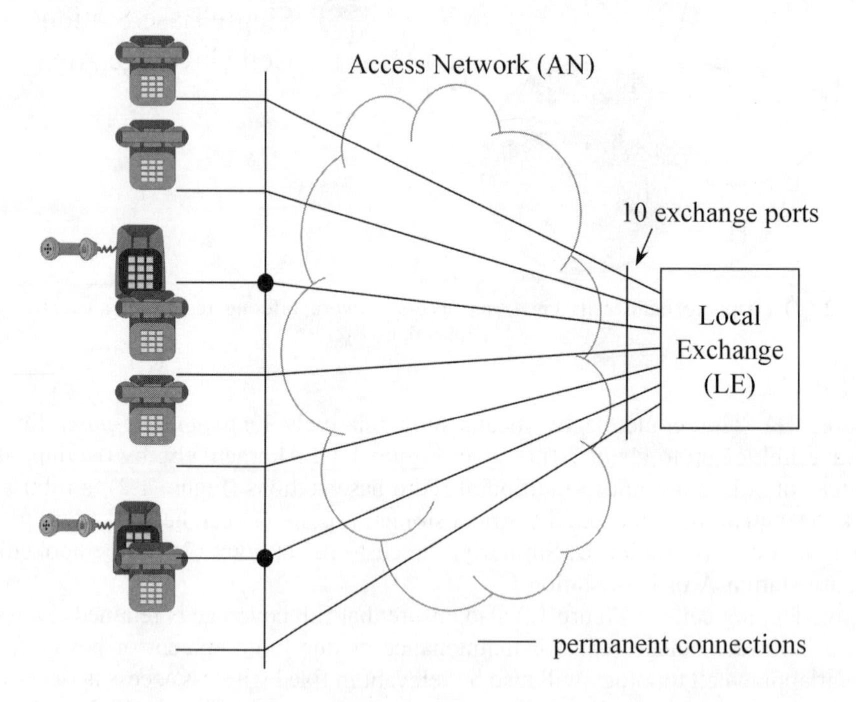

Figure 1.3 Basic copper-cable wireline access network without concentration function

premises to the remote local exchange location, where each individual customer line is connected to a separate exchange port. All the switching takes place in the local exchange.

Typically, only a small percentage of the customer lines (telephones in Figure 1.3) are actually in use at any one time (Figure 1.3 illustrates two out of 10 in use — 20%. Actual typical usage of residential telephones is much lower than this — 0.5% during the daily *busy hour* [the hour of greatest telephone activity]). So in the example of Figure 1.3, a maximum of two access lines and two exchange ports are in use at any one time, but 10 of each must be provided — an overprovision of 80%. The overprovision in a 'real' network could be even higher than this (99.5%—for 0.5% traffic in the busy hour). Lines are mostly not used!

Actually, most real access networks are a little more complex than we show in Figure 1.3. Figure 1.4 shows how in reality, a complicated star-cable topology is created using a system of streetside boxes used as cable-consolidation points which allow individual cables (single pair, two-pair or ten-pair cables) to be collected onto *distribution* and *junction* cables of higher capacity (e.g. 100 pair, 200 pair, 600 pair or fibre optic).

By using cascaded cables and streetside boxes, the operator of the network in Figure 1.4 is able to consolidate small capacity cables into larger cables. Thus, the cables provided into customer premises are typically *drop wires* of either 1-copper-pair, 2-pairs or 10-pairs. After the first consolidation at a streetside cabinet (typically within a few hundred feet), the connections are carried by *distribution cables* of 20, 50 or 100 pairs. Further stages of consolidation may result in very large capacity cables (e.g. 600 pair). However, 600 pair cable is about the maximum size (approx. 8 cm diameter) which can be installed in normal conduit.

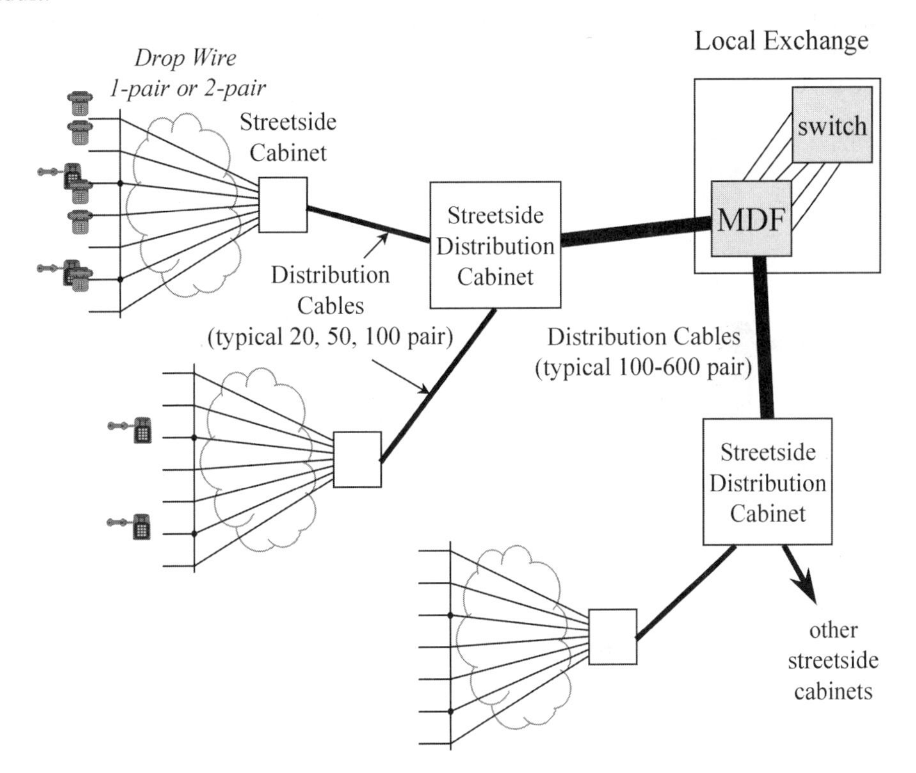

Figure 1.4 Typical copper-cable based wireline access network of a public telecom network

Consolidating the individual drop wires into ever larger distribution cables does have the benefit of reducing the number of cables and conduits which are necessary within the access network as a whole. This is clearly to both economic and operational advantage. However, it does not get away from the basic flaw which we discussed in Figure 1.3 — that most of this capacity is unused for most of the time, and most of the exchange ports also go unused. Still, the exchange ports are provided on a one-per-customer basis.

A more efficient access network is illustrated in Figure 1.5. The example is directly comparable with that of Figure 1.3. Again, there are 10 customers in total, and again only a maximum of two lines are actually ever in use. The difference is that the *access network* of Figure 1.5 includes a *concentration* function. Because of the concentration capability, only two connections are required in the access network, and only two ports are necessary at the exchange. The concentration capability relies on the ability of each of the customers to use *any* of the available connections across the access network (and thus to *any* of the ports at the exchange). In simple terms, the concentration function is a simple *switch* or *bus* function. Radio is an ideal medium for concentration, since all of the remote end customer sites inherently can use any of the available radio channels. We therefore only need to provide as many radio channels as are actually *used* at any one time, provided we have a *signalling* methodology for identifying reliably which of the customers is using the line (and thus the exchange) at any one time. Such signalling is discussed in Chapter 10. (Examples are V5.1 and V5.2 or north American TR-303.)

A *fixed wireless access* network can be realised with a structure as simple as that of Figure 1.5. Compared with the 'pure copper' infrastructure of Figure 1.4, there are considerable savings in the costs and operational problems associated with multiple physical cables, streetside cabinets, conduits and cabling ditribution frames (e.g. the *MDF* or *main distribution frame* at the local exchange). Instead, a radio base station at the local exchange site is cabled directly to the switch using a *concentrated* cable interface (i.e. high

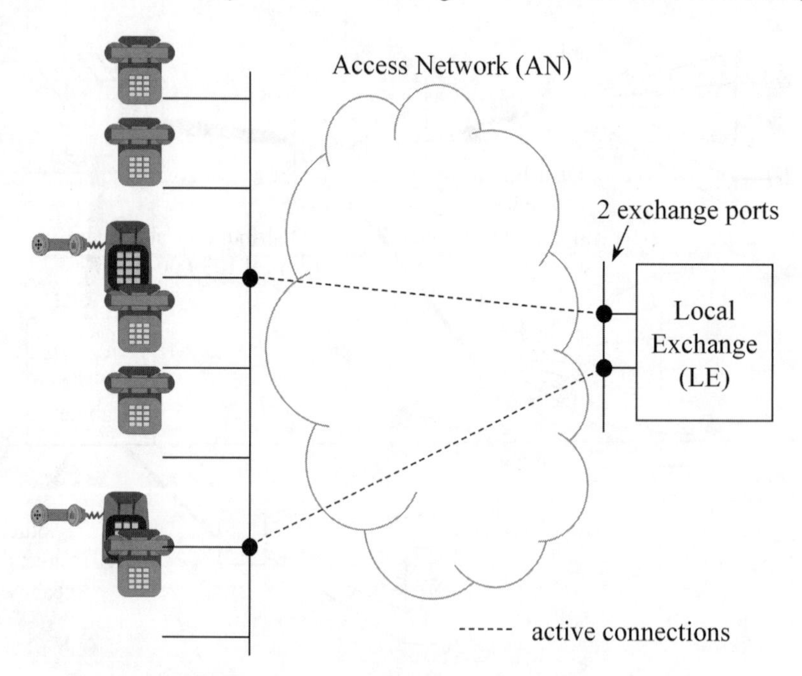

Figure 1.5 Access Network (AN) with concentration function

bitrate connection) and appropriate signalling (e.g. V5.1 or V5.2 as we shall discuss in Chapter 10), and two radio channels connect to the active telephones.

It is not to say that the concentration function illustrated in Figure 1.5 cannot be achieved within wireline networks, but the effort and cost is somewhat greater. It is common nowadays for example, to replace the distribution cables of Figure 1.4 by fibre optic cables. This has the benefit that far more connections can be carried by a single cable (and thus conduit) than was previously the case with 600-pair copper distribution cable (8 cm diameter). With a single pair of fibre optic cables (diameter only a few millimetres) running at 155 Mbit/s, the equivalent of nearly 2000 copper telephone connections can be carried. To achieve the conversion from electrical signals on copper cable into *multiplexed* (i.e. multiple connections sharing the same cable) optical signals, an *active* electronic device is installed in the streetside cabinet (in the space previously occupied by a copper cable patch panel). Such devices are also normally designed to include the basic switching capability required for the concentration function we discussed in Figure 1.5. This type of technology goes under the generic name of *Fibre To The Kerb* (*FTTK*).

At the local exchange building a *head end* device is usually needed to reconvert the optical signals to the normal cable interface format required by the switch. In comparison with a fixed wireless access solution, a fibre to the kerb solution retains the need for a significant conduit and streetside cable box infrastructure, within which power and electronic equipment must be installed. For an establised operator with available capacity in existing (and maybe already written-off) conduit and streetside cabinet infrastructure, fibre to the kerb is likely to be the most economic means of augmenting access network capacity. But where the conduits and streetside boxes must be newly established (because capacity has run out or because a new competing operator has no existing infrastructure), fixed wireless access is likely to be the most economic and quickest means of providing service.

1.9 Fixed Wireless Access is More Reliable than Mobile Telephony!

One way of realising a fixed wireless access network is to use mobile telephone network technology, using permanently installed 'mobile handsets' as fixed endpoints. Such an approach has been followed by licensed 'mobile operators', who have attempted to tap the market for 'fixed' telephony using their licences for 'mobile telephony'. It has also been the approach of investment schemes aimed at developing improved national infrastructure in developing economies (e.g. United States investment schemes in the far east and European Union investment [e.g. the PHARE projects] in central Europe).

The limitation of using mobile technology for fixed wireless access is that it provides only for a 'basic' telephone service of a quality which may be poor by fixed network standards (poor transmission caused by radio fading, resulting in line 'dropouts' during conversation). Such an approach fails to capitalise on the potential services, bitrates and system quality which can be achieved with systems designed specifically for fixed wireless access. The fact is that most of the technology (and thus much of the cost and most of the problems) associated with mobile telephone networks is required to overcome the *mobile* nature of the end customer. The system must permanently track his whereabouts in order to direct incoming calls to him, and must be able to maintain a connection by *handover* of radio connections from base station to base station even during a call. Not only this, since the customer's mobile radio unit does not know the exact directional location of the nearest base station, it must use an *omnidirectional* antenna (in other words, must transmit in all possible directions, in order to make sure that the signal will arrive). Such omnidirectional transmission limits the range of the system and increases the problem of radio *interference*.

In comparison with mobile telephone network systems, fixed wireless access systems are usually designed:

- to incorporate only stationary end-users (though some 'portability' or 'slow moving' mobility may also allow the customer unit to be relatively easily moved from one location to another). The restriction to stationary end devices dispenses with the need for the complicated tracking and handover procedures, as well as easing some of the problems of radio and interference planning;
- to incorporate higher performance, more *directional* customer site antennas which lead to high system availability and longer potential range (i.e. greater coverage area of the base station);
- to operate in radio bands specifically designated for fixed wireless access. These bands have the advantage of greater radio spectrum availability and thus give the scope for higher bitrate services to be offered to customers by means of fixed wireless access.

1.10 Applications of Fixed Wireless Access for Public Telecommunications Services

The commonest applications of fixed wireless access systems in public telecommunications networks are for:

- rural telephony access networks;
- basic rate *ISDN* (*Integrated Services Digital Network*) and $n \times 64$ kbit/s leaseline service access networks;
- high bitrate *Internet* access;
- high bitrate connection of business premises ('wireless-to-the-building', for single or multiple tenant office blocks);
- broadband and multimedia service access networks;
- wireless access to 'interconnection' sites (*points-of-presence*, or *POPs*), where equipment may be 'collocated' at the ex-monopoly operator's local exchange site in order to permit *unbundled* access to the copper line plant (we discuss this later in the chapter).

As we shall discover during the course of this book, there are a number of different types of *fixed wireless access* systems, running from simple analogue solutions for *Wireless Local Loop* (*WLL*) connections in telephone networks, through digital radio systems for basic ISDN service, to high bandwidth and high bitrate *broadband wireless access* and *broadband wireless access* systems.

In general, one can say that the simpler systems operate at lower radio frequency bands (around 2 GHz) and offer low bitrates to customers but have relatively long range, while the broadband systems operate at higher frequencies (e.g. 10 GHz, 24 GHz, 26 GHz,

28 GHz, 38 GHz, 42 GHz), offering much higher bitrates to customers but having more restricted range. The PMP band at 3.5 GHz was initially viewed by operators, regulators and manufacturers alike as ideal for basic rate ISDN speeds. In the meantime, it is increasingly being used for broadband services. The trend towards offering broadband services is partly due to the market demand for high bitrate Internet-based services, and partly due to the better economics of broadband (higher revenues for a similar capital investment).

1.11 Economics of Fixed Wireless Access

It is not possible for us to determine here in absolute terms the economic case for fixed wireless access. We are, however, able to discuss the elements of investment and other costs which must be considered when planning a fixed wireless access network, and can also compare the investment and outlay profiles of alternative means.

We shall consider both of two possible starting points for an operator considering the installation of fixed wireless access:

- a 'green field' access network installation, where the public telecom operator is simply interested to determine the most economic technology for establishing a network in an as yet unserved area (or in an area where a new 'competing' operator is determined [for service reasons] to own his own access network rather than use lines leased from his competitor);
- a new 'competing' public telecom operator in a deregulated telecommunications market, who compares the cost of access line provision against the cost of leaselines which he may alternatively rent from his longer-established (typically ex-monopoly) competitor.

In our first example of a 'green fields' site, the focus of cost analysis will be upon direct comparison of the capital outlay required to establish the network to cover a given area. In this case, the benefit of the fixed wireless access solution is that a single radio base station gives an immediate *coverage area* of many square kilometres. The coverage is achieved for the capital outlay in equipping a single operations site and installing a single radio station of minimal capacity. Subsequent extension of capacity at the base station and investment in customer site equipment can be programmed to run in step with the acquisition of customers for the network.

By comparison, the initial investment required for a cable-based network (e.g. fibre or copper) is relatively large and the coverage area is often extremely restricted. At fully-loaded investment costs between about $50 and $150 per metre. Laying a new fibre network around all the streets of a major city (typically with upwards of 5000 km of road) is a very heavy investment! For this reason, most of the new 'city' operators who have emerged in recent years to build fibre networks have concentrated on cabling only the city centre areas where there is a high density of high revenue potential customers. Even well-provided-for cities often have less than 100 km of road actually cabled for fibre. This gives a connection potential for only a small portion of even the business premises, and provides no real access network capability for 'residential customer' access services.

The investment profiles of fixed wireless access and cable-based (e.g. fibre network access) are very different (Figure 1.6).

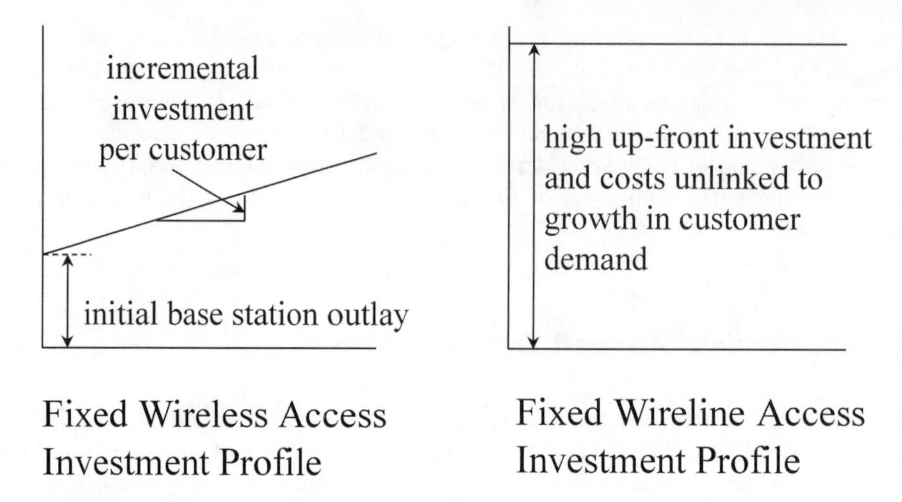

Figure 1.6 Comparative investment profiles of wireless and wireline access networks

The grave difference in the upfront investment in wireless, compared with 'wireline', networks in a 'green field' investment is starkly illustrated by the business case calculations of a real network operator (who raised his financing at IPO [initial public offering — on the NASDAQ] based on these calculations). The operator calculated that the investment needed to establish a single conduit and fibre cable line of 4 km length was about the same as the investment to create a radio coverage area of nearly 80 km^2!

In our second economic analysis we consider instead an 'operating cost' comparison of the investment in a fixed wireless access system (by a new 'competing' carrier) against the alternative of using 'leaselines' from the established (ex-monopoly) operator. This type of analysis requires more careful application of accounting procedures. In particular, the result of the analysis is sensitive to the depreciation period chosen for the financial 'writing-off' of the capital costs of equipment over time. (The longer the depreciation period chosen, the more favourable is a capital investment in an own infrastructure when compared with the continuing ongoing payment for rental of services. In essence, the longer the depreciation period of the capital investment, the more expensive the leaseline alternative looks, because of the mounting leasing charges.)

Figure 1.7 illustrates the cumulative costs of operating fixed wireless access systems of different bitrates and system types (point-to-point microwave, as well as point-to-multipoint radio systems loaded with differing numbers of customers) compared against the cost of leaselines. The comparison is based upon operating costs within the German environment for new operators. It seems likely that the regulatory framework for wireless access already established in Germany will also be adopted (maybe slightly adapted) in various European and other countries, so that similar results are likely in other countries.

The costs included in the comparison of Figure 1.7 are the capital outlay in radio equipment (depreciated over five years), mounting and installation costs, frequency fees, base station roof right acquisition and maintenance/running costs (set at 12% of equipment price).

These costs are compared against the cost of leaselines. The type of leaseline chosen for the comparison is the type *Carrierfestverbindung* (*CFV*). CFVs are a type of leaseline

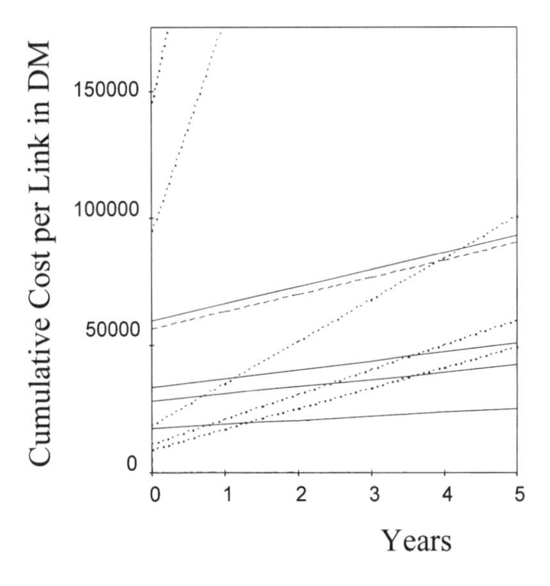

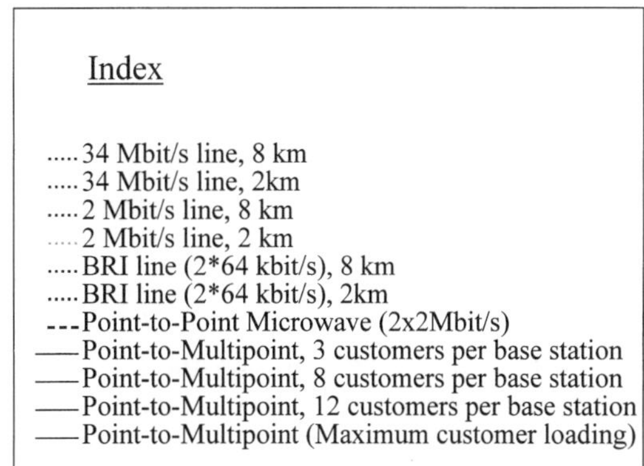

Figure 1.7 Comparison of total operating costs of fixed wireless access versus leaseline as an alternative

offered by Deutsche Telekom (the former monopoly carrier) only to licensed public carriers. CFVs are intended to provide for 'last mile' connections of customers to the switches of the new 'private' carriers competing against Deutsche Telekom. Thus they are 'access network' leaselines. Typical cost of CFV leaselines is far less than the cheapest type of 'normal tariff' leaselines (*Standardfestverbindung* (*SFV*)).

The cost analysis of Figure 1.7 helps us draw a number of conclusions, including:

- Point-to-point microwave is a lot cheaper than a 34 Mbit/s leaseline. (Typically, the cost of a 34 Mbit/s microwave link is about 10–20% above the cost of a 2×2 Mbit/s microwave link.) Even when only a single 2 Mbit/s leaseline (of 8 km length — a modest range for PTP microwave radio) is replaced with double the capacity via a microwave link, the payback is within four years;
- Point-to-multipoint radio 'breaks even' over a period of time which depends upon the number of customers connected to each base station (i.e. according to the customer

density) and to the bitrate required by the customer. The higher the bitrate required, the sooner the breakeven point. (From Figure 1.7, a point-to-multipoint system loaded at a density of 12 customers per base station and providing for 2 km long lines as 'replacements' for 2 Mbit/s leaselines pays in after about two years. However, should the required bitrate per customer be only 2×64 kbit/s, then the payback is more like $3\frac{1}{2}$ years.)

However, the economic analysis will not always draw the same conclusions; and it is surprisingly not always the equipment costs which are the critical determining factor, although these are always an important element of the costs. The most significant factors determining the outcome of economic analysis are discussed later in the chapter, but we should perhaps, before leaving the subject of leaselines, also mention the growing trend towards *unbundling* of access services.

1.12 Critical Factors Affecting the Economic Viability of Fixed Wireless Access

The factors which most affect the outcome of an economic analysis which compares fixed wireless access against other possible access network solutions are:

- The price paid for radio spectrum and frequency usage (this can be a very substantial sum, particularly if the radio spectrum is allocated by public auction).
- The amount of radio spectrum available (limited spectrum may limit overall network capacity).
- The density of customers and the total coverage area (affects the amount of equipment needed).
- The range of the radio system (this is governed by the frequency band of operation and the *climate zone*—we discuss these factors in Chapter 7).
- The capital cost of equipment, the cost of a base station, but even more important is the marginal cost of each remote customer radio terminal device.
- The effort required for planning and preparing radio links (site survey costs, *line-of-sight* confirmation (from the customer site to the base station), as well as steelwork and cabling installation prior to radio equipment installation can add significantly to the initial capital outlay (sometimes as much again as the price of the equipment itself).
- Costs associated with finalising installation details with customer office building landlords and pacifying local authorities and other regulatory administrators (such matters can sometimes unduly delay installation work).
- Lastly, the relative attractiveness of a radio solution depends naturally upon the costs of the alternatives. These may be unusually high or low due to particular local conditions.

In Germany the regulator currently (January 1999) has published expected radio spectrum charges for *Point-to-Multipoint* (*PMP*) radio systems which are to include a one-time charge of 17.5 million Deutschemarks for a nationwide licence plus an annual fee of 4700 DM per transmitter. In other countries, notably the USA, and more recently Switzerland, there is a practice of auctioning off the spectrum to the highest bidder. Either one of these principles may be adopted by individual national regulators, but one thing is clear: the spectrum will not be cheap (see Appendix 12).

On the other hand, the radio frequency charges for *point-to-point* (*PTP*) microwave links are very reasonable. Most European countries are generally in line with the German charges — a one-time fee of 430 DM plus a further annual fee of 350 DM, irrespective of the link bitrate.

1.13 Alternative Technologies for the 'Local Loop'

An academic might list five different technologies which pose an alternative means of telecommunications transmission in the *local loop* to *fixed wireless access*, and which therefore might be considered to be competing technologies. These are listed below. In reality, however, as we discuss next, the practicality of the real operators' situation rules out most of the 'alternatives'. The competing technologies are:

- owned copper network infrastructure;
- *unbundled* copper infrastructure, probably used with modern high bitrate technology (xDSL);
- *cable modems* used in association with a local cable TV network (usually coaxial cable);
- A fibre access network;
- A satellite access network.

We consider each of the 'competing' technologies in turn.

For a new operator without existing assets, the establishment of a new copper network infrastructure would require a huge capital investment programme with very long payback period — unacceptable to the financial investment community. For new operators building new infrastructure fibre is the preferred technology, because of its much higher capacity and its much smaller conduit size needs. A single pair of small fibre optic cables replace a number of large diameter 600-pair copper distribution cables.

Of course, for an ex-monopoly operator with an existing, and probably written-off, copper network infrastructure there may be little motivation to consider any type of alternative technology — up until the point in time where the existing conduit or cable capacity runs out in a given area, or it proves unsuitable for the carriage of a particular new high bitrate service.

Unbundled access (discussed in detail later) is likely to be the favoured means for the new operators to gain control of direct copper cable access to their customers by using *local loop* lineplant of the ex-monopolist. Where this is possible and economic, undoubtedly the new operators will start to offer local switching services. Where it remains uneconomic for offering local basic telephone service, the new operator faces three options: either to restrict himself to offering *long distance* telephone service (leaving the ex-monopolist with a *de facto* local switching monopoly), to use a different local loop technology (e.g. to purchase a cable TV network or fixed wireless access network), or to try to improve the economics of the *unbundled access* by using new technology (e.g. xDSL) to provide much higher bitrates over the unbundled plant.

xDSL is the generic name given to a new class of technology intended for use in conjunction with local loop copper pairs to enable very high speed *Digital Subscriber Lines* (*DSLs*) to be realised. The 'x' represents any of various different but similar technologies, specifically 'x' may stand for 'H' (*HDSL — High Speed Digital Subscriber*

Line) or for 'A' (*ADSL—Asymmetric Digital Subscriber Line*). Both HDSL and ADSL provide for much higher bitrates than telephone or *ISDN* (*Integrated Services Digital Network*) access lines.

HDSL provides for 2 Mbit/s duplex (2 Mbit/s in both directions, *downstream* from the exchange to the customer, and *upstream* from the customer to the exchange). ADSL, on the other hand, provides much more *downstream* capacity (typically 8 Mbit/s) than *upstream* capacity (typically 512 kbit/s). ADSL was conceived for video distribution and 'online' services access, where the customer only uses the upstream channel to order programmes or point to *worldwide web* pages which he wishes to receive or download.

The problem that xDSL poses when used in conjunction with existing copper lineplant (whether *unbundled* or not) is that it requires a much greater dynamic range than the telephony for which most copper lineplant was designed for. The much higher frequencies which have to be carried for xDSL lead to problems of *interference*. Furthermore, the quality of the lineplant must be very good in order to support xDSL, and there must be no amplifiers or filters installed in the line. This is a 'tall order' for a copper lineplant network, which in some cases has been in the ground for years, and for which the planning records might either have been lost or may no longer be reliable, due to plant degeneration (wear and tear). For these reasons it is unlikely that, at least initially, the ex-monopoly operators will have the resources to cope with the demands of *unbundling* and *xDSL* simultaneously on a nationwide basis. We can expect xDSL initially only to be available in certain cities or city areas.

Cable modems provide a potential similar service potential to that of xDSL, but for carrying public telecommunciations services across established coaxial cable TV networks. Of course, they only come into question where there is an existing coaxial cable TV network. Even a cable TV operator will not install new coaxial cable today. The preferred technology is fibre.

The problem with new fibre networks, of course, and as we have discussed several times in this chapter, is the huge upfront investment associated with digging conduits and laying cables. There are, in addition, a number of practical difficulties and maybe additionally costs associated with gaining rights of way for laying cables. Even for public licensed operators, who may have a 'right of use' of public streets and other thoroughfares without charge, there are likely to be restrictions about where, when, whether and how often streets may be dug up. Historic cities can pose particular problems due to their building restrictions.

Satellite is an effective and highly economic means of providing *broadcast* transmission (where the same signal is to be received by many end customers simultaneously—as in the case of broadcast television), but is very expensive when used for *point-to-point* connections which are the basis of most public telecommunications. The long delay times (0.5 seconds) incurred by signals sent via satellite also lead to transmission quality problems. Satellite is therefore only economic for broadband applications (e.g. television or high speed business leaselines via *VSAT* (*Very Small Aperture Terminal*) systems).

1.14 Unbundled Access—A Serious Competitor to Fixed Wireless Access Networks?

Unbundling (or *unbundled access*) is the term given to the leasing of 'raw' cable lineplant by new carriers from the established 'monopoly' carrier. It has come about in the United

States, in the European Union and in other countries as the result of regulator decree. It is considered by some to remove the need for wireless access. So let us explain why wireless will remain.

The political idea which has established unbundling is the argument that it is unfair for the ex-monopoly carrier to gain undue benefit in a deregulated market from his ownership of the existing, already written-off access network cable infrastructure, which probably was financed with taxpayers money.

Unbundling attempts to create a 'fair' situation in which the ex-monopoly operator is forced to allow the new competing operators to use only a portion of the service he normally offers his end customers. Put simply, the new operators can choose to lease only the physical cable, without the normal line terminating and other transmission devices which make up a normal 'leaseline' (conditioned for data or voice transmission). Instead, the new operator can use his own equipment directly on his competitors copper or other cable, to offer his own service to end customers—all without the expense of having to dig new conduits and lay new cables, and without the limitations of his competitor's technology.

Figure 1.8 shows the basic principle of unbundling. It illustrates how the new carrier gets access to the copper cable infrastructure (the *local loops* which connect the end customers) of the ex-Monopoly operator by diverting the lines at the *MDF* (*Main Distribution Frame*) of the ex-Monopolist. In this way, the new carrier is even able to provide local switching services to the end customer. The *de facto* monopoly on local switching which had, before unbundling continued to exist, has been removed.

Unbundled access undoubtedly increases the range of services which may be offered by the new carriers and hugely increases their business potential. With it, the new carriers can more effectively and more economically attack the local switching service monopoly. In the future there is also the potential to upgrade the technology used on the local loop and to offer end customers newer and much higher bitrate services over the same physical

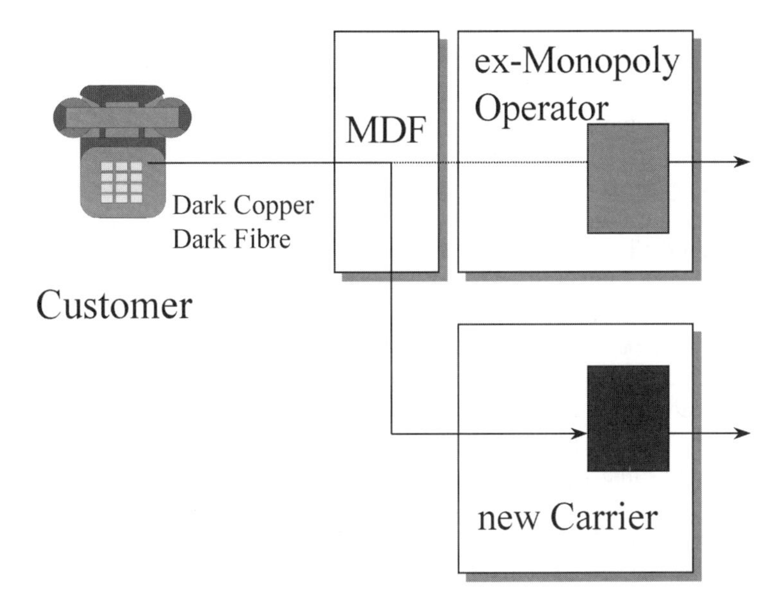

Figure 1.8 The principle of *unbundled* access

lineplant, thereby greatly increasing the value received by the customer for a given price. However, since unbundling is still in its early stages of realisation, it will take some time before a number of practical and operational matters are resolved.

While deregulators have been quick to *determine* the price (i.e. dictate the price) which ex-monopoly operators may charge new carriers for the various service elements (ranging from the copper cable itself, to the various installation and service/maintenance check work which must be carried out), it is unclear in practice how effectively the planning and operational procedures will run. Certainly, it is not conceivable that all the operators can decide which technology they will use over a particular cable in isolation from the other operators using the cable, and not expect to *interfere* with one another. When a new cable is being planned and laid, how much the new operators will be able to determine the specification, quality and capacity of cables is unclear, as is their ability to quickly respond to the maintenance call-out and restoration needs of their customers during normal service.

The ex-monopoly players will certainly have scope to present all manner of arguments to the regulator about how particular desires of the new operators cannot be met as quickly as they might like. This is likely to mean some years yet of regulatory and legal wrangling between the various players (regulators, ex-monopolists and new network operators) before a stable situation is reached. Until then, exactly what is possible with unbundling (i.e. which technology can reliably be used, for what, in which areas and at which overall cost) is likely to remain unclear. A fixed wireless solution during this period will provide greater independence and certainty for the new operators — if only as a second alternative for the really 'hardened' *unbundled access* addicts.

1.15 Objections to the Use of Radio

There are two common 'objections' to the use of fixed wireless radio systems:

- Concern about the effects of radio transmission (*electro-smog*).
- Concern about the security of transmission and the possibility of overhearing or interception of messages by third parties.

Fixed wireless systems sometimes use highly directional antennas (e.g. parabolic antennas) which appear to have a threatening effect upon the population. People feel that they are being directly irradiated by such devices. The truth is that the total output radio power of such systems rarely exceeds 100 mW. This is about one-twentieth of the output radio signal power of a handheld mobile telephone, and many thousand times weaker than either a mobile telephone base station, or worse still, the transmitter aerial of a broadcast TV network. According to current safety standards, such small radiation levels do not require any special precautions (e.g. fencing off areas directly neaighbouring main transmitter stations). Simply put, the level of radiation caused by a transmitter (either base station or customer terminal) of a fixed wireless access system is considerably less than that of a mobile telephone, and since it never comes into close proximity with the body, therefore represents a much lower risk than mobile telephony.

Tales of overhearing mobile telephone conversations are widespread. Most of these tales, however, relate to old style *cordless telephones* and analogue mobile telephones. Such devices were relatively easy to eavesdrop because they used relatively simple modulation (FDMA — discussed in Chapter 4) and made no provision for encryption. Modern fixed wireless access systems use digital modulation, which lends itself more easily to signal encryption (e.g. using the *DES* (*Defence Encryption System*) and modern overhearing-immune modulation schemes (e.g. *spread spectrum* modulation techniques — see Chapter 4).

1.16 Commercial Benefits of WLL Deployment in Public Networks

To conclude this chapter, we summarise the commercial benefits of deploying *Wireless Local Loop* (*WLL*) systems in the development of *access networks* in modern public telecommunications networks:

- WLL gives rapid 'market access' to a greatly expanded customer potential. This is made possible by the ability to connect customers in large geographical areas directly to an operator's network.
- An 'owned infrastructure' such as that made possible by WLL ensures greater 'customer loyalty', and thus lower turnover of customers. A customer considers more carefully before he changes his access line provider. Access network owners and operators can therefore expect to have less turn-over of customers than the *long distance* telephone companies, from which customers can defect at the time of their next phone call, simply by dialling the *access code* of an alternative long distance operator.
- Modern WLL offers the possibility to offer a full range of modern ISDN and broadband data services without having to try to resolve the problems associated with a copper access network designed and installed years ago expressly for the sole purpose of telephony.
- WLL offers the fast realisation of a large geographical *coverage area* for an access network with minimum investment. The initial investment in minimum capacity base stations provides for coverage. Further investment in capacity and subscriber terminals can be managed to coincide with take-up in business and revenues.
- WLL offers the potential for rapid realisation of a *local loops* to new customers, thereby circumventing the sometimes prolonged lead times associated with cabling and providing the operator with an important differentiator against his competition.
- Compared with cable infrastructure, the capital investment in WLL is less heavily front-loaded, giving a potential better return on investment.
- The simple network architecture of WLL requires only base stations in prominent locations, thereby saving the thousands of conduits and streetside cabling cabinets associated with cable lineplant. This simplifies the planning, eases operations and minimises investment in long depreciation time building assets.

2

Radio Communication, The Radio Spectrum and its Management

This chapter lays out the basic terminology of radio communication, and explains the worldwide practices used to manage the radio communications spectrum. In particular, it explains the most frequently used terms as well as the operating practices, standards and governing bodies associated with radio communications and public telecommunications networks, with particular focus on the radio methodologies and spectrum relative to fixed wireless access systems, including both, 'terrestrial point-to-point radio' and 'terrestrial point-to-multipoint'.

2.1 Radio Communication

Radio has long been a means of transporting information through the air by electromagnetic means. It is physical scientific fact, that if you supply sufficient power to a high frequency electromagnetic wave (i.e. an electric current), it has the ability not only to propagate through good conducting materials like metals, but also through poorer conductors. At frequencies above about 100 kHz (kilo Hertz, or thousand cycles per second), electromagnetic radiation is capable of propagation direct through the air. This type of electromagnetic wave propagation is called radio. By coding the radio signal (a technique known as *modulation*), transport of useful communications signals, such as television, audio, telephone or data signals, becomes possible. This is known as radio communication.

2.2 The Radio Communications Spectrum

In principle, all electromagetic signals with *frequencies* higher than about 100 kHz (equivalent to a *wavelength* of 3000 metres — the relationship between frequency and wavelength is governed by the equation $v = f\lambda$, where v is the velocity of light, 3×10^8 m/s, f = frequency and λ = wavelength). The lower frequency radio bands (up to about 100 MHz) were first used for 'public broadcasting'. These bands include *Long Wave* (*LW*), *Medium Wave* (*MW*) and *Short Wave* (*SW*), *Very High Frequency* (*VHF*) and

Ultra High Frequency (*UHF*). These are the bands we are familiar with tuning in to with our audio radio sets at home to receive national broadcast radio and television programmes.

Radio propagation is also possible at much higher frequencies than those above. However, in general, it is the case that the higher the frequency of the radio signal, the lower the range of the possible radio communication and the greater the complexity of the equipment required to achieve radio *transmission* (the generation and 'launching' of radio waves into the air). This has meant that, over time, radio engineers started with the first radios at relatively low frequencies. However, as these frequencies became congested with ever increasing numbers of people and purposes for radiocommunication, there has been a pressure on the radio engineers to develop new radio equipment for use in higher frequency ranges.

The total range of frequencies at which radio communication is possible is called the radio *spectrum*. For telecommunications purposes, the spectrum can be thought of as extending at the low frequency end from about 100 kHz (100 thousand cycles per second) to about 300 GHz (300 billion cycles per second). In particular, the part of the spectrum corresponding to frequencies above about 1 GHz are used in conjunction with *microwave radio* and *millimeter wave radio* equipment. This is the most common type of radio equipment used in 'fixed' public telecommunications network applications such as fixed wireless access.

2.3 Spectrum Management

To ensure that the radio spectrum is not used 'willy-nilly' by all sorts of people, all competing destructively to destroy the potential of communication for everyone by cluttering up the atmosphere with *interfering* radio signals, an international organisation associated with the United Nations is responsible for radiocommunications and spectrum management. This organisation is *ITU-R* (*International Telecommunications Union—Radiocommunications Sector*). ITU-R is a subdivision of the *ITU* (*International Telecommunications Union*). (Note: The ITU-R was previously known as CCIR (Consultative Committee for Radio Communications, and some people still refer to it by this acronym.)

ITU-R sets out in a series of *ITU-R recommendations* and *reports* the radio planning and operating procedures to be observed internationally in radio communication. It also determines and documents the various parts of the spectrum (correctly called *bands*) which may be used for given purposes. It administers the specification of technical radio standards, the registration of radio users and the allocation of radio bandwidth to ensure that interference of signals is minimised. ITU-R carries out these functions in coordination with two other ITU bodies, *IFRB* (*International Frequency Registration Board*), and *WARC* (*World Administrative Radio Council*).

Bands are allocated by ITU-R on an international basis for such uses as:

- terrestrial fixed network communication;
- maritime communication;
- navigational radio;
- satellite communication;
- radio astronomy;

- public broadcast radio and television;
- mobile communication;
- amateur radio.

The band structure is adjusted as necessary every four years at the *World Administrative Radio Council* (*WARC*), at which all ITU (International Telecommunications Union) members may attend. The members of the ITU are governments of member countries, and it is the responsibility of the members to ensure that the ITU rules and regulations are complied with within their territories.

The ITU spectrum plan divides the world into three regions (Regions 1, 2 and 3 as illustrated in Figure 2.1). These correspond roughly with America, Europe/Africa and Asia/Pacific. All individual countries follow one of the three regional plans. Within each *region*, regional and national agencies deal with the detailed spectrum planning and management. There are two influential bodies in setting the regional plans. These are:

- *CEPT* (*Conference of European Post and Telecommunications administrations*). This is a conglomerate of European Regulatory agencies and PTTs (national Post, Telephone and Telegraph companies) as well as public telecommunications operators.
- *FCC* (*Federal Communications Commission*) of the United States of America.

These two bodies in effect set the spectrum plan for the European and American regions respectively, but they also find influence in region 3.

The regional bodies (e.g. CEPT and FCC) perform a more detailed sub-division of the spectrum than ITU-R. Based on the band allocations of ITU-R, the regional bodies further split the spectrum into a larger number of smaller bands, each allocated for a particular use. However, these 'uses' may not be easily recognisable to the non-expert. One such defined use, for example, is that for *Public Land Mobile Networks* (*PLMN*) (this includes the

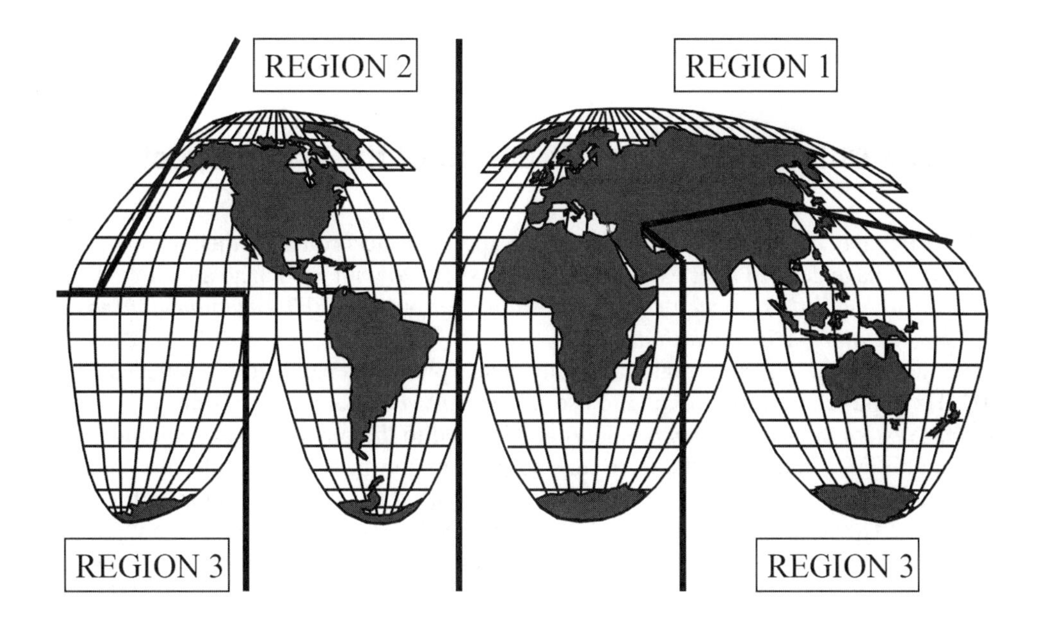

Figure 2.1 ITU-R regions

bands allocated for cellular telephone networks, including *GSM* (*Global System for Mobile communications*)). Another use might be 'terrestrial radio'.

Once a given radio spectrum has been earmarked for a given purpose in a given region, regional technical standards bodies (principally *ETSI* (European Telecommunications Standards Institute) in Europe and the *FCC* (*Federal Communications Commission*), *American National Standards Institute* (*ANSI*) or the *IEEE* (*Institute of Electrical and Electronics Engineers*) in North America) lay down the detailed radio spectrum *masks* and other specifications which ensure that individual radio transmission and receiving devices use the spectrum efficiently, do not disturb neighbouring systems and, as far as necessary, are compatible with one another. One of the important roles of the *regional* bodies is to resolve the procedures and band allocations necessary to ensure the most efficient and interference-free use of radio spectrum even in border regions between countries.

The actual use of radio spectrum, the licensing of its use and the detailed registration of authorised users is regulated on a national basis. Typically, the transmission or receipt of a given radio frequency is regulated by national government bodies, to whom individual users must apply for registration. In the UK, for example, this is carried out by the *Radio Communication Agency* of the *DTI* (*Department of Trade and Industry*). In Germany the task is conducted by *RegTP* (*Regulierungsbehörde für Telekommunikation und Post*), while in the United States it is the responsibility of the *FCC* (*Federal Communications Commission*) (see Appendix 13). Only after the division of the spectrum by these bodies do we recognise the radio system applications in very specific terms:

- longhaul *point-to-point* (*PTP*) radio (up to 100 km);
- shorthaul *point-to-point* radio (up to 20 km);
- *point-to-multipoint* (*PMP*) radio, including *Wireless Local Loop* (*WLL*);
- satellite *uplinks* and *downlinks*;
- broadcast services (including television, public broadcast radio, etc.);
- mobile telephone networks (DCS, GSM, PCS, PHP, etc);
- military usage.

2.4 The Differing Regional Structure of the Radio Spectrum

Both CEPT and FCC have set out the entire spectrum according to their own different needs. Both have defined a series of *bands*, and associated with each band an allowed usage and a *recommended frequency raster*. These are set out in CEPT *recommendations* and FCC *specifications* and *regulations*.

The fact that the different regional bodies agree plans independently is reflected in the 900 MHz band. Harmonisation of the use of the 900 MHz band across Europe for *GSM* (*Global System for Mobile communications*) has led to the possibility of *roaming* (use of the same GSM cellular telephone in any country in Europe). Meanwhile, the 900 MHz band is allocated in the USA by the FCC for *unlicensed* use in conjunction with cordless telephones. The result: roaming to the USA with a 900 MHz cellular phone is not possible. For this type of roaming, *dual band* or *triple band* mobile telephones are necessary (i.e. ones which operate at 900 MHz and/or 1800 MHz in Europe, as well as at 1900 MHz in the USA).

Let us next consider an example of a few of the *bands* allocated by FCC and by CEPT for *terrestrial radio communication*. In fact, we shall consider the bands used for *point-to-point (PTP)* and *point-to-multipoint (PMP)* radio systems, as used in public telecommunications networks. Specifically, we shall consider the 24 GHz, 25 GHz, 26 GHz and 28 GHz *bands*.

Each band commonly becomes known under a shorthand name (e.g. 26 GHz band), but this shorthand terminology is often inconsistent and confusing, as we can see when we consider the actual spectrum covered by each of the bands. We also shall see how some of the bands allocated in the US have the same name as bands allocated in European countries at the same frequency, but equipment which is approved to work in the US band is often incompatible with equipment allowed to be used in the European band at the same frequency. Thus, 28 GHz in the USA is not the same band as the CEPT 28 GHz band. Even within the regions, the terminology may differ. To illustrate some of the inconsistencies we illustrate in Table 2.1 the spectrum between about 24 GHz and 29 GHz as used in the US, UK and continental Europe for public telecommunications purposes.

There are other inconsistencies in the naming of the *bands*, as Figure 2.2 illustrates. Figure 2.2 illustrates some of the most commonly used European public network radio bands. Amongst other inconstencies in naming: first, not all of the 'bands' cover the same *bandwidth* of spectrum; so that the 10 GHz band includes only 500 MHz of bandwidth while the 38 GHz band has 2.5 GHz of bandwidth available. Secondly, the name of the band is often not the centre frequency within the band. It would be more logical to call the 26 GHz band the 25.5 GHz band, and indeed this band is sometimes known in the UK as the 25 GHz band, as we saw in Table 2.1. The 10 GHz band shown in Figure 2.2 is sometimes also known as the 10.5 GHz band. The naming is more slang than it is tight terminology, so if you are unfamiliar with a particular band, check the specifications in detail!

Table 2.1 Examples of the inconsistencies in band structures and names

Frequency band (name)	USA	UK	Continental Europe
24 GHz	Allocated for fixed wireless operation (PTP and PMP)	No 24 GHz band but some of the spectrum is used as the 25 GHz band	No 24 GHz band but some of the spectrum is used as the 25 GHz band
25 GHz	No 25 GHz band, but some of the spectrum is used as the 24 GHz band	24.50–26.50 GHz, allocated for PTP operation	Same spectrum normally known as the 26 GHz band
26 GHz	No 26 GHz band, but some of the spectrum is used as the 24 GHz band	Same spectrum normally known as the 25 GHz band	24.50–26.50 GHz, allocated mostly for both PTP and PMP operation
28 GHz	27.5–30.0 GHz plus 31.0–31.3 GHz Allocated for LMDS (Local Multipoint Distribution Service, a type of PMP operation)	27.50–29.50 GHz, not yet allocated, but considered for PMP operation	27.50–29.50 GHz, not yet allocated, but considered for PMP operation

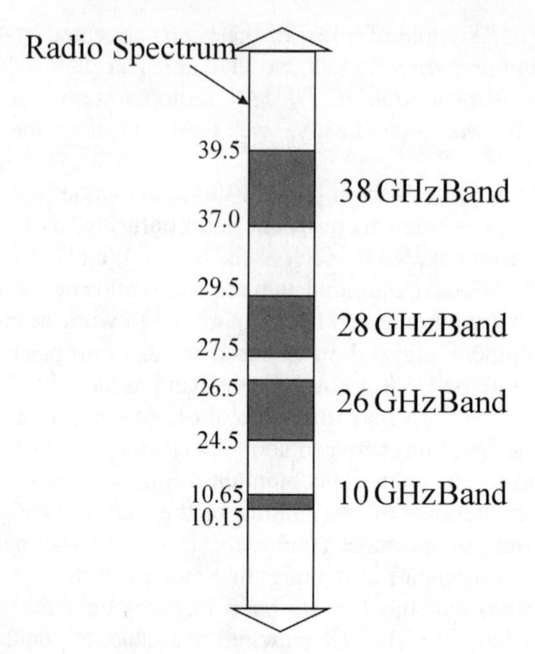

Figure 2.2 Example CEPT bands used in public fixed radio telecommunications networks

Unfortunately, the band naming is also confused by the different regional band structure defined in the USA by the FCC. Thus, for example, the 24 GHz band used in the USA as we saw in Table 2.1 is also from about 24 GHz to 26 GHz, but this band is structured differently from the CEPT 26 GHz band. Perhaps helpfully, though, the conventional name for the FCC band is 24 GHz.

So what's the best advice to the novice who wishes to ensure he is referring to the correct band? To make sure that he has checked not only the name of the band (e.g. 26 GHz), but also the minimum and maximum frequencies (24.5–26.5 GHz) and the band structure (CEPT or FCC). By checking all these facts, one can be sure that a given radio equipment is suitable for use in a given country or allocated frequency band.

A comprehensive listing of the bands used for fixed wireless applications in public telecommunications networks, along with a summary of the relevant band *raster* (explained later in this chapter), appears in Appendix 1.

2.5 One-Way Radio Communication and Introduction to Radio Channels

Within the domain of a given radiocommunications regulatory agency (usually on a national basis), a given radio *band*, as we have seen, is allocated to provide *spectrum* for a given application. Since different applications make different demands on the quantity of spectrum required and the nature of the use of the radio signal, each radio band is structured differently, and has a different allocated *bandwidth*.

All radio applications may be split into one of three types of communication: *simplex* communication; *duplex* communication; and *half duplex* communication. We next discuss the nature of these different communication types, so that we can understand the structure and *raster* applied to radio bands. We first discuss simplex communication, and then go on to duplex communication in the next section.

Simplex is the the simplest type of radio communication. In simplex communication, there is a one-way communication between the sender or *transmitter* and the *receiver*. Simplex communication is nowadays mostly limited to *broadcast* applications. The sender or *transmitter* (shorthand *Tx*, also marked 'A' in Figure 2.3) can only talk, the receiver (shorthand Rx, also labeller 'B') can only listen. There is not an opportunity for full conversation; there is no opportunity for B of Figure 2.3 to talk to A.

In a simplex radio system a radio *channel* is used for the communication. The amount of information which may be sent from transmitter to receiver is dependent upon the bandwidth of the radio channel (i.e. to the amount of spectrum allocated to the radio channel).

A radio band is usually sub-divided into a number of channels (there may be a very large number of channels in a given band). Each radio channel is defined in terms of its centre frequency and its bandwidth. The structure of the channels within the radio band is defined by the band *raster*. The raster determines how much *bandwidth* is allocated to each channel and how much *spacing* (i.e. unused bandwidth) should be left free between *adjacent channels*. The raster is usually documented as either a matrix of *centre frequencies* or as the formula for calculating the centre frequencies. The centre frequency of a given individual channel is the frequency right in the centre of the allocated spectrum or bandwidth. Each channel within the raster is spaced at equal intervals and each channel usually has the same bandwidth. Figure 2.4 illustrates the raster of channels defined in the UHF (ultra high frequency) band for television transmission.

We can see from Figure 2.4 that the *UHF band* is split into a number of numbered channels, each having a bandwidth suitable for transmission of a television signal. These are the channels used for public television broadcasting. But beware! These raster channel numbers do not correspond with the numbers used by the broadcasters (e.g. Channel 4, Channel 5, BBC1, BBC2, ITV, ABC, CBS, Channel 7, etc.). The raster channel numbers correspond to the actual radio channels used for transmission, as we discuss next.

The raster channel numbers used to broadcast a given TV channel vary from one part of a country to another. Thus the public broadcasting channel known as 'Channel 4' might be transmitted in the south of England on UHF channel 27 and in the north of England on UHF channel 23, etc. Using different UHF channels at different but neighbouring transmitting stations avoids the possibility of radio *interference*. Such an approach thus

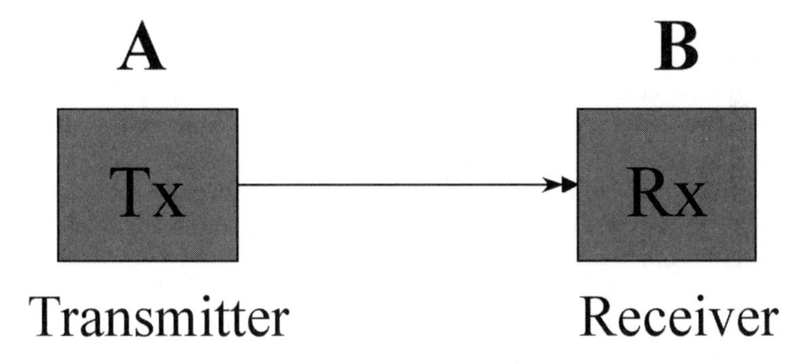

Figure 2.3 *Simplex* radio system (one-way communication)

54 MHz	Channel 2
	Channel 3
	Channel 4
	Channel 1 (A-8)
	Channel 5 (A-7)
84 MHz	Channel 6 (A-6)

Channel number (n)	Picture channel frequency (-1.75 MHz to + 4.75 MHz)	Channel name
7-13	132 MHz + 6n MHz	
14-22	36 MHz + 6n MHz	A to I
23-36	78 MHz + 6n MHz	J to W
37-59	78 MHz + 6n MHz	AA to WW
60-82	78 MHz + 6n MHz	AAA to WWW
83-94	78 MHz + 6n MHz	AAAA to LLLL
95-99	-480 MHz + 6n MHz	C57 to C61
100-110	48 MHz + 6n MHz	MMMM-WWWW
111-125	48 MHz + 6n MHz	AAAAA-WWWWW

Figure 2.4 UHF raster for cable television (EIA IS-6 May 1983)

enables a television broadcaster to use multiple broadcasting sites at different geographical locations to make sure that households throughout a country can receive his programme, without suffering picture quality problems caused by interference. The subjects of interference and frequency *re-use* (i.e. planning the geographical re-use of the same raster channels at *non-adjacent* transmitting stations) we shall return to in Chapter 9. Next, we return to the subject of duplex communication.

2.6 Two-Way (Duplex) Radio Communication

For permanent two-way communication, duplex communication is necessary. Duplex communication allows communication in both directions at the same time. It requires two transmitters and two receivers, one transmitter and one receiver at both ends of the communication link, as shown in Figure 2.5.

There are various technical ways of achieving duplex communication. The most common form of full duplex communication, and the form we shall discuss first, employs what is known as *Frequency Division Duplexing* (*FDD*).

In *Frequency Division Duplexing* (*FDD*), the radio bandwidth allocated to a duplex channel is actually equivalent to two separate simplex channels, one simplex channel for each of the two directions of transmission, one for the transmission A-to-B and one for the receipt direction A-from-B.

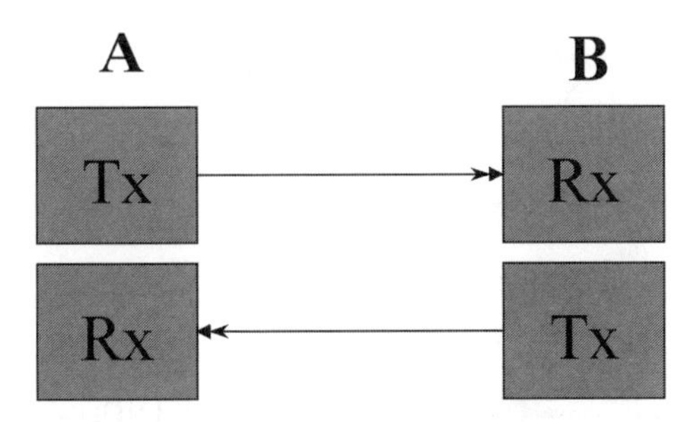

Figure 2.5 Full duplex radio communication system

The total radio bandwidth required for a duplex system is twice that needed for the equivalent simplex system. Instead of using, for example, a single 7 MHz channel at a given frequency (we are here referring to the centre frequency of the channel), we need two separate 7 MHz channels at different frequencies for our duplex system. Furthermore, these channels should not be *adjacent channels* within the band, otherwise we might suffer interference between the two directions of transmission, so there is usually a defined *spacing* (called the *duplex spacing* or *duplex separation*) between the two channels used for a given duplex communication.

Where a given radio band is foreseen by the regional or national radio planning agency to be used for duplex communication, it is normal practice within a given radio band for the spectrum management body (e.g. CEPT or FCC) to define as part of the duplex channel raster not only the *channel spacing*, but also the *duplex spacing*. The duplex spacing or *duplex separation* is the fixed difference in the values of the frequencies (i.e. *centre frequencies*) between the transmit and receive radio channels. The duplex spacing is chosen to maximise the usage of the spectrum available within the band, but simultaneously minimise the possibility of radio interference between the transmit and receive directions.

Let us consider the 26 GHz CEPT radio band again in order to illustrate a *duplex channel raster* and the concept of *duplex channel spacing*. CEPT *recommendation TR 13-02* sets out the channel raster for this band, which runs from 24,549 MHz to 26,453 MHz (a total bandwidth of 1904 MHz). The recommendation stipulates that the duplex spacing shall be 1008 MHz and that individual channel bandwidths shall be either 3.5 MHz or an integral multiple of 3.5 MHz (e.g. 7 MHz, 14 MHz, 28 MHz, 56 MHz up to 112 MHz). From this information, the centre frequencies of each of the channels can be calculated, as can the frequencies of the pairs of frequencies (duplex pairs) which shall be used to create duplex (i.e. two-way) radio communications links. The raster is illustrated in Table 2.2 and Figure 2.6.

From Table 2.2 the split into lower and upper sub-bands can be seen, where the lower sub-band runs over the spectral range 24,549 MHz to 25,445 MHz and the upper band runs over the range 1008 MHz higher, i.e. from 25,557 MHz to 26,453 MHz. The channels run consecutively on a 3.5 MHz *raster* or integral multiple of this raster (7 MHz, 14 MHz, etc), and the channels are numbered from the lowest frequency. The channels use duplex pairs of frequencies, separated by 1008 MHz, one channel from each of the lower and upper bands for the two directions of radio transmission. Thus a 'duplex channel' is actually two separate 'channels'. Unfortunately, the double use of the word 'channel' can thus be somewhat confusing. Maybe the terminology 'channel 1 (upper band frequency)' and 'channel 1 (lower band frequency)' is a little clearer. This reserves the use of the word 'channel' when applied to duplex radio transmission to the 'pair' of frequencies. Figure 2.5 illustrates the structure.

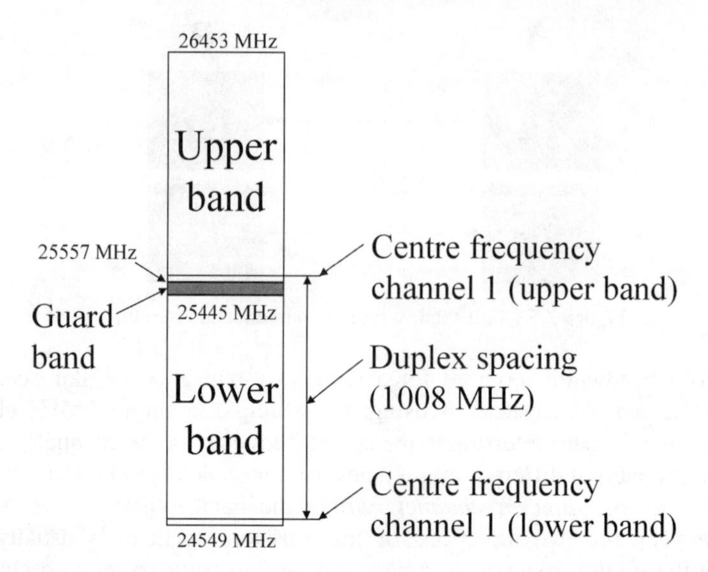

Figure 2.6 Band structure and channel raster of CEPT 26 GHz band (24.5–26.5 GHz)

From Table 2.2 and Figure 2.6, it is also clear that there is an unused block of 112 MHz which separates the lower and upper sub-bands. This unused block is to guard against possible interference of channels. The spacing of 112 MHz is chosen to coincide with the largest possible foreseen channel bandwidth of 112 MHz. The significance of this choice will become clear later in this chapter when we discuss the subject of channel spectral *masks*.

Appendix 1 details the bands and their channel rasters as defined by the various regional radio regulations agencies (CEPT, FCC).

2.7 Half-Duplex Communication and Time Division Duplexing

The third form of radio communication after simplex and duplex communication is *half duplex* communication. In half duplex communication, two-way communication is possible, but simultaneous two-way transmission is not possible. Only one party can 'talk' at any one time, as only one radio channel is available. The same radio channel is used for both transmit and receive directions of transmission, and simply shared over time. This is the sort of transmission used in *Citizens Band* (*CB*) radio; you 'press to speak' and punctuate sentences not only with full stops, but also with the word 'over': 'How are you, over' . . . 'Fine thanks, over'.

Half-duplex radio is thus a 'halfway house' between simplex and duplex communication. Like simplex radio systems, there is only one radio channel, but by providing both a transmitter and a receiver at both ends of the link, the same channel may be used for both directions of transmission by a 'taking in turns' routine.

Half duplex communication is a simple form of *Time Division Duplexing* (*TDD*). In TDD the same radio channel is used for both directions of transmission, but shared over time between transmission and reception. It is used in some forms of digital radio (for

Table 2.2 Duplex channel raster of 26 GHz band, showing calculation of the limit frequencies (minimum and maximum) of the lower and upper sub-bands as well as the channel centre frequency calculations

	(3.5 MHz raster 3.5 MHz duplex per channel)	7 MHz raster (7 MHz duplex per channel)	14 MHz raster (14 MHz duplex per channel)	28 MHz raster (28 MHz duplex per channel)
Total number of channels in the band (maximum value of n)	256	128	64	32
Minimum frequency in the band	24,549 MHz	24,549 MHz	24,549 MHz	24,549 MHz
Maximum frequency of lower sub-band	25,445 MHz	25,445 MHz	25,445 MHz	25,445 MHz
Minimum frequency in upper sub-band	24,549 MHz + 1008 MHz (duplex spacing) = 25,557 MHz	24,549 MHz + 1008 MHz (duplex spacing) = 25,557 MHz	24,549 MHz + 1008 MHz (duplex spacing) = 25,557 MHz	24,549 MHz + 1008 MHz (duplex spacing) = 25,557 MHz
Maximum frequency of band	26,453 MHz	26,453 MHz	26,453 MHz	26,453 MHz
Formula for calculating centre frequency of channel n (lower subband channel)	24,549 MHz plus $(n^{-\frac{1}{2}}) \times 3.5$ MHz	24,549 MHz plus $(n^{-\frac{1}{2}}) \times 7$ MHz	24,549 MHz plus $(n^{-\frac{1}{2}}) \times 14$ MHz	24,549 MHz plus $(n^{-\frac{1}{2}}) \times 28$ MHz
Formula for calculating centre frequency of channel n (lower subband channel)	25,557 MHz plus $(n^{-\frac{1}{2}}) \times 3.5$ MHz	25,557 MHz plus $(n^{-\frac{1}{2}}) \times 3.5$ MHz	25,557 MHz plus $(n^{-\frac{1}{2}}) \times 3.5$ MHz	25,557 MHz plus $(n^{-\frac{1}{2}}) \times 3.5$ MHz
Total bandwidth in band	896 MHz (= 256 × 3.5 MHz duplex)	896 MHz (= 128 × 7 MHz duplex)	896 MHz (= 64 × 14 MHz duplex)	896 MHz (= 32 × 28 MHz duplex)
Separation of lower and upper bands	25,557–25,445 = 112 MHz	25,557–25,445 = 112 MHz	25,557–25,445 = 112 MHz	25,557–25,445 = 112 MHz

example, TDD is used in the DECT (digital European cordless telephony) radio standard). We describe TDD in detail in Chapter 13.

2.8 Bandwidth — Determines the Capacity of a Radio Channel

So far we have spoken of the subdivision of the radio spectrum into bands, and the structuring of these bands into individual channels for given communications applications. We have also learned that each channel is assigned a given bandwidth. So what exactly is the significance of the bandwidth?

The bandwidth of a radio channel is the quantity of radio spectrum resource available for communication. The more bandwidth, the more complex a signal may be transported. More bandwidth means higher fidelity of audio signals, better signal resolution of video signals and higher bitrates of digital (e.g. data) signals.

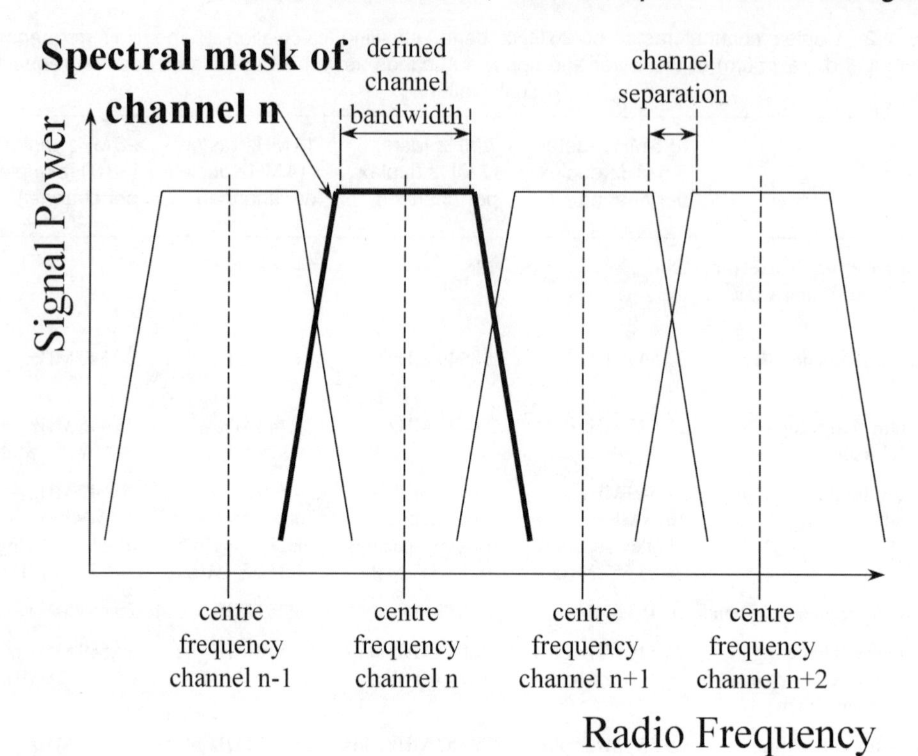

Figure 2.7 The *spectral mask*

A telephone channel requires 4 kHz of bandwidth. A broadcast television channel requires around 70 MHz. Digital signals require of the order of 1 Hz bandwidth for each 1 bit/s of bitrate (i.e. approx. 1 bit/s per Hertz), depending upon the *modulation* scheme used. *Higher modulation* schemes, as we shall discuss in chapter 3 achieve more bit/s per Hertz (e.g. 2 bit/s per Hertz, 4 bit/s per Hertz or 16 bit/s per Hertz), but at the cost of reduced system range and increased susceptibility to interference.

2.9 The Spectral Mask

To allow the possibility of more than one network operator and more than one type of manufacturer's equipment sharing a given radio band without interference to one another, it is necessary for the radio regulation agency, in association with a given technical standards body, to define the permissible *spectral mask* of radio equipment which may be operated within the band. In the US the radio specifications and spectral masks are defined by the *FCC* (*Federal Communications Commission*) or *IEEE* (*Institute of Electrical and Electronics Engineers*). In Europe the technical standards are laid down by *ETSI* (*European Telecommunications Standards Institute*).

Figure 2.7 illustrates the spectral masks of a number of *adjacent channels* within a given radio *band*. The aim of the spectral mask is to ensure that the radio *emissions* from a given radio transmitter operating in one channel do not disturb or *interfere* with the operation of a second system operating nearby on an adjacent channel. The spectral mask defines the maximum power which may be emitted at all possible frequencies within the radio channel itself, and, perhaps more importantly, limits the power which is allowed to be transmitted at frequencies outside the channel, particularly at frequencies fringing the neighbouring channel.

As we see from the mask of channel n in Figure 2.7 (drawn in bold), the centre frequency lies in the middle of the flat part of the mask. The flat part of the mask indicates that a radio transmitter operating in this channel may transmit at the maximum power any radio frequency within the defined bandwidth of the channel. At frequencies beyond the bandwidth limits of the channel (the sloping part of the mask), the transmitter is not allowed to emit strong signals. Actually, it is not in the interest of the radio system operating at channel n to waste its output power on 'unwanted' emission outside the normal channel bandwidth, but the problem is that it is difficult to design radio systems which are able to produce maximum power across the full channel bandwidth without creating extraneous transmission at neighbouring frequencies.

To limit the interference to the *adjacent* $(n-1)$th and $(n+1)$th channels, a small portion of spectrum is allocated to provide a *spacing* between the channels. This is indicated in Figure 2.7. The steeper the sides of the radio mask, the less spectrum within the band need be wasted in the spacing of channels.

The significance of the mask is that it is used in the *homologation* or *approvals testing* procedure of radio systems. Measurements are made to ensure that the transmitter conforms with the mask, before a given radio system is *approved* for usage.

Figure 2.8 illustrates a typical spectral output of a real radio system when compared against the spectral mask. Such plots are made as a matter of course by technical approvals testing houses.

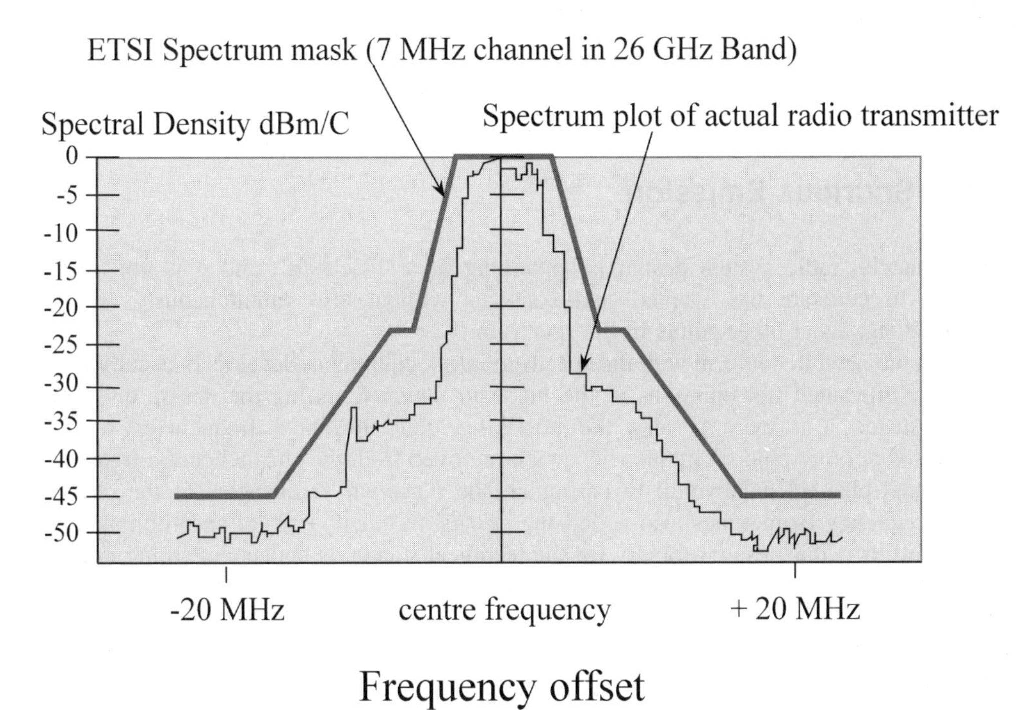

Figure 2.8 Typical channel spectrum mask and spectrum plot of actual radio equipment

Note how the spectrum mask of Figure 2.8 is defined relative to the centre frequency of the channel and the maximum power of the single strongest frequency. Note also how the real spectrum mask has 'shoulders' at around 7 MHz offset from the centre frequency. It is no coincidence that this corresponds to twice the frequency offset (3.5 MHz) of the nominal channel bandwidth. By designing the radio for a 'flat patch' from 0 up to 3.5 MHz offset, a 'flat patch' at double and other multiples of this offset is almost inevitable. (Note: a multiple of a given frequency is called a *harmonic* frequency. Eliminating undesired harmonic frequencies is one of the greatest challenges in radio design.)

Note also how in Figure 2.8 the output of the real radio transmitter, when measured, appears to produce output across a wide spectral range at a weak signal level about 50 dB weaker than the maximum power. This emission partly comes from the transmitter itself, but is partly due to so-called *background noise*. Such weak broad spectral signals are always present in the atmosphere, and must be taken into account in the design of practical radio systems. Thus design engineers seek to ensure that the 'wanted' signal is much stronger in power than the noise, so that it can be clearly distinguished, correctly received and decoded.

The relative strength of the 'wanted' signal to the noise is usually referred to as the *signal-to-noise* (S/N) or *carrier-to-noise* (C/N) ratio. Radio design engineers will usually specify for their equipment the required minimum S/N (or C/N) for reliable reception of the signal at the receiver. Where the signal does not arrive at the receiver with adequate power relative to the noise, the received signal, once decoded may be poor or even indistinguishable. Such degradation of the *carrier signal* into the noise may be caused by radio signal *fading* (e.g. loss of signal strength caused by bad weather), by interference of other nearby radio systems, or by trying to operate the system too far beyond its *range*. A well-designed radio which can operate even at relatively low S/N and good radio network planning are essential prerequisites for reliable operation. We shall return to this subject in Chapters 7, 8 and 9.

2.10 Spurious Emission

Unfortunately, radio system design is something of a 'black art', and it is not always possible to generate the 'wanted' radio signals without also simultaneously creating unwanted signals at other points in the spectrum.

Given the need to conform with the spectrum mask, equipment developers usually try to eliminate unwanted transmissions in the *adjacent channels* during the design and early testing stages. But there is also the possibility that unwanted frequencies will be transmitted at other random frequencies much removed from the channel centre frequency and normal channel bandwidth. In particular, the *harmonic* frequencies of the channel centre frequency (sometimes also called the *carrier frequency*) are often problematic in this regard. It is thus also customary for the technical standards bodies governing radio to limit these so-called *spurious emissions*. The approvals testing bodies are typically required to measure across a wide spectral range, up to three or more times the carrier frequency to check for such spurious emissions.

Some technical radio standards require not only that spurious emissions are not transmitted by the radio, but also that the radio receiver equipment is *immune* to interference from the (defined) spurious emissions of other equipments.

2.11 Authorised Uses of Radio and the Regulations Governing Use

Over the last few years, fixed wireless access systems (in particular *point-to-point* (*PTP*) and *point-to-multipoint* (*PMP*) millimetre and microwave radio systems) have gained significantly in importance as major corporations and new network operators alike have sought to use radio techniques to bypass the transmission resources of the old monopoly network providers. Table 2.3 illustrates the range capabilities of the new frequency bands which have come into use, and for reference, the uses allocated to each of the bands by ITU-R.

In most deregulated national public telecommunications markets there are now a plethora of new public operators or carriers fighting for market share in both the public telephone switching business and the data communications and value added services sectors. At first, many of these companies attempted to build long distance fibre networks and compete for long distance telephone service minutes. However, as the competition has become tougher and the profit margins tighter, there has become a growing number of carriers turning to fixed wireless networks as a fast means of realising both longhaul backbone infrastructure as well as access networks for direct connection of customers.

Table 2.3 Frequency bands allocated by ITU-R for public network and corporate microwave radio links, either point-to-point or point-to-multipoint

Frequency band	Countries of application	Application and range
1 GHz	Worldwide	Cellular mobile radio
2 GHz	Worldwide	cellular mobile radio DECT some point to point microwave
3 GHz	Worldwide	3.5 GHz rural radio
4 GHz	Worldwide	high capacity point to point (e.g. STM-1) satellite communications
6 GHz	Worldwide	satellite communications
7–8 GHz	Worldwide	longhaul point to point microwave (up to 50 km)
10 GHz		US: cellular base station backhaul UK: point to multipoint wireless local loop
11 GHz	Worldwide	medium capacity (E3, T3)
12, 14 GHz	satellite communication	direct broadcast by satellite
13, 15 GHz	Worldwide	short to medium haul point to point (25 km)
18, 23, 26 GHz	Worldwide	short haul point to point (10–15 km)
28 GHz	Worldwide	point to multipoint, multimedia applications
38 GHz	Worldwide	short haul (5–7 km)
above 40 GHz	Worldwide	Extremely short haul (1–3 km)

Before installing a radio network, corporates or public network operators have to fulfil number of regulatory conditions, usually comprising most or all of the following:

- Operators must be in possession of relevant public operator licences (e.g. satellite operating licence, mobile telephone network operating licence, public network infrastructure licence, etc.) or operate under the auspices of some sort of user *class licence* (which may allow even non-public network operators (e.g. corporate network operators) to operate certain restricted classes of radio equipment.
- In some countries there is a need to apply for a radio operating licence.
- Operators usually need to apply (separately from their licence application above) for an allocation of radio spectrum. For *point-to-point* (*PTP*) system operation it has become normal to have to apply to the national radio regulatory agency for a channel allocation on a link-by-link basis. (For each link a separate application. The channel which may be used is determined by the radio planning department of the radio regulatory agency, which keeps records of all users within a given band.)
- Usually only approved (i.e. *homologated*) radio equipment may be operated. (This restriction does not always apply to the ex-monopoly public operators, who have been used to years of deciding for themselves what is suitable, safe and non-intrusive to other radio users. Only equipment which meets the local radio raster (e.g. *CEPT* in Europe and *FCC* in USA) and conforms with local technical specifications (*ETSI* in Europe and *FCC* in USA) is likely to get a homologation certificate, as we discussed earlier in the chapter.
- There may be further safety standards to be observed at main radio transmitter stations. In particular, where a given radio transmitter exceeds a given output power it may be necessary to take precautions at the site to guard against accidental over-exposure of personnel to the transmitted radiation. It may, for example, be necessary to fence off an area around the main transmitting antenna to prevent personnel unknowingly exposing themselves for long periods to the radiation. For lower power transmitters or radio station sites such precautions may be unnecessary.
- Finally, it may be necessary to gain other operating permissions or building permissions of local authorities and loandlords before installing and operating radio equipment.

2.12 Satellite Transmission Management

Satellite transmission is a specialised form of fixed wireless communication, and so like other types of radio transmission it requires proper frequency management and careful siting of both ground based antennas (called *earth stations*) and the *orbit vehicles* (called *satellites*). Satellite systems are used widely in public telecommunications networks for *point-to-point*, *point-to-multipoint* and *broadcast* applications.

A major benefit offered by satellite transmission over cable is the fact that a single earth station in one country and a single satellite in space may be used to provide transmission links to not just one, but a number of other countries. The signal is sent up by the transmitting earth station to the satellite (the *uplink*). The satellite responds by amplifying and re-transmitting a downpath radio signal over the whole coverage area. Any earth station within the coverage area can pick up the signal, and select from it the information intended for that particular destination. Satellites are therefore excellent media for widescale signal broadcasting, a fact well demonstrated by the operation of satellite TV networks by a number of worldwide system operators (Astra and Eutelsat in Europe, Orion in USA, etc). However, satellite systems can also be effectively used for Point-to-Point (PTP) communication. As point-to-point communications media they are best suited (i.e. have their greatest advantage over other alternative wireline transmission means) when they are employed either to carry high bandwidth signals (such as video or TV signals) or to reach geographically remote locations.

To-date, most of the satellites used for telecommunications have been *geostationary*. By this we mean that they orbit around the earth's equator at a speed equal to the spin speed of the earth itself. They therefore appear to be geographically stationary above a point on the earth's equator, which allows them to be tracked easily by earth station antennas. Any given earth station will either be able to 'see' a satellite for 24 hours a day, every day (dependent on the satellite's longitudinal position), or will never be able to see it.

More recently, however, a number of new telecommunications systems based on *orbiting* satellites have been introduced. The advantage of such orbiting satellites is that they fly at lower altitude above the earth's surface, so that the signal path from the earth's surface to the satellite and back is shorter, the signal *propagation time* therefore also. Examples of such orbiting satellite systems are those of *Iridium*, *Globalstar*, *ICO* and *UMTS* (Universal Mobile Telephone Service, a system intended to extend mobile telephone networks to global coverage).

The world's leading organisation for the development and exploitation of satellite communications is called *INTELSAT* (*International Telecommunications Satellite consortium*). INTELSAT is a jointly owned consortium, set up in 1964 by the world's

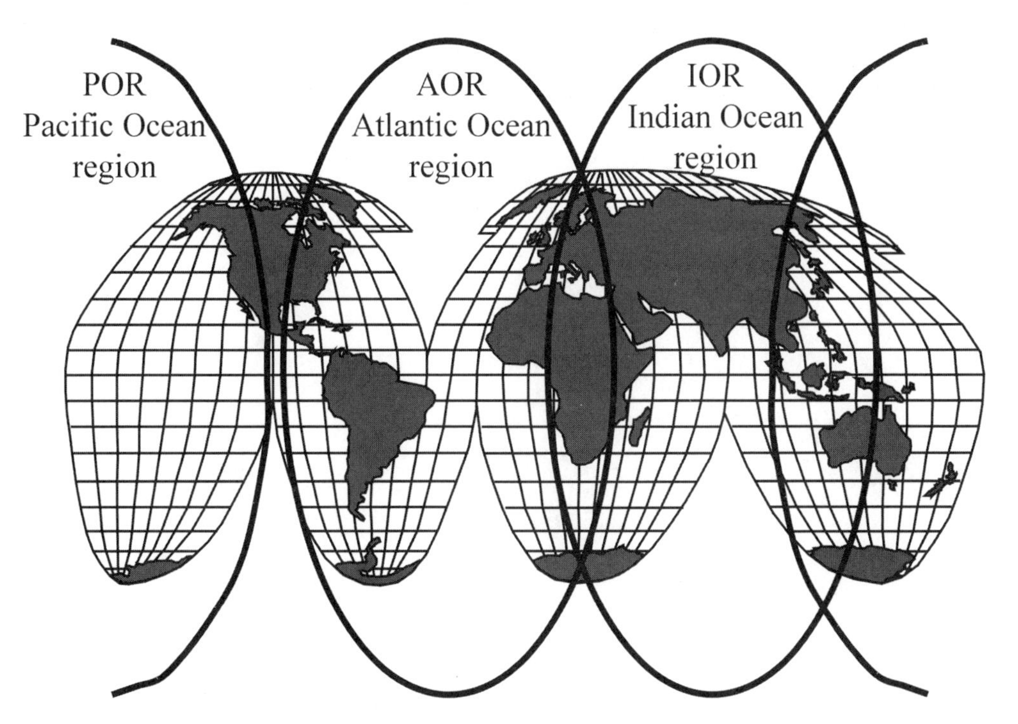

Figure 2.9 Ocean coverage regions of the INTELSAT satellite system used for interlinking the world's national public telecommunications networks

leading public telecommunications operators to provide a means for worldwide transmission. It is not the only satellite body. Others include *INMARSAT* (*International Maritime Satellite organisation*), *EUTELSAT* (the European equivalent of INTELSAT), and a number of privately owned and run companies (e.g. COMSAT in the USA), which operate satellites for a range of national telecommunications purposes (e.g. direct broadcast by satellite, telephone circuit access to remote areas, or most recently, the *Universal Mobile Telephone Service*, *UMTS*.

The geographical coverage area, or *footprint*, of a satellite in orbit (i.e. the area on the ground in which earth stations are able to work to it) is affected by its geographical position above the equator, and by the design and power of its antennas. The position above the equator is usually stated as the geographic longitude. The antennas may either be directed at a small area of the earth's surface in a *spot beam*, or broadcast to all the hemispherical zone on the earth's surface that the satellite can see. The latter type of antenna needs to be much larger, and it demands radio signals of considerably greater power. The INTELSAT system comprises a number of geostationary satellites, positioned in clusters over the middle of the world's three great oceans, the Atlantic, the Pacific and the Indian Ocean, and give the coverage areas shown in Figure 2.9. Between them, the satellites allow most countries to have telecommunications with almost any other country in the world.

European countries have access to the Atlantic and the Indian Ocean INTELSAT satellites, while countries in North America can view the Pacific and Atlantic satellites (from west and and east coasts, respectively).

Allocation of radio bandwidth and planning for new satellites in the Intelsat system are carried out annually at the *Global Traffic Meeting* (*GTM*), held in Washington, DC, United States. At the meeting, all constituent operators and users of the INTELSAT system declare and specify their forward five year bandwidth requirements. The technical standards to be used in conjunction with the INTELSAT system are defined directly by the *ITU-R* (*International Telecommunications Union — Radio communication sector*).

3

Point-to-Point (PTP) and Point-to-Multipoint (PMP) Wireless Systems and Antennas

As we shall discover in this chapter, the most important types of systems used in public telecommunications networks for 'fixed wireless access' and Wireless Local Loop (WLL) are 'point-to-point' and 'point-to-multipoint' systems. We discuss the types of applications for which such systems are suited and, in particular, we describe how the characteristics of the antennas used determine the point-to-point or point-to-multipoint nature of the system. In the following chapters, we then go on to discuss the different types of radio signal modulation used in PTP and PMP systems.

3.1 Characteristics of Wireless Systems

Wireless systems are characterised by the applications for which they were designed, varying technically in the *modulation* techniques used to code the radio signal itself, and in the nature of the antenna, particularly its *directionality*. However, these technical differences in the radio hardware design are themselves only a reflection of the specific purpose for which the equipment is intended to be used.

In Chapter 2 we introduced a number of terms which will help us to define the nature of a communication we wish to transport via radio. We learned how a duplex radio connection allowed for full two-way communication between two end-points. Such communication is desirable for private conversations or similar conversations between two parties. Where such communication between two fixed end-points takes place on a frequent or permanent basis, it may make sense to establish a permanent point-to-point radio link. On the other hand, the conversations between two specific parties may not be that frequent, and it may make sense instead to allow a number of users to establish temporary links to any other user within a pool on an *any-to-any* basis.

In addition to the *symmetric* forms of communication, most of which are link or point-to-point oriented, there are also many *asymmetric* forms of communication. The simplest types of asymmetric communication are simplex and half duplex communication. A simplex communication system is the most efficient where only one direction of communication is required (e.g. for a video surveillance system, where the signal is carried only one way, from camera to control room). Even many types of conversational and computer communication, although two-way communications, are far from symmetric. Often, far more information needs to transported in one direction than in the other. (An Internet user browsing the *World Wide Web* (*WWW*) usually downloads far more information to his PC than he sends to the network. The information sent to the network comprises little more than the commands to request the download of the next web page.)

Wireless systems which take account of the asymmetric nature of a given communication can be designed to be much more efficient in their use of radio spectrum than comparable symmetric systems. As examples of asymmetric wireless systems, broadcast and point-to-multipoint (PMP) wireless systems have been designed to use radio spectrum highly efficiently for certain asymmetric types of communication.

In broad terms we can classify wireless systems in three categories of applications:

1. *Point-To-Point* (*PTP*) systems intended to provide line communication (i.e. a *link*) between two fixed end-points.
2. *Any-to-any* systems intended to provide for communication directly between any two or more of a number of mobile systems or other undefined end-points.
3. *Point-To-Multipoint* (*PMP*) systems intended to allow a number of users to share the facilities provided by a single *base station* or to allow for broadcast of the same signal simultaneously to a number of remote stations.

We disccuss these various types of systems in turn.

3.2 Point-to-Point (PTP) Radio Systems

Point-to-point (PTP) radio systems are commonly used in public telecommunications networks for trunk network and other line-type connections. A PTP radio system provides a link much like a wireline trunk, connecting two (usually fixed) end-points and providing a fixed bandwidth or bitrate on a permanent basis. Unlike broadcast radio systems the desired transmission is along a single line of transmission, and it is therefore usual to optimise the system for the transmission along this line (or link).

A point-to-point link is usually equipped with two high-*gain* antennas (e.g. a parabolic reflector antenna — see Figure 3.1). By *high gain* antenna, we mean an antenna which focusses or concentrates the output radio power of the transmitter in a particular direction.

A typical point-to-point wireless system is designed like the simple system of Figure 3.1. The system, comprises two radios (transmitter and receiver) and two antennas, one radio and one antenna (together sometimes referred to as a radio *terminal*) at either end of the *link*. The antennas are high gain antennas (in this case parabolic reflector antennas). These are designed to ensure that nearly all of the signal generated by the radio is transmitted along the *pole* of the antenna. The *pole* of the antenna is an imaginary line representing the main direction of radio transmission and reception of the antenna. (In some types of antennas there is actually a metal conductor along the *pole* — as it were a metal 'pole' along the pole, but do not be confused by the double meaning.)

In an ideal system, the entire radio signal generated by the transmitter from the left-hand radio of Figure 3.1 arrives in the antenna of the right-hand radio. If this were possible, the radio signal power generated could be kept to a minimum, along with radio interference. In reality, however, point-to-point system operators face a difficult dilemma between two conflicting interests:

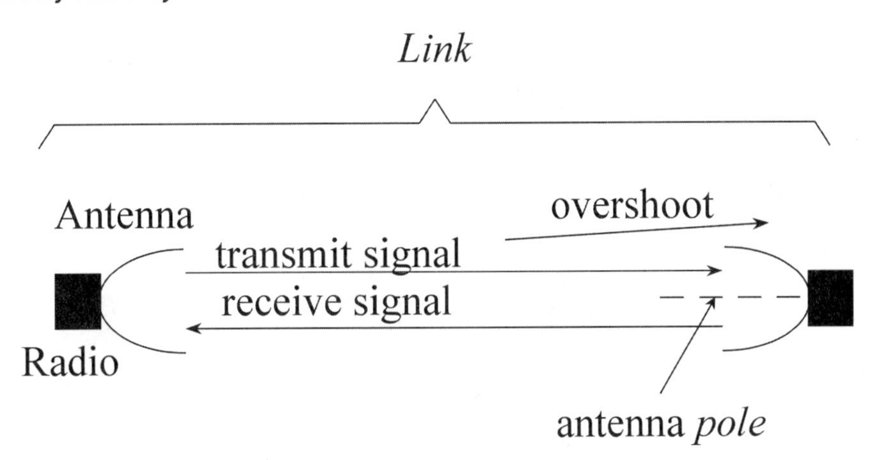

Figure 3.1 A simple point-to-point wireless system

- The signal is subject to attenuation in the atmosphere between the transmitting and receiving antennas. The weather, particularly rain, leads to signal *fading*. As a result we must transmit a stronger signal, not just the minimum power for reaching the far end of the link, but with some extra power to cover a *fade margin*. (We discuss in more detail in Chapter 7 how we plan radio links against the negative effects of fading.)
- Sending a strong signal increases the likelihood of *overshooting* the intended receiving antenna. An unrequired strong signal may travel much further than the intended receiver, and disturb an antenna in quite a wide area beyond. The potential interference with other neighbouring wireless systems due to overshoot may be quite extensive, since even a small *beamwidth* antenna has a wide area of *illumination* even at modest distances. The arc of illumination of a 2.5° beamwidth at 20 km is, for example, 872 metres!

The radio frequency bands allocated to PTP system usage (i.e. in the GHz bands) are not easily able to propagate through obstacles or diffract around them, so that a *Line Of Sight* (*LOS*) is usually necessary between the transmitting and receiving antennas. This is determined at the time of system installation, when an LOS-check (line-of-sight check) is performed (Chapter 8). The need for a line-of-sight between the end-points, and the effort associated with the *LOS-check* during installation of a high frequency PTP *link* is inconvenient, and adds to the cost of the system. On the other hand, this extra effort makes possible the realisation of high bandwidth point-to-point links using high frequency radio bands which are unsuited to general broadcast uses because of these practical difficulties. In practice, a line-of-sight can be achieved between almost any two end-points — it is only a question of how tall the radio masts need to be at either end of the link.

3.3 Any-to-Any Radio Systems

Most any-to-any radio systems are designed to provide temporary *point-to-point* communication between any two end-points within a large pool of devices. Figure 3.2 illustrates the principle. It illustrates a *HIPERLAN* wireless network connecting Personal

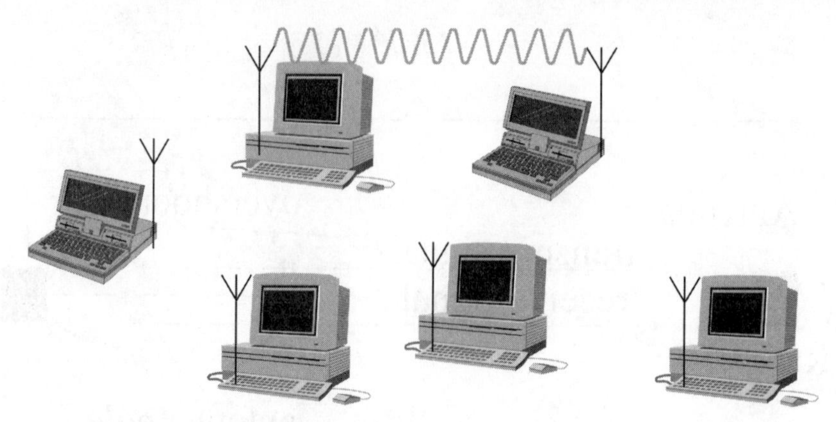

Figure 3.2 The principle of an *any-to-any* communication system

Computers (PCs) in an office environment. In this case, the radio spectrum (actually a single radio channel) is used as a 'shared medium'. Provided the radio channel is not already in use, any PC may use the channel to communicate with any other PC in the network on a direct point-to-point basis.

Any-to-any type wireless systems are an efficient and effective means of providing full connectivity between a relatively large number of low-usage users. The fact that each possible pair of users has only very little information to communicate would mean that a permanently allocated point-to-point system would be both uneconomic and wasteful of radio spectrum. By allowing a large number of users to share the radio spectrum, much more efficient use is possible. However, this efficiency in the use of the spectrum is not without its price. The problem with any-to-any radio systems is that the direction of radio transmission is not known (as it is in point-to-point radio systems, so omnidirectional antennas have to be used. In comparison with a directional point-to-point antenna, an omnidirectional antenna has a much lower gain, reducing the quality and reliability of any-to-any systems in comparison with point-to-point wireless systems. Not only this, the omnidirectional transmission leads to greater potential for radio interference, and a higher security risk associated with possible undesired interception and overhearing of the radio signals by third parties.

3.4 Point-to-Multipoint (PMP) Radio Systems

Point-to-Multipoint (*PMP*) radio systems comprise a base station and a number of remote stations as illustrated in Figure 3.3. In such systems, the radio transmission is not symmetric as in point-to-point radio systems. In PMP systems, you must consider separately the *downlink* direction (from the base station to the multiple remote stations) and the *uplink* direction (from the remote stations to the base station).

In the *downstream* direction (i.e. from the base station to the remote terminals or subscriber stations), the same signal is broadcast to all the subscribers in the *cell* or *sector* area. To achieve this a wide angle antenna is used. In the diagram of Figure 3.3 a single *omnidirectional* antenna is being used to broadcast the *downstream* signal to all points of the compass equally. The result is a *cell* area (the grey-shaded circular area). Alternatively,

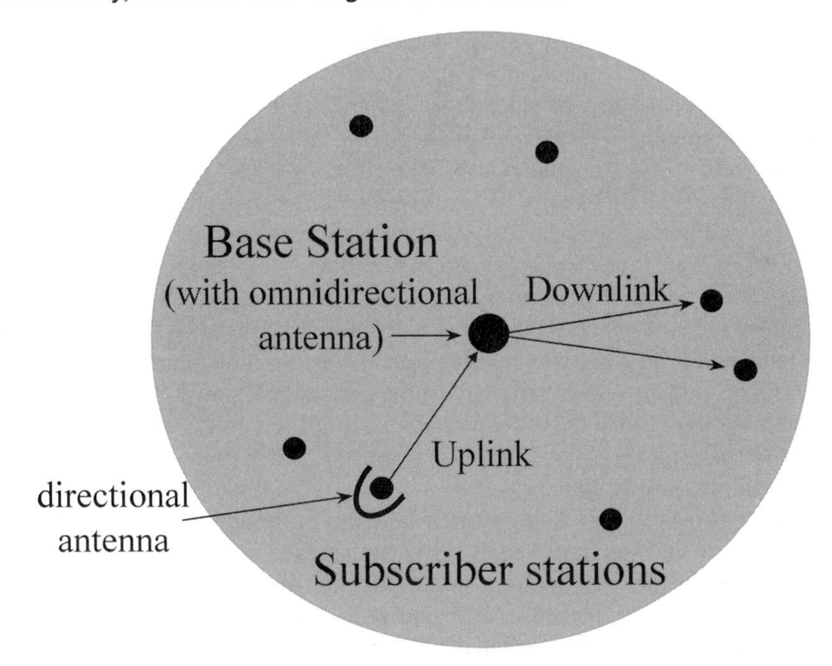

Figure 3.3 Point-to-Multipoint (PMP) radio system

a sector antenna could have been used with a slightly more directional characteristic (e.g. covering just a 90° sector portion of the cell. In this case, four sectors and four *sector antennas* would be necessary to provide full 360° coverage).

The radio *modulation* in the *downstream* direction and the *multiple access* scheme of a *point-to-multipoint* system (we discuss these in Chapters 4 and 5) must be so designed that a given communication broadcast from the base station is received by only the correct (i.e. desired) remote subscriber terminal, without interference of other communications from, other subscribers or base stations.

In the *upstream* direction, the available radio spectrum has in some way to be shared between all the active terminals. This is the job of the *multiple access* scheme. Various different alternative schemes are available. We discuss these in Chapter 5.

As for the antenna used at the remote station, it is normal in *fixed* wireless systems (where the antenna is permanently mounted and is not mobile) to use a highly directional antenna (as shown schematically in Figure 3.3). The use of a *directional* antenna restricts the likelihood of unwanted interfering reception from radio sources other than the base station. In addition, during transmission, interference caused to other subscriber terminals can be minimised by directing all the transmission at the base station.

3.5 The Directionality, Gain and Lobe Diagram of an Antenna

The *gain* of an antenna is a measure of the 'focusing' capability of the antenna. It is the ratio of the signal strength of the antenna along its *pole* (i.e. along the main direction of transmission) relative to the power which would be transmitted along this direction if instead an *omnidirectional* or *isotropic* antenna were used instead. (An omnidirectional or *isotropic* antenna has no 'focusing power': instead it transmits in all directions of space equally — it is the perfect sort of antenna for broadcasting.)

The gain of an antenna is usually expressed as a ratio, and quoted in units called *dBi* (*decibels* relative to an isotropic antenna*). The focusing effect of an antenna and its gain are normally illustrated by means of an *antenna diagram* (otherwise called the *lobe pattern*) diagram. Figure 3.4 illustrates the lobe pattern of both an *isotropic* antenna and that of a typical PTP antenna. The grey pattern (circular area) illustrated in Figure 3.4 is the *antenna diagram* of the *isotropic* (*omnidirectional*) antenna. The centre point of the circle is the imaginary point at which the antenna is located. The distance from this point to the line of the lobe pattern in a given compass direction (i.e. to the circumference of the circle in this case) represents the relative strength of the radio signal transmitted by the antenna in this compass direction. Since the antenna radiates in all directions equally, the line of the lobe pattern remains at a constant distance from the centre point of the diagram for all compass directions — so we have a circle.

In contrast, a point-to-point radio system antenna is designed to try to maximise the radiation along a given direction. Specifically, the antenna is designed to try to focus the entire output signal of the radio along a single direction known as the pole of the antenna, since it is only the receiver placed at a distant point along this line which is intended to receive the signal. Focusing the signal in this way has two beneficial effects. First, the signal strength at the point of reception is increased, making for better reception. Second, fewer other users are able to intercept, cause interference to, or overhear the conversation or other communication taking place on the link.

The second lobe diagram illustrated in Figure 3.4 is that of a directional point-to-point antenna. The antenna is oriented with its pole in a direction approximately 50° from north for transmission across a link in this direction. As in the case of the isotropic antenna lobe diagram, the distance from the centre point of the diagram to the line (shown in black) shows the relative signal strength radiated in the given direction. Thus, the signal strength along the main pole is significantly greater than that radiated by the isotropic antenna. However, the antenna does not achieve the ideal pattern — all the power radiated along the pole direction. There are some *null points* in the antenna pattern (another name for the lobe pattern) corresponding to directions in which there is no transmission, but other points (called the *sidelobes*) corresponding to directions in which quite strong undesired transmission takes place.

The aim of the antenna designer is to try to maximise the *gain* of the antenna (i.e. the transmission along the pole direction) while simultaneously minimising the unwanted sidelobes. It is not possible to eliminate the sidelobes. There is, for example, usually quite a strong sidelobe of the antenna corresponding to radiation in the reverse direction (out of the 'back' of the antenna). This particular sidelobe can cause problems in the radio frequency planning of a network, and so its strength is usually quoted in the specification of an antenna as the *front-to-back* ratio.

As annotated on Figure 3.4, it is usual to plot the *antenna diagram* (or *lobe pattern*) on a decibel (dB) scale, showing the performance of the antenna relative to an isotropic antenna being fed with the same power of radio signal. The units are therefore in dBi. The gain of a

*It is usual to quote the gain of an antenna in the GHz band relative to an *isotropic* antenna, but the gain of antennas intended for use in the lower frequency (MHz) bands is sometimes quoted relative to a *dipole* antenna.

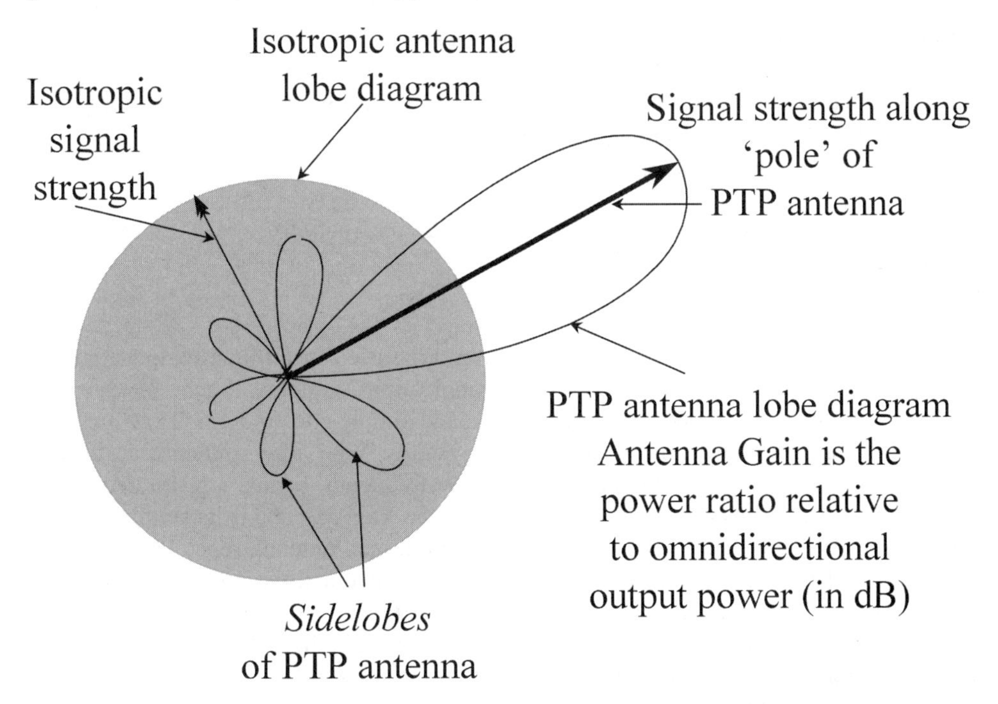

Figure 3.4 Antenna lobe patterns of *isotropic* and *point-to-point* antennas

particular antenna is the maximum value of the antenna pattern, the power transmitter along the antenna's pole. The dBi value is calculated using the following simple formula:

$$\text{Antenna gain in dBi} = 10\log_{10}\frac{(\text{radiated signal strength in Watt/m}^2 \text{ at a given distance along the antenna pole})}{\text{radiated signal strength in W/m}^2 \text{ of an } \textit{isotropic} \text{ antenna at the same distance}}$$

3.6 High Gain Antennas Used in PTP Applications

The most common types of high gain antennas used in conjunction with point-to-point wireless systems are parabolic reflector antennas, though at some lower frequencies, *Yagi* (or *array*) antennas are also used.

Parabolic antennas such as illustrated in Figure 3.5 are the most commonly used type of antennas in millimetre wave and microwave point-to-point radio (e.g. in the bands upwards of about 7 GHz for terrestrial applications, and also at 4 GHz and 6 GHz in association with satellite radio applications. Such an antenna works like a concave mirror. Whereas a torch is designed to have a light source (a bulb) at the *focus* point of a concave mirror (or reflector) and to shine the entire light along a single direction, so a parabolic microwave antenna has a microwave radio source at its focus and aims to reflect all the radiation along its single *pole* direction.

Figure 3.6 shows the typical basic construction of a point-to-point radio terminal. The diagram illustrates the radio transmitter and receiver and the parabolic antenna, interconnected with waveguide. The radio signals are generator in the radio unit and fed by waveguide to the focal point of the parabolic reflector antenna, where they launch into free space. The parabolic-shaped antenna acts only to reflect the radio signals along a path parallel with the pole of the antenna.

The gain of an antenna depends upon three main factors:

- The physical form of the antenna (its shape and construction).
- The size of the antenna.
- The frequency of the radio *carrier signal.*

Parabolic-type antennas are the most common type used in millimetre and microwave radio systems employed in public telecommunications systems. Such antennas offer relatively high gain from a relatively small size and simple construction. They are not the only type of high gain antennas used in radio systems. Yagi array antennas are the most common type of high gain antennas used in low frequency bands, e.g. for reception of television (terrestrial broadcast). The advantage of the Yagi antenna is its simple and cheap construction. We will discuss a number of other high gain antenna types later.

Generally, the larger the physical dimensions of an antenna, the greater its gain. In the case of the parabolic reflector antenna illustrated in Figures 3.5 and 3.6, the area of the reflector which is perpendicular to the *pole* of the antenna is the critical factor. Since most parabolic antennas are circular in form, their area is proportional to the square of their diameter. Thus, an antenna of 60 cm diameter has a collecting area (the reflector surface) four times greater than that of a 30 cm diameter antenna. The strength of the resulting signal collected in the waveguide of the receiving antenna (see Figure 3.6) will thus be four times stronger (in decibels, 6 dB stronger). In other words, the doubling in diameter of the antenna affords it an additional gain of around 6 dB. Doubling the radio frequency of operation will also add 6 dB of gain.

Figure 3.5 Parabolic antenna used in *point-to-point* wireless (Reproduced by permission of Netro Corporation)

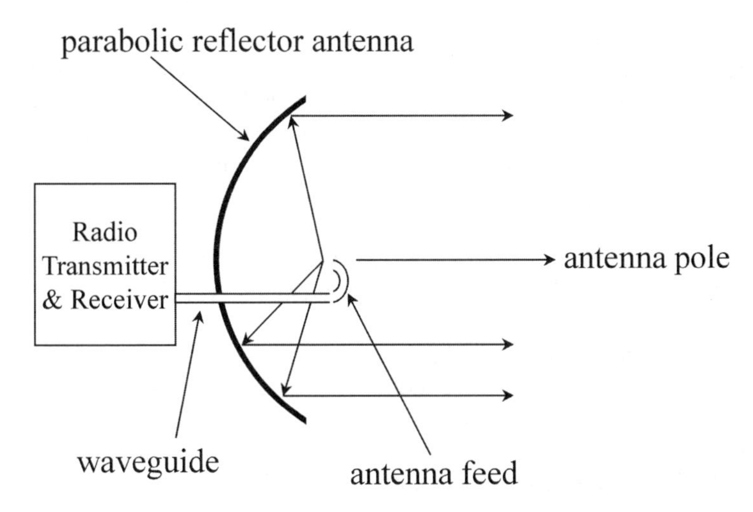

Figure 3.6 Basic construction of *point-to-point* radio terminal

The frequency of the radio carrier signal is also critical to the effectiveness of a given size and form of antenna. The greater the *wavelength* of a given radio carrier signal, the greater the size of the antenna needed to receive a given signal strength. Thus, the lower frequencies (the larger wavelengths) require larger antennas. In practical terms, we note that broadcast radio and television antennas in the kHz and MHz ranges (up to about 100 MHz) require huge mast antennas to generate a strong enough signal for broadcast. Meanwhile, mobile telephone system antennas for 900, 1800 or 1900 MHz are much smaller (typical 1–2 metres in length at the base station, and maybe 10 cm length in the handset).

A typical 30 cm parabolic antenna designed for use in the 26 GHz band has an antenna gain around 35 dBi, while a similar sized antenna designed for use in the 38 GHz will have a gain around 39 dBi. Doubling the diameter of the antennas to 60 cm will increase their gains respectively to 41 dBi and 45 dBi, while 120 cm antennas would have gains around 47 dBi and 51 dBi.

The gain of a parabolic antenna can be estimated from the following formulae:

$$\text{Antenna gain} = 66.7 \times D^2 f^2 \qquad \text{[absolute value]}$$
$$= 18.2 + 20\log_{10}(D) + 20\log_{10}(f) \quad \text{dBi}$$

where D = the diameter of antenna (m), and f = the frequency of operation (GHz).

3.7 Point-to-Point Antenna Diagram and Gain

Figure 3.7 illustrates a typical antenna diagram of a real point-to-point antenna. Note how in Figure 3.7 the antenna has a very high gain along its pole and a very narrow *aperture*. This is the sharp, narrow point at the 0° antenna *pole*. The aperture is the width of the transmitted radio 'beam' in degrees. The aperture corresponds to an imaginary hole in an imaginary sphere (see Figure 3.8) which surrounds the antenna. Through the aperture we imagine the entire radiated power of the antenna to be transmitted, and this imaginary 'hole' we normally define in terms of the vertical and horizontal angles of opening (correctly called the *beamwidth* — as shown in Figure 3.8). In reality, of course, not all the power is transmitted in a single direction, and we need a convention for defining the area of the aperture. The convention is to define the aperture to include all angles of radiation of the antenna for which the signal strength is greater than 3 dB below the maximum value (i.e. no more than 3 dB weaker than the signal strength along the pole of the antenna). The edges of the aperture are thus defined by the '3 dB points' of the antenna diagram. In Figure 3.7 these points are about 1° left and right of the antenna pole. (The total beamwidth is thus 2°.)

The *Radiation Pattern Envelope* (*RPE*) of an antenna is laid down by radio technical standardisation bodies. This envelope defines the maximum allowed radiation of the antenna at angles offset from the pole direction (the main direction of transmission). The RPE is designed to ease the task associating with the radio frequency planning of wireless networks, by limiting the strength of disturbing radio signals which might *interfere* with the operation of other neighbouring systems.

The typical gain of a point-to-point antenna lies in the range 28 dBi to 50 dBi, and the typical beamwidth (and aperture) is very small (0.5° to 7° in both horizontal and vertical directions for parabolic antennas, maybe up to 15° for slightly lower gain antennas which are intended to be easier to *align* during installation).

As we have seen, the gain of an antenna is a measure of the ability of the antenna when transmitting to 'focus' the output signal of the transmitter along a single direction, the pole

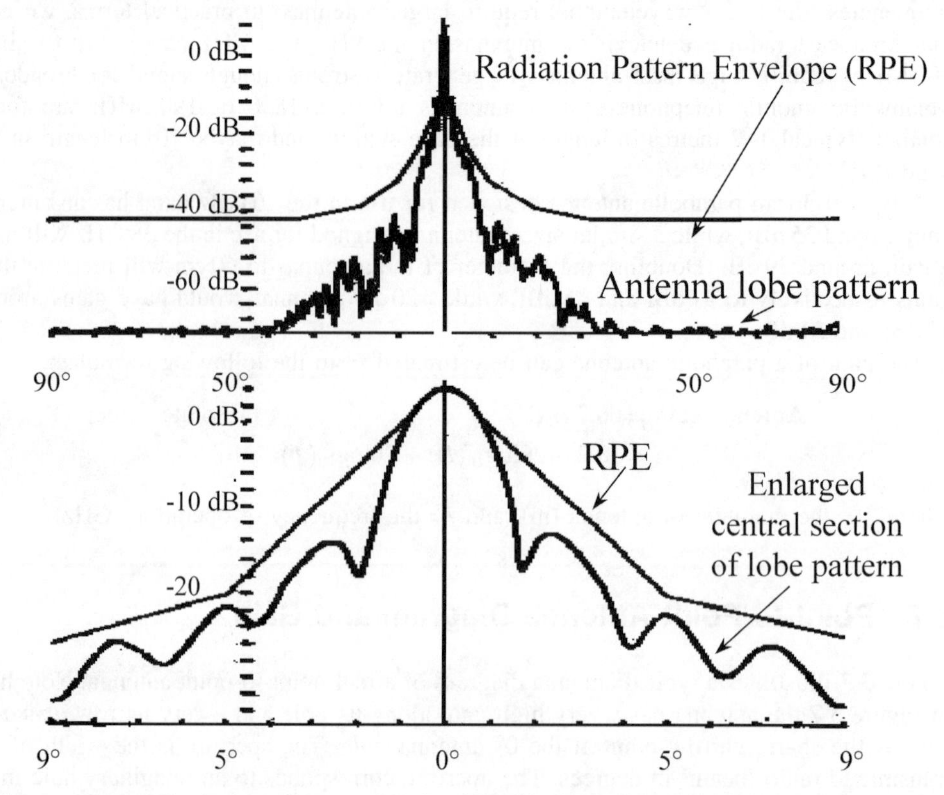

Figure 3.7 Antenna diagram of high gain parabolic point-to-point radio system

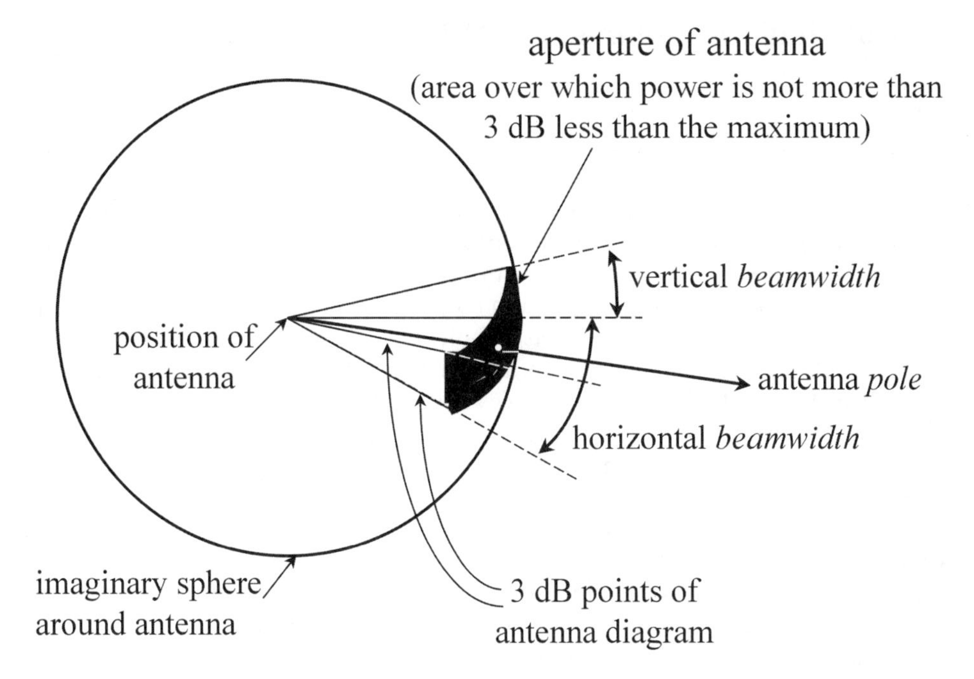

Figure 3.8 The aperture of an antenna

or *axis* of the antenna. Maybe surprisingly for the newcomer to radio, an antenna has the same gain when receiving as it does when it is transmitting. The gain when receiving is the ability of the antenna to pick out a distant signal from a single arriving direction, to exclude signals emanating from other directions, and to maximise the strength of the desired received signal in order that the radio receiver can perform an accurate decoding to recover the original user data or analogue (audiovisual) signal.

The fact that the antenna should work with an equally high gain whether receiving or transmitting may not at first appear obvious. But at least maybe the following rationalisation may help. Let us return to the antenna shown in Figure 3.6. When the same antenna is used for reception, it is again the area of the antenna which is presented to the incoming radio signal which is important. The larger the area of the antenna, the greater the amount of radio signal energy we will be able to reflect into our receiving head at the focus of the reflector.

For the best performance of a single link one should use highly directional *high gain* antennas at both ends of the link in order to maximise the signal strength along the direction of transmission, and to maximise the effectiveness of transmission.

Highly directional high gain antennas are also often used at the remote stations in point-to-multipoint systems. At the remote stations the only desired reception is of the signal from the *central station* or *base station*. If the remote stations are *fixed* (stationary, i.e. part of a *fixed wireless access* system), then the base station signal will always be received from a single fixed direction. In addition, the desired transmission from the remote station to the base station will also be in a single direction. So the use of a directional antenna at the remote station is appropriate.

3.8 Base Station Antennas in PMP Systems

At the base station of a PMP system, lower gain antennas have to be used. The lower gain of the antenna reflects the need for the antenna to transmit to and receive from a large number of remote stations spread throughout the *sector* of the antenna.

The simplest type of base station antenna is an *omnidirectional* (or *isotropic*) antenna. We already discussed in Figure 3.4 the antenna diagram of such an antenna. In practical terms the simplest realisation of an omnidirectional antenna is a *dipole* antenna. This usually comprises a simple wire, a metal post or a mast. The length of the dipole is matched to the frequency of the radio carrier signal (a typical length, e.g. used in mobile telephone networks at 900 MHz and 1800 MHz, is of the order of one half of the wavelength — the longer the wavelength, the lower the frequency and the larger the antenna needed). Figure 3.9 illustrates an actual base station dipole antenna (as used in a cellular mobile telephone network), as well as showing the antenna diagram in both horizontal and vertical elevations.

We should note from the vertical elevation antenna diagram of the dipole antenna, shown in Figure 3.9, that this type of antenna is strictly not an omnidirectional antenna, as there is only weak radiation in the directions vertically upwards and vertically downwards. Cellular telephone users may have wondered how it can be that the signal strength is sometimes poor when you are standing directly underneath the base station! Nonetheless, slang convention has it that a dipole antenna is an omnidirectional antenna — radio network designers are generally most interested to transmit across the horizontal surface of the earth rather than into space!

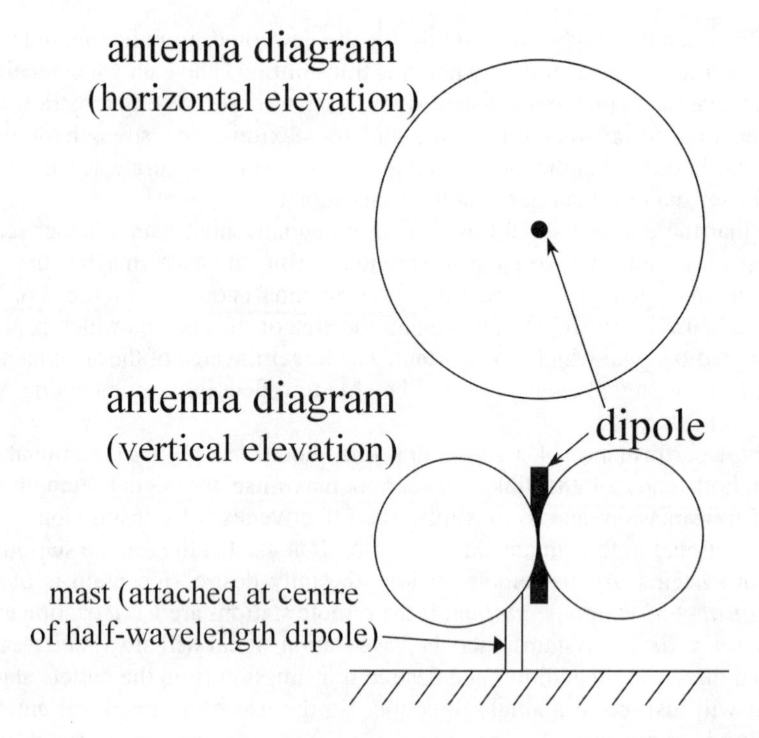

Figure 3.9 An omnidirectional base station dipole antenna and its antenna diagram

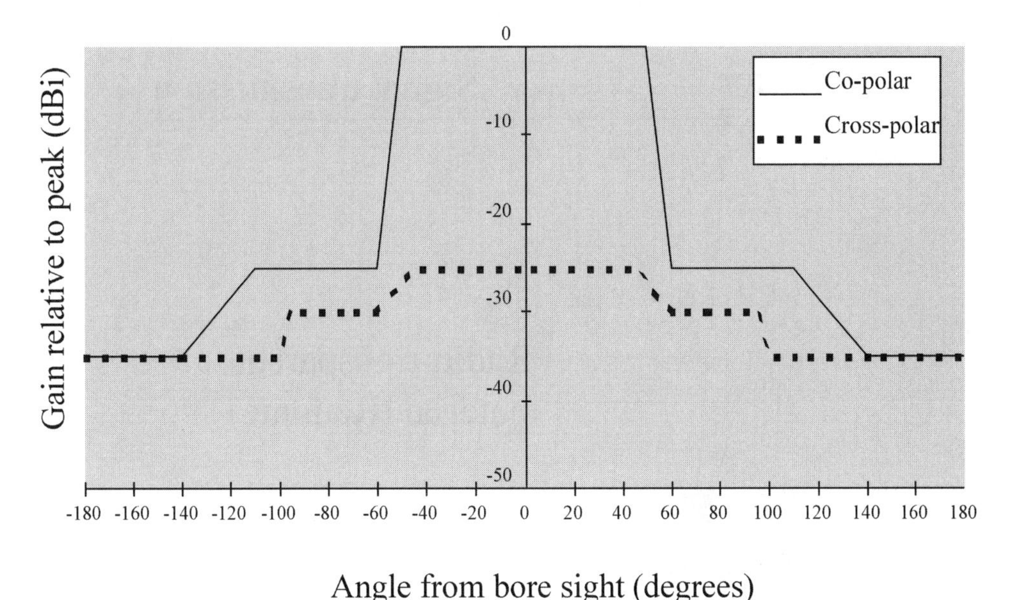

Figure 3.10 Typical antenna diagram of a sector antenna of a PMP system

Figure 3.10 illustrates a typical *Radiation Pattern Envelope* (*RPE*) of an alternative type of base station antenna — a *sector antenna.* The designer of a real sector antenna will try to match this envelope, but the actual antenna diagram will be a wavy-line approximation to the envelope (as we saw in Figure 3.7), with gain always slightly lower than the envelope itself.

The sector antenna of which Figure 3.10 is the antenna diagram is not intended to have a single pole direction of transmission, but instead to radiate at a relatively constant signal strength across a sector of 90° (45° either side of the pole, but now more appropriately called the *bore sight*).

A sector antenna may be constructed in a number of ways: either using a reflector with a beam splaying effect, using an array of mini-antennas (somewhat like the *Yagi* antenna), but purposely designed to have a wide main lobe) or by means of a horn antenna, as we shall describe in the final section of this chapter.

3.9 Other Types of Antennas

We have discussed so far a few of the common typpes of antennas used in fixed wireless systems, including parabolic antennas, array antennas and dipole antennas. In our final section on the subject of antennas, let us also briefly review some of the other main types of antennas used in fixed wireless systems and their relative strengths.

Horn Antenna

As its name suggests, a horn antenna is physically shaped like a horn. In some types of horn antennas (Figure 3.11a), the shape of the horn is very precisely manufactured to act as a

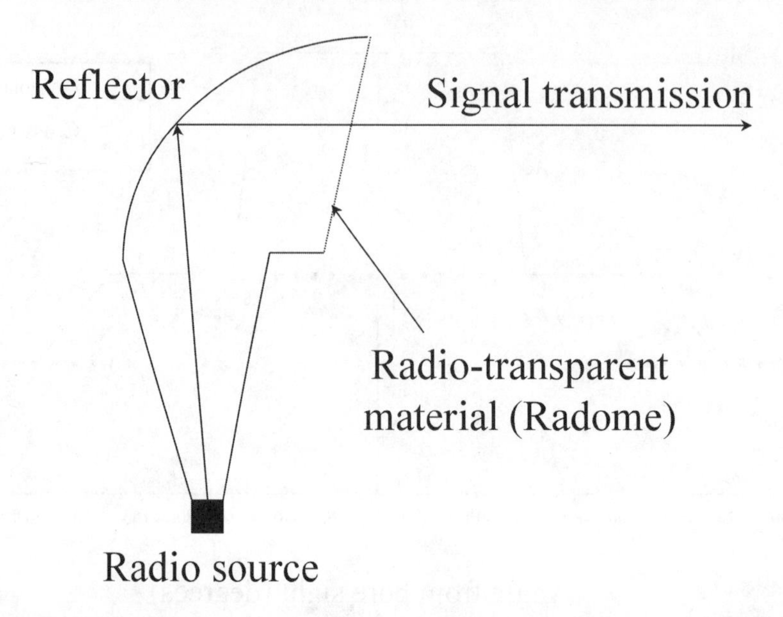

Figure 3.11 (a) High gain horn antenna used in long haul point-to-point radio links; (b) Photograph of three horn antennas of long haul point-to-point microwave links

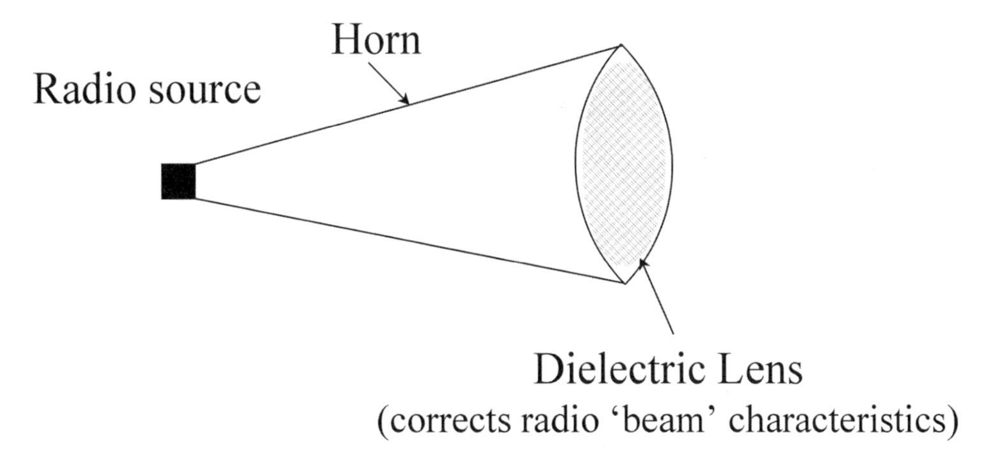

Figure 3.12 Lens-corrected horn antenna

reflector antenna. These type of antennas are sometimes used as an alternative to parabolic antennas in relatively low frequency (e.g. under 7 GHz) long haul point-to-point systems and you may have seen such antennas on microwave radio towers (Figure 3.11b). The principle of such antennas is similar to that of a parabolic antenna.

Horn antennas such as those shown in Figure 3.11 are usually formed to protect against entry of rainwater or other precipitation. This is most simply achieved by making the top part of the antenna extend beyond the lower 'lip' of the aperture. But sometimes, in addition, a radio-transparent material is used to seal the antenna from the weather. Such a cover over the antenna aperture is called a *radome*. But while a radome obviously protects against the intrusion of water into horn, parabolic or other types of antennas, it unforunately also provides a surface on which ice can accumulate. The accumulation of ice (e.g. during the winter) can be just as great a (if not a greater) problem than the intrusion of water to the performance of the radio system. Ice build-up is sometimes combatted by means of heating the radome.

Horn antennas of the type illustrated in Figure 3.11 are not widely used in wireless access networks, but another type of horn antenna is sometimes used. This is the lens corrected horn. In a lens corrected horn antenna, the horn itself may be of a relatively simple and easily manufactured nature (e.g. the antenna of Figure 3.12 is a simple wedge-shaped horn intended to act as a *sector antenna* of a given wide horizontal and relatively narrow vertical aperture angle. The lens (usually placed at the exit from the 'horn' is used to correct the radio beam output from the antenna. The lens is made from a *dielectric* material which acts to focus or disperse radio waves in the same way that a glass lens disperses or focuses light waves.

Planar Antenna

Because of public sensitivity and local authority objections to the mounting of large and unsightly antennas on the outside façades of buildings, manufacturers have increasingly sought to produce smaller and less obtrusive antenna designs. As a result an increasing

Figure 3.13 Construction of a planar antenna (Reproduced by permission of Netro Corporation)

number of *planar* or 'flat plate' antennas have appeared. The idea is that, to neighbours and passers-by, the presence the antenna should go entirely unnoticed.

As Figure 3.13 illustrates, *planar antennas* may have a pleasing look to the eye, but the technical performance of such antennas (measured in terms of their antenna gain) is not always very satisfying. A different antenna type of similar physical dimensions will often have a better gain.

As shown in Figure 3.13, the internal construction of a planar antenna comprises an *array* of small dipole-like antennas. Each individual mini-antenna within this array has an omnidirectional transmitting and receiving characteristic, as we discussed previously in the section on dipole antennas. However, by adding together the received signals of all of the mini-antennas in the array, we strengthen the signals received from a particular direction (i.e. those transmitted to or received from the pole of the antenna (usually a point in the middle of the plate, extending along a line perpendicular to the plate). Meanwhile, the signals received by the antennas from different directions tend to cancel one another out (i.e. the signals from these directions contributed by two or more of the mini-antennas *interfere* with one another to weaken one another). The physical spacing and lay-out of the *array* of mini-antennas is critical to ensure this strengthening of signals along the *pole* of the antenna and simultaneous weakening of signals from other directions.

4

Radio Modulation

'Modulation' is the means by which a user signal is prepared for carriage by radio. The 'user signal' is 'modulated' onto the radio 'carrier' signal. The modulation may either be some form of superimposition or a means of encoding. Audio and broadcast radio systems use *Amplitude Modulation (AM)* and *Frequency Modulation (FM)* as appear on the dial of a standard radio receiver. Modern digital radio systems, on the other hand, use encoding techniques as modulation schemes. These modulation schemes include *On-Off keying (OOK)*, *Frequency Shift Keying (FSK)* and *Quadrature Amplitude Modulation (QAM)*. Modulation at the transmitting end of a radio link and demodulation at the receiving end is carried out by the radio 'modem' (*mo*dulator/*dem*odulator). We discuss these terms in this chapter.

4.1 Radio Modulation and the Radio 'Modem'

Modulation or encoding is the act of conversion of a communications signal for transport over a radio medium. *Modulation* of the radio carrier signal takes place within the radio transmitting device prior to signal transmission and involves 'imprinting' the user information (analogue, hi-fidelity or data signal) onto the radio *carrier* frequency.

At the receiving radio device, the carrier signal is *demodulated* to recover the original user signal, information or other communication. The device which carries out the modulation of the transmitted (Tx) signal and the demodulation of the received (Rx) signal is called a radio *modem.*

Figure 4.1 illustrates the typical component layout of a high frequency wireless system, such as are used in point-to-point and point-to-multipoint public wireless networks. Each radio terminal comprises three basic components, the high frequency radio and antenna (sometimes together called the *Outdoor Unit (ODU)*) and the radio modem (sometimes called the *Indoor Unit (IDU)*).

The modem receives the end-user signal (the analogue signal, hi-fidelity signal, video or data signal) from the end user device and *modulates* this signal onto a radio *carrier* channel. Initially, the conversion may be from an analogue or digital signal into a *baseband* radio channel of a bandwidth equal to that of the available radio channel bandwidth, but with a centre frequency which does not match with the centre frequency of the intended radio transmission channel. Instead, the initial modulation (if to a baseband channel) is to a frequency range from 0 Hz to a frequency corresponding to the channel bandwidth. As it were, the user signal is modulated onto the first 'base' channel of the given bandwidth, hence the term *baseband.*

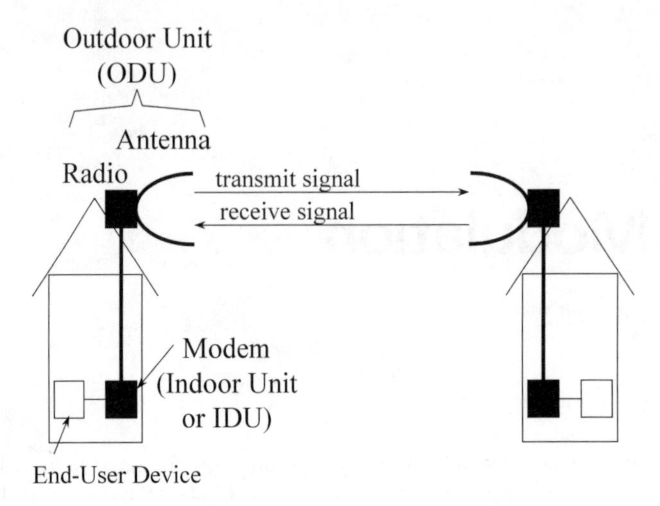

Figure 4.1 Basic components of a simple radio system

The baseband signal is usually conveyed to the outdoor unit by means of a coaxial cable connection (or a *waveguide*) to the *Outdoor Unit* (*ODU*), where it is simply frequency shifted (*upconverted*) into the appropriate radio channel range for transmission (Figure 4.2).

The *Radio-Frequency* (*RF*) carrier signal used for the upconversion is produced by a high quality oscillator. The RF carrier is mixed with the modulated baseband signal to produce the *Radio-Frequency* (*RF*) *signal*. The RF signal is filtered again to prevent possible interference with other radio waves of adjacent frequency. Finally, the signal is amplified by a high power amplifier and sent to the antenna, where it is converted into radio waves. A short connection to the antenna (using *waveguide*) ensures that not too much of the radio signal strength is lost before transmission into *free space* by the antenna.

Sometimes the user signal is modulated onto an *Intermediate Frequency* (*IF*) by the modem, rather than left as a *baseband* signal, and carried as an IF signal between the indoor and outdoor units. The modulated IF signal also has a bandwidth equal to that of the destination radio channel, but as the name suggests, the centre frequency of the IF signal is 'intermediate' between *baseband* and the *RF* centre frequency. There may be one of two reasons why the radio designer chose to use IF rather than a baseband signal. First, the type of modulation may be easier to perform using an intermediate frequency. Secondly, there may be benefits in running IF rather than baseband over the indoor-to-outdoor connection cable. These benefits may include a lower loss characteristic and reduced susceptibility to

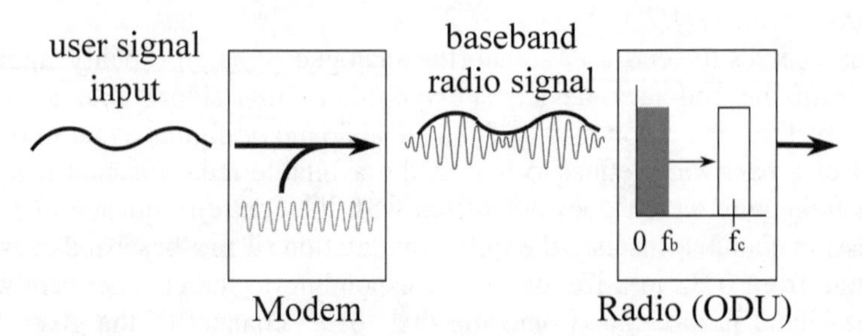

Figure 4.2 Modulation of baseband signal followed by frequency shifting by the outdoor radio unit

interference. (Baseband signals, for example, can be prone to interference from nearby power cables which emit relatively high electromagnetic fields at around 50 Hz).

At the receiving end, the *demodulation* process works in reverse. The antenna collects the maximum available radio signal energy it can. This is *downconverted* to the modulated baseband signal by the outdoor radio unit and passed to the receiving *modem* or *indoor unit* for demodulation to recover the original user signal.

The radiowaves are received by the antenna and are reconverted into electrical signals. A filter removes extraneous and interfering signals before demodulation. As an alternative to a filter, a *tuned circuit* could have been used with the antenna. A tuned circuit allows the antenna to select which radio wave frequencies will be transmitted or received.

After demodulation, the information signal is processed so as to recreate the original audio signal as closely as possible. The signal is next amplified using an amplifier with *Automatic Gain Control* (*AGC*) to ensure that the output signal volume is constant, even if the received radio wave signal has been subject to intermittent *fading*. Finally, the an analogue output might be adjusted to remove various signal distortions called *group delay* and *frequency distortion*, by devices called equalisers. Alternatively, a digital signal might be checked and *corrected* for errors caused by such radio fading, if the radio system (as many digital ones do) includes *Forward Error Correction* (*FEC*). We will discuss FEC in more detail in Chapter 6.

4.2 The Basic Methods of Modulation

The modulation of the radio carrier signal may follow either an analogue or a digital regime. Analogue modulation has three basic forms: *Amplitude Modulation* (*AM*), *Frequency Modulation* (*FM*) and *Phase Modulation* (*PM*, also known as *quadrature modulation*).

Digital modulation can be either be by 'on/off' carrier signal modulation (i.e. by switching the carrier on and off), or by other methods such as *Frequency Shift Keying* (*FSK*), *Phase Shift Keying* (*PSK*) or *Quadrature Amplitude Modulation* (*QAM*). These are really just variants of the analogue modulation methods. We discuss the various modulation methods in turn, commencing with the analogue methods.

Amplitude Modulation

Modems employing *amplitude modulation* alter the amplitude of the carrier signal, so that the trace outline of the carrier signal amplitude (Figure 4.3a) matches the original analogue signal. For amplitude modulation to work correctly, the frequency of the carrier signal must be much higher than the highest frequency of the information signal. (This ensures that the 'peaks and troughs' of the carrier signal of Figure 4.3a are more frequent than the 'peaks and troughs' of the information signal (Figure 4.3b), so enabling the carrier signal to 'track' and record even the fastest amplitude changes in the information signal.)

Amplitude modulation is carried out simply by using the information (end-user) signal to control the power of a carrier signal amplifier. Demodulation can be carried out by filtering the carrier signal with a baseband filter.

In the digital version of amplitude modulation, the amplitude of the carrier frequency is varied between a given amplitude and zero. These two amplitude states correspond

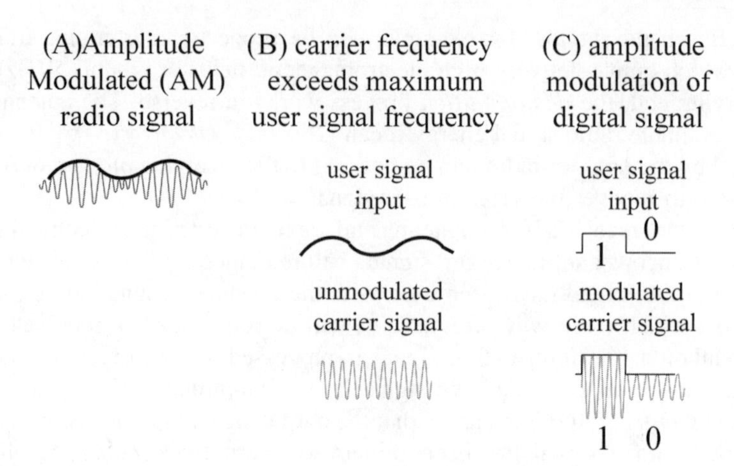

Figure 4.3 Amplitude modulation

effectively to 'on' and 'off', respectively to values '1' and '0' of the modulating digital bit stream. Alternatively, two different, non-zero values of amplitude may be used to represent '1' and '0'. A digital amplitude modulation scheme is illustrated in Figure 4.3c.

Frequency Modulation

In *Frequency Modulation* (Figure 4.4), it is the frequency of the carrier signal that is altered to carry the signal content of the modulating bit stream. The amplitude and phase of the carrier signal are left largely unaffected by the modulation process.

Frequency modulation is achieved simply by mixing the information and carrier signals. The *interference* of the two signals during mixing leads to *inter-modulation*, creating *sidebands* (products of *inter-modulation*) near to the *carrier* frequency. A given frequency f in the original information signal creates *intermodulation products* at frequencies $f_c - f$ and $f_c + f$, where f_c is the carrier frequency. An original information signal occupying the band from frequency f_1 to f_2 is thus mapped into two sidebands $f_c - f_2$ to $f_c - f_1$ (lower sideband) and $f_c + f_1$ to $f_c + f_2$ (upper sideband). The *sidebands* thus mirror the signal spectrum of the original information signal. Figure 4.5 shows an example of frequency modulation of a baseband voice telephone channel (in the baseband 300 Hz to 3400 Hz) onto a carrier of 8000 Hz. This is the basis of *Frequency Division Multiplexing* (*FDM*) which we will discuss later in the chapter.

It is normal to filter the signal following modulation to remove both the baseband and one of the sideband signals (*single sideband mode*), and in some cases, the carrier signal is also removed prior to transmission (*suppressed carrier mode*). This leaves just a single sideband (or a single sideband and the carrier signal). This signal is amplified for radio transmission. The benefit of the suppressed carrier mode is that the transmitter power is not 'wasted' simply sending the carrier signal. Also, sending just one sideband is sufficient, since it contains all the necessary information to recreate the baseband signal.

At the receiver, demodulation of an FM signal is achieved by once again mixing the carrier frequency with the received sideband signal. This again produces two 'sideband-like' *intermodulation* products. One of these products is a reconstruction of the original

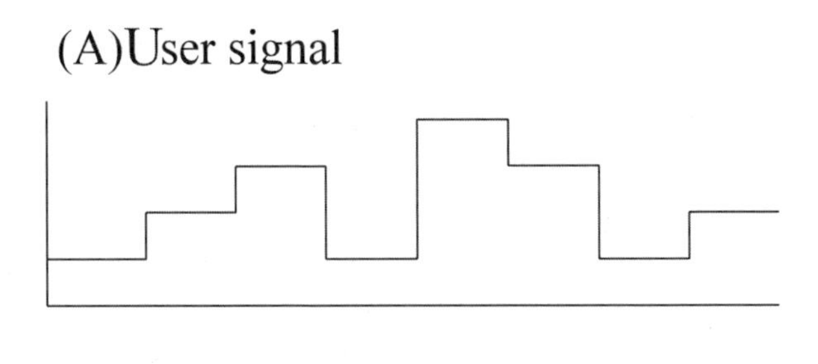

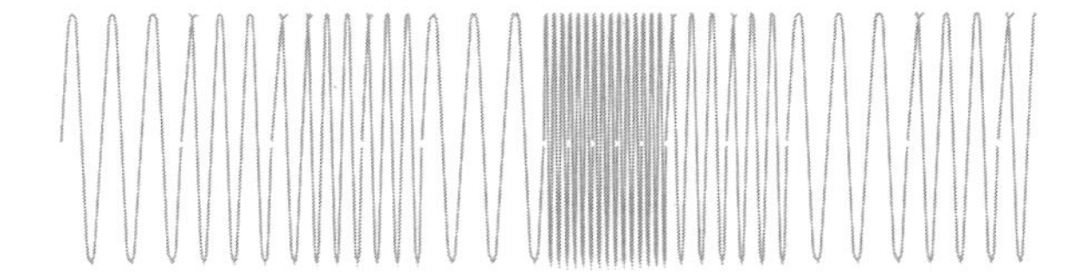

Figure 4.4 Frequency modulation

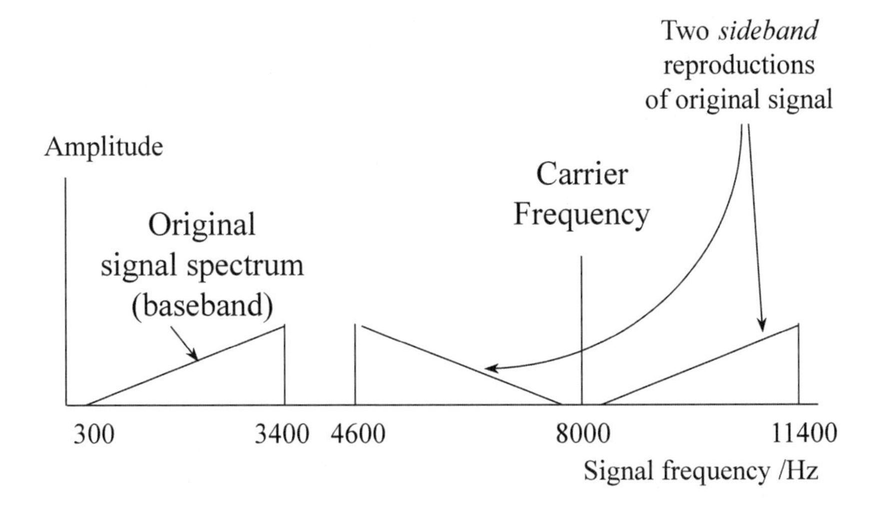

Figure 4.5 Spectrum content of a Frequency Modulated (FM) signal (narrowband frequency division multiplexing)

baseband signal. The other is a high frequency intermodulation product which is removed using a filter.

One of the disadvantages of the *suppressed carrier mode* of operation is that an accurate oscillator and signal generator is required at the receiver in order to create the carrier signal necessary for demodulation. Thus, particularly in one-way broadcast radio systems, where the aim is often to keep the cost of receivers as low as possible, the suppressed carrier mode is unusual. Conversely, where an oscillator and carrier signal generator is available at both ends of the radio link (e.g. in a duplex public network link — where the carrier signal is needed for transmission anyway), the suppressed carrier mode is feasible.

Modems using frequency modulation for encoding of digital data (i.e. digital radio modems) are either *OOK* (*on-off keying*) or *FSK* (*frequency shift keying*) modems.

The simplest form of digital *frequency modulation* is *OOK* (*on-off keying*). An OOK radio simply modulates the radio carrier signal by switching it 'on' and 'off' according to the value '1' or '0' of the data 'bit' to be carried (Figure 4.6a). Morse Code is a type of OOK.

Frequency Shift Keying (FSK) is a slight advance on OOK, in which radio carrier signals of two or more different frequencies (both located within the 'bandwidth' of the channel) are used to represent different '1' or '0' bit values of a digital bitstream. The simplest form of FSK is 2FSK in which two different frequencies are used to represent the bit values '0' and '1' (Figure 4.6b). The advantage of FSK over OOK is that a signal is always being sent, so the receiver can always tell that the transmission is active.

More complicated versions of FSK include 4FSK (Figure 4.6c). 4FSK was commonly used in previous generation digital radio equipment before the emergence of *QAM* (*quadrature amplitude modulation*, which we discuss later in this chapter). In 4FSK, four different frequencies are used to represent two digital *bits* at a time, one frequency each to represent the consecutive bit combinations 00, 01, 10 or 11. By coding two bits at a time, the *baud rate* of the transmitted radio signal may be reduced. The *baud rate* or *symbol rate* is the rate at which the radio receiver must be able to distinguish the *symbols* of the incoming signal. The higher the *baud rate* of the signal, the greater must be the *agility* of the radio receiver and transmitter. The significance of the baud rate is that the maximum

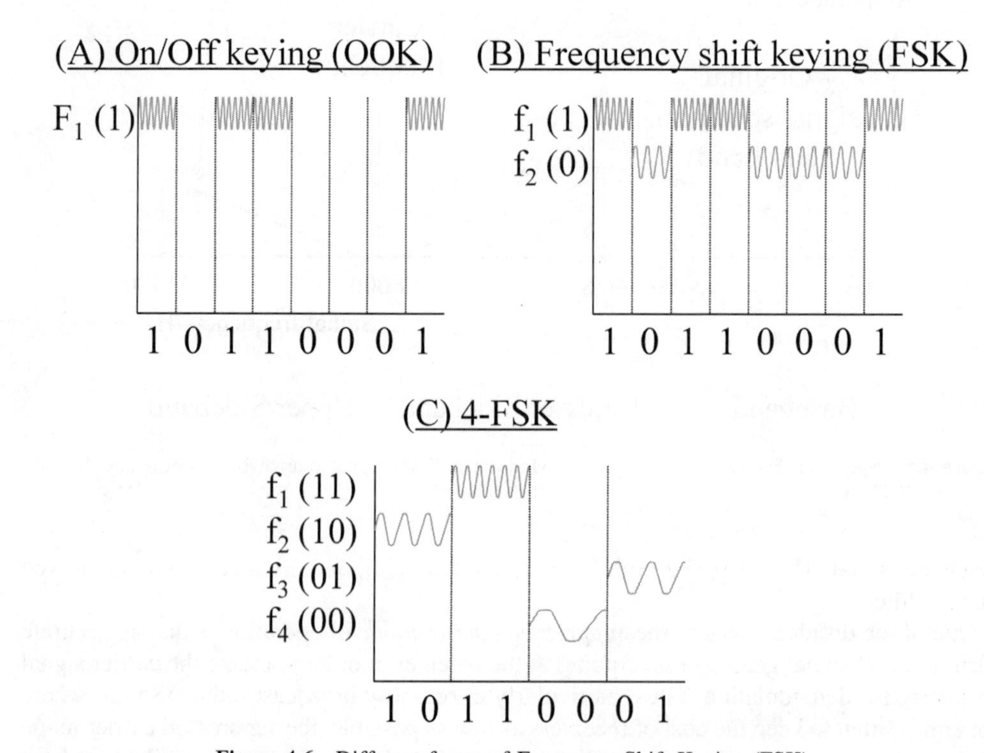

Figure 4.6 Different forms of Frequency Shift Keying (FSK)

baud rate which may be carried is roughly equal to the *bandwidth* of the radio channel (*carrier*) being used.

Phase Modulation

In *phase modulation*, the carrier signal is advanced or retarded in its *phase* cycle by the modulating signal (end-user provided information). Phase modulation has appeared with the advent of digital radio systems, and Figure 4.7 illustrates the coding of a radio carrier signal to carrier digital information.

Under digital phase modulation, the radio carrier signal (at the beginning of each new bit) will either be allowed to retain its phase or will be changed in phase. Thus, in the example shown the initial signal phase represents value '1'. The change of phase by 180° represents next bit '0'. In the third bit period the phase does not change, so the value transmitted is '1'. Phase modulation (often called *phase shift keying*, or *PSK*) is conducted by comparing the signal phase in one time period to that in the previous period, thus it is not the absolute value of the signal phase that is important, rather the phase change that occurs at the beginning of in each time period.

The advantage of phase modulation is that radio systems using it are relatively less prone to the interference of *noise* and other disturbing signals. As a result, it is possible to get away with a weaker signal arriving at the receiver, and yet still recover the original digital signal without errors. Radio systems using phase modulation are thus less prone to link *outage* (unavailability). For a similar length of link, such radios exhibit a higher *availability*. Or seen another way . . . the same *availability* of link can be achieved over greater link *range*.

Quadrature Amplitude Modulation

The most modern digital radio systems employ *Quadrature Amplitude Modulation* (*QAM*). In QAM, the frequency of the radio carrier signal is left unchanged during

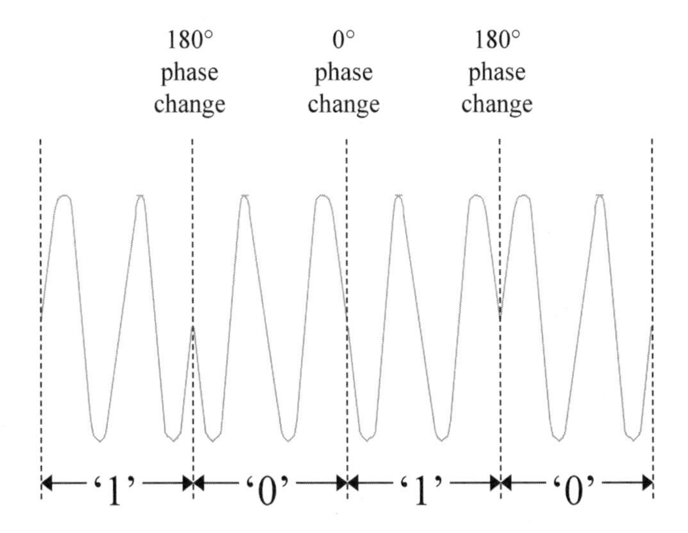

Figure 4.7 Phase modulation of a digital signal

modulation, but the various bit patterns within the digital stream are instead coding onto the carrier by means of signal *amplitude* and *phase* (or *quadrature*) changes. Very sensitive QAM radios may be built. This is the advantage of QAM. Just like FSK, there are also different variants of QAM, allowing *higher modulation* such as 4-QAM, 8-QAM, 16-QAM, 64-QAM, etc. The value 4, 16, 64 or whatever represents the number of different *quadrature/amplitude* combinations used in the *modulation* scheme. Thus, in 16-QAM modulation, 16 combinations are available, allowing 4 bits (equivalent to binary value 16) to be carried for each *symbol*. 64-QAM carries 5 bits per *symbol*, and so on.

The significant advantage of *higher modulation* schemes is the increased number of bits per *symbol* which may be transmitted. Thus, while 4-QAM is able to carry about 1 bit per Hertz of radio bandwidth, 16-QAM achieves about 2 bits per Hertz and 64-QAM 4 bits per Hertz. However, there is a penalty for higher modulation. This is that a more sensitive radio receiver is required, so either the equipment is more expensive or the system range is restricted. We consider the subject of higher modulation (also called *multilevel transmission*) next.

4.3 High Bit Rate Modems and Higher Modulation or Multilevel Transmission

The transmission of high bit rates can be achieved in modem design in one of two ways. One is to modulate the carrier signal at a *high* bit rate (equal to the bitrate of the modulating signal. The other way is to use *higher modulation* (also called *multilevel transmission*).

Now the rate (or frequency) at which we modulate (i.e. change) the carrier signal is called the *baud rate*, and the disadvantage of modulating a high bit rate with an equally high baud rate is the high baud rate that it requires. The difficulty lies in designing a modem and radio capable of responding to the line signal changes fast enough (i.e. *agile* enough). However, this is not the only limitation. Because the baud rate is limited by the bandwidth of the radio channel available (the baud rate may not exceed the bandwidth), there is a maximum bit rate we can carry with this type of modulation.

Fortunately, *higher modulation* (*multilevel transmission*) offers an alternative in which the baud rate is lower than the bit rate of the modulating bit stream. The lower baud rate is achieved by encoding a number of consecutive bits from the modulating signal and representing them by a single signal state. The method is most easily explained using a diagram. Figure 4.8 illustrates a bit stream of 2 bits per second (2 bit/s being carried by a modem which uses four different line signal states. The modem is able to carry the bit stream at a *baud rate* of only 1 per second (1 Baud)). Specifically, Figure 4.8a uses four different frequencies (*4-FSK modulation*) for the four different combination values of 2 bits, while Figure 4.8b uses four different signal phases (*4-PSK* or *QPSK* [*Quarternary Phase Shift Keying*]).

The modem used in Figure 4.8a achieves a lower baud rate than the bit rate of the data transmitted by using each of the line signal frequencies f1, f2, f3 and f4 to represent two consecutive bits rather than just one. It means that the line signal always has to be slightly in delay over the actual signal (by at least 1 bit as shown), but the benefit is that the receiving modem will have twice as much time to detect and interpret the datastream represented by the received frequencies. *Multi-level transmission* is invariably used in

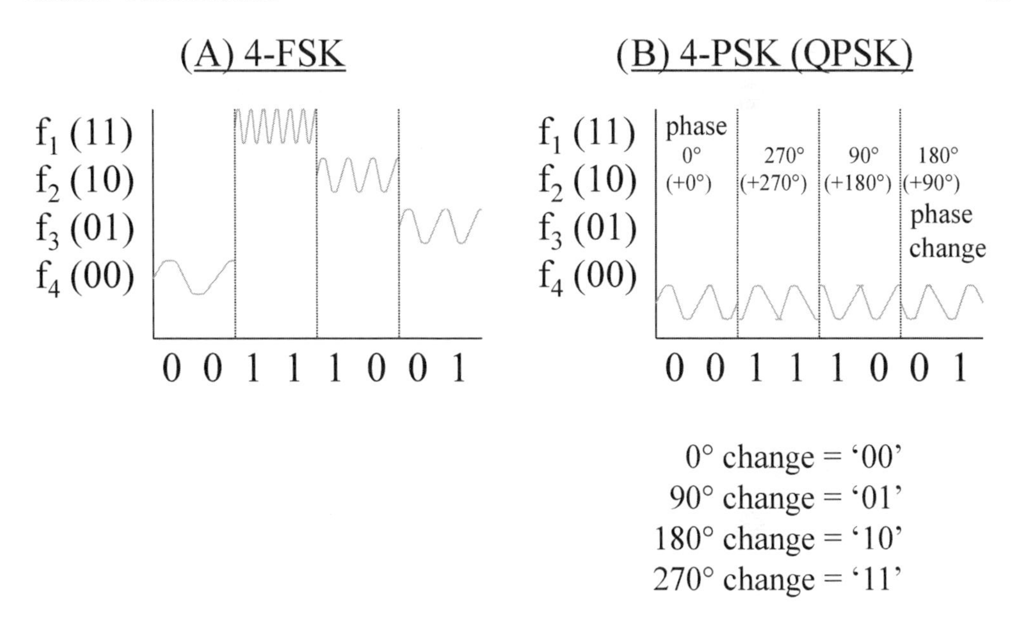

Figure 4.8 Higher modulation (multi-level transmission)

the design of very high bit rate modems. Figure 4.8b achieves the same effect with four different signal phase changes.

4.4 Modem 'Constellations'

At this point we introduce the concept of *modem constellation* diagrams, since these assist in the explanation of more complex *Phase-Shift-Keyed* (*PSK*) and *Quadrature Amplitude Modulation* (*QAM*) modems. Figure 4.9 illustrates a *modem constellation* diagram composed of four dots. Each dot on the diagram represents the relative phase and amplitude of one of the possible line signal states generated by the modem. The distance of the dot from the origin of the diagram axes represents the amplitude of the signal, and the angle subtended between the X-axis and a line from the point of origin represents the signal phase relative to the signal state in the preceding instant of time.

Figures 4.9b and 4.9c together illustrate what we mean by *signal phase*. Figure 4.9b shows a signal of 0° phase, in which the time period starts with the signal at zero amplitude and increases to maximum amplitude. Figure 4.9c, by contrast, shows a signal of 90° phase, which commences further on in the cycle (in fact, at the 90° phase angle of the cycle). The signal starts at maximum amplitude, but otherwise follows a similar pattern. Signal phases, for any phase angle between 0° and 360°, could similarly be drawn. Returning to the signals represented by the constellation of Figure 4.9a, we can now draw each of them, as shown in Figure 4.9d (assuming that each of them was preceded in the previous time instant by a signal of 0° phase). The phase angles in this case are 45°, 135°, 225° and 315°.

We are now ready to return to the more complicated but nowadays common digital modulation technique known as *Quadrature Amplitude Modulation* (*QAM*). QAM is a

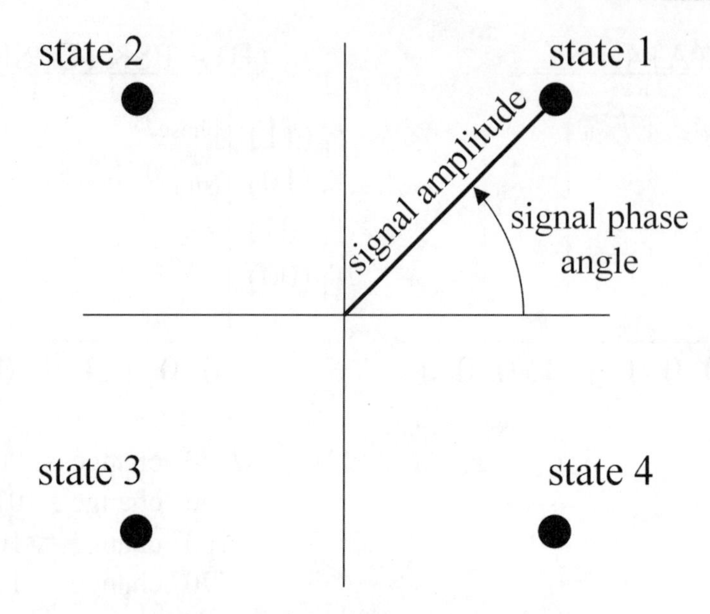

Figure 4.9 A *modem constellation* diagram

technique using a complex hybrid of phase (or *quadrature*) as well as *amplitude modulation*, hence the name.

Figure 4.10 shows a simple eight state form of QAM in which each line signal state represents a 3-bit signal (values nought to seven in binary can be represented with only three bits). The eight signal states are a combination of four different relative phases and two diferent amplitude levels. The table of Figure 4.10a relates the individual 3-bit patterns to the particular phases and amplitudes of the signals that represent them.

(A) bit combinations and signal attributes

Bit combination	Signal amplitude	Phase shift	Typical Signal
000	Low	0°	
001	High	0°	
010	Low	90°	
011	High	90°	
100	Low	180°	
101	High	180°	
110	Low	270°	
111	High	270°	

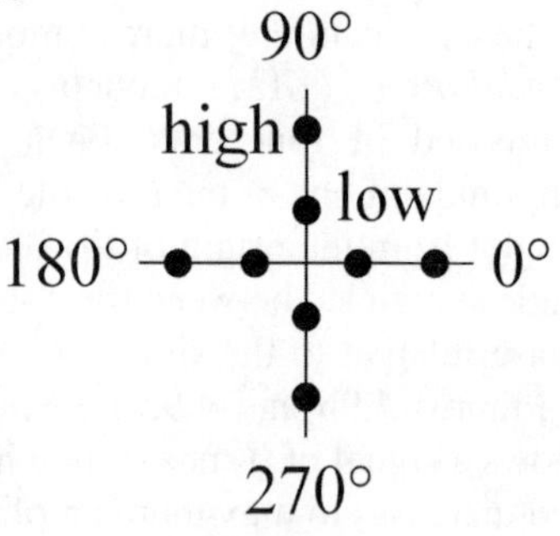

90°
high
low
180° 0°
270°

(B) Modem constellation diagram

Figure 4.10 *Constellation* diagram of a *Quadrature Amplitude Modulation* (*8-QAM*) modem

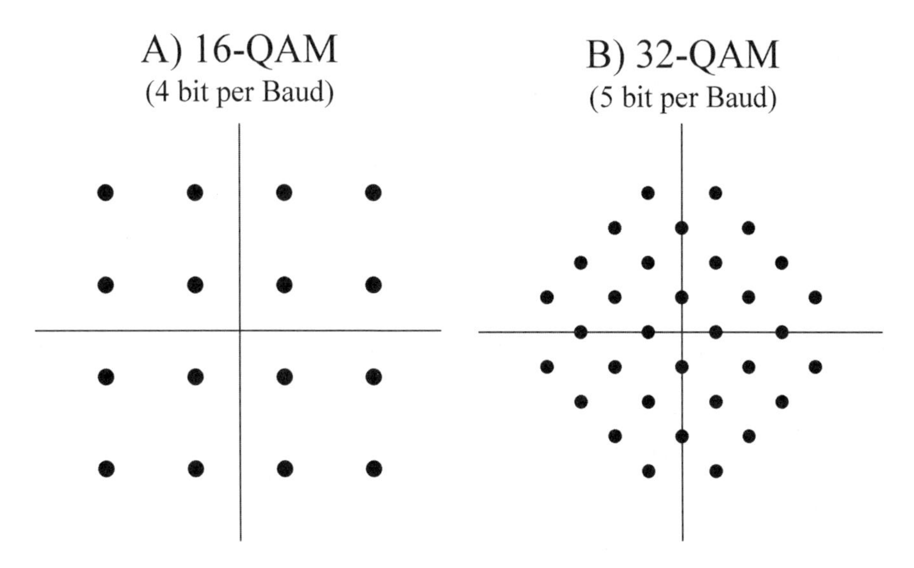

Figure 4.11 Constellation patterns of possible 16-QAM and 32-QAM modulation schemes

Figure 4.10a also illustrates typical line signal patterns that would be used to represent the various signals in the table. Each signal is shown assuming that the preceding signal had been one of 0° phase. In reality, as we saw in Figure 4.9, the same signal state is not used consistently to convey the same 3-bit pattern, since the phase difference with the previous time period is what counts, not the absolute signal phase.

Figure 4.10b shows the constellation of this particular modem.

To finish off the subject of modem constellations, Figure 4.11 presents, without discussion, the constellation patterns of a couple of very sophisticated modems using very high level modulation. As in Figure 4.10, the constellation pattern will allow the interested reader to work out the respective 16 and 32 line signal states. Even more dedicated students might like to conceive for themselves possible 64-QAM and 128-QAM constellation patterns. Such radios do already exist!

4.5 Frequency Division Multiplexing (FDM)

Frequency Division Multiplexing (*FDM*) is a means of concentrating multiple analogue circuits over a common physical connection. It was used widely in the analogue era of telecommunication — before glass fibres and digital techniques were available. It was also, and still remains, a main principle of both analogue and digital radio systems.

FDM provides us the basis on which the radio spectrum is divided up into a large number of radio channels, so enabling its use to be shared between many users, even if they try to communicate simultanneously.

Since FDM requires that we 'shift' baseband signals into other radio frequency ranges corresponding with our *raster* of radio channels (as we discussed in Chapter 2), it is inextricably linked with *frequency modulation* as we already discussed in conjunction with Figure 4.5, but it is worth re-capping the discussion with FDM in mind. To simplify our

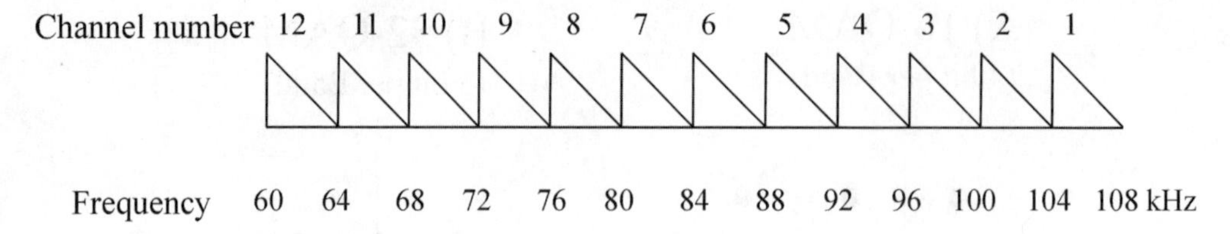

Figure 4.12 FDM of 12 channels into a *basic group* structure

initial discussion, we shall consider the FDM system previously widely used in analogue public telecommunications networks.

FDM systems formed the backbone of *analogue* public telephone networks until the mid 1970s, when *digital* transmission and *Pulse Code Modulation* (*PCM*) took over as the cheaper and higher quality alternative. FDM was commonly used on the long distance and international links between *trunk* (*toll*) telephone exchanges. The transmission lines were typically either coaxial cable or analogue radio.

The lowest constituent bandwidth going to make up an FDM system designed for the carriage of multiple telephone channels has a channel bandwidth of 4 kHz. This comprises the 3.1 kHz needed for the speech itself, together with some spare bandwidth on either side which goes to buffer the channel from interference by its neighbouring channels. The speech normally resides in the bandwidth between 300 Hz and 3400 Hz.

To prepare the baseband signal which forms the basic building block of an FDM system, each of the individual input speech circuits are filtered to remove all the signals outside of the 300–3400 Hz band. The resulting signal is represented by the baseband signal (schematically a triangle) which we illustrated in Figure 4.5.

The next step, as we know, is the shifting of the baseband signal by modulating it with one of a number of standard carrier signals. The frequency of the carrier signal used will be equal to the value of the frequency shift required. In Figure 4.5, as in standard FDM systems for telephony use, a carrier signal frequency of 8000 Hz is being used. This creates the two *sideband* images of the signal, the *lower sideband* and the *upper sideband.* As we discussed in conjunction with Figure 4.5, since all the original information from the baseband signal is held by each of the sideband images, only one of them is actually transmitted. The other one is filtered out. However, unlike Figure 4.5, in FDM systems things do not stop here.

In an FDM system a number of different carrier frequencies are used, one each for a number of different telephone channels. This has the effect of spreading the different user telephone channels throughout the available cable or radio spectrum, so enabling many users to conduct private conversations at the same time. Figure 4.12 illustrates how 12 telephone channels can be carried on separate radio channels, each with a bandwidth of 4 kHz, but with differing *carrier* and channel *centre frequencies.*

While FDM is nowadays falling into disuse in cable networks, as analogue networks are replaced with digital networks based on optical fibre and copper line systems, it remains an important technique in radio for subdision of the radio spectrum and the individual radio bands into *rasters* of radio *channels.* In addition, the associated technique FDMA (Frequency Division Multiple Access) is a very effective way of sharing radio bandwidth for *multiple access* between multiple communicating endpoints, as we shall discover in Chapter 5.

5

Multiple Access Schemes for Point-to-Multipoint Operation

We have learned how a radio modem is used to convert analogue or digital communications into a form suitable for radio transmission by 'modulation' of a *Radio-Frequency* (*RF*) carrier signal. An equipment configuration based on transmitters modulating a given signal and a remote receiver and demodulator acting as the partner to decode the same signal is the basis of point-to-point wireless communication. However, for point-to-multipoint wireless communication, in which a number of remote stations share a common radio spectrum, further technical provisions are necessary to allow for 'multiple access' to the spectrum. We discuss in this chapter the three most important technical realisations of 'multiple access', together with their relative strengths and limitations: FDMA (Frequency Division Multiple Access), TDMA (Time Division Multiple Access) and CDMA (Code Division Multiple Access).

5.1 Multiple Access Schemes used in PMP Radio Systems

In a *point-to-multipoint* (*PMP*) radio system it is usual in the *downstream* direction to broadcast the same radio signal to each of the remote stations in a sector (or cell) and rely upon the *downstream* (i.e. remote) stations to determine which part of the signal is relevant to them. Either all downstream stations receive and decode the signal (in the case of a broadcast signal intended for all remote stations), or alternatively, each remote station decodes only that part of the downstream signal intended for them.

In the *upstream* direction, things are rather more complicated, since the channel bandwidth of the upstream channel has to be shared between, and transmitted on (without interference), by multiple users. For this purpose a *multiple access* scheme has to be applied, to ensure that the remote stations do not simultaneously transmit on the same frequency at the same time. This would only result in interference.

Three principal types of multiple access scheme are used in modern digital radio systems. These are *FDMA* (*Frequency Division Multiple Access*), *TDMA* (*Time Division Multiple Access*) and *CDMA* (*Code Division Multiple Access*).

In FDMA, the bandwidth of the available spectrum is divided into separate channels, each individual channel frequency being allocated to a different active remote station for its transmission.

In TDMA the same upstream channel frequency is shared by all the active remote stations, but each is only permitted to transmit in short bursts of time (correctly called *slots*), thus sharing the channel between all the remote stations by dividing it over time (hence *time division*).

In CDMA all the remote stations use the same channel frequency for their transmission and all may send simultaneously. Separation of the individual signals is achieved by a special coding technique, hence the term *code division*.

We discuss each of these multiple access techniques and their relative strengths in turn.

5.2 FDMA (Frequency Division Multiple Access)

FDMA (*Frequency Division Multiple Access*) is based on *FDM* (*Frequency Division Multiplexing*—as we discussed in Chapter 4). The radio base station (typically at the network local exchange) uses different frequency *sub-bands* or *channels* to communicate with each of the remote stations individually. The allocation of the different channels or sub-bands to the individual remote stations may be either on a permanent basis or on a temporary (*on-demand*) basis, according to the need at a given point in time. Thus, for example, a total available radio bandwidth of 100 kHz could be split into 4×25 kHz individual channels, as shown in Figure 5.1.

When the channels of a multiple access system such as FDMA are permanently allocated for all-time, the system is said to be a *PAMA* (*Pre-Assigned Multiple Access*) system. The allocation and use of the spectrum over time in a PAMA/FDMA system is shown in Figure 5.2.

As well as PAMA (*Pre-Assigned* or *Permanently Assigned Multiple Access*) systems, there are also *DAMA* (*Demand Assigned Multiple Access*) systems, which allocate the radio channel *on-demand*. By allocating the frequencies on-demand, they can be shared between many end users according to the calling patterns or traffic profiles. In this case, the exact number of channels needed in a DAMA-system depends upon the maximum number of *simultaneous* connections needed by the users as a whole. This value is typically much less than the number of connected end-users. In contrast, a PAMA system would need to

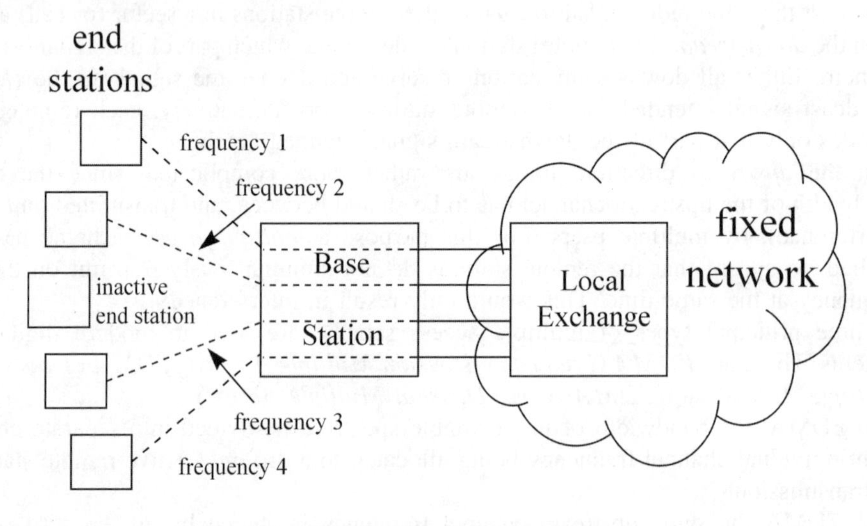

Figure 5.1 The principle of *frequency division*

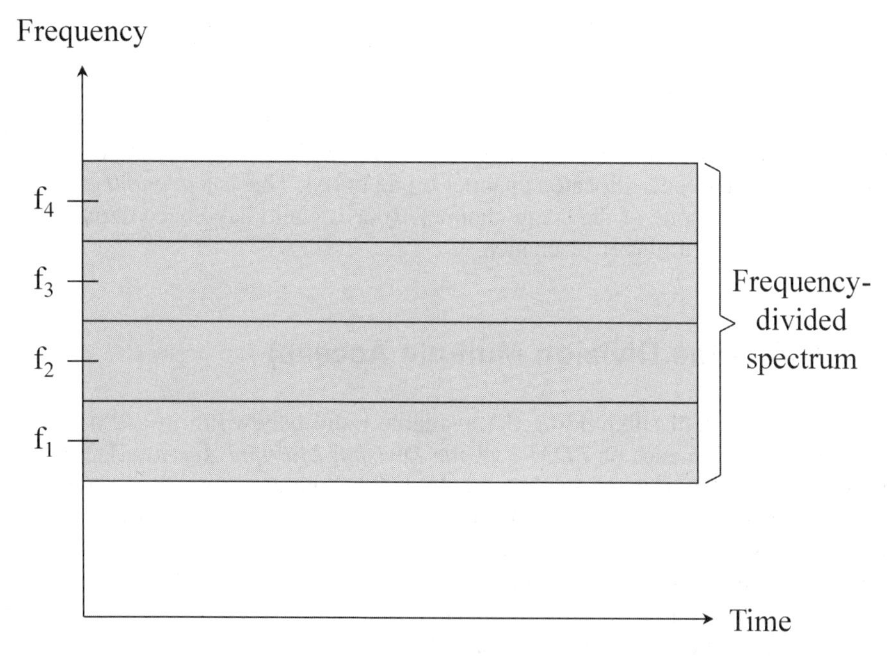

Figure 5.2 The assignment of spectrum for a PAMA system based on FDMA

provide as many radio channels as there were end-users in total (and most of the time the individual channels would not be in use).

Figure 5.3 illustrates how a DAMA/FDMA system re-allocates the different channels and frequencies over a relatively long time period. The diagram shows how a small number of channels are being shared by a large number of users on a demand-assigned basis.

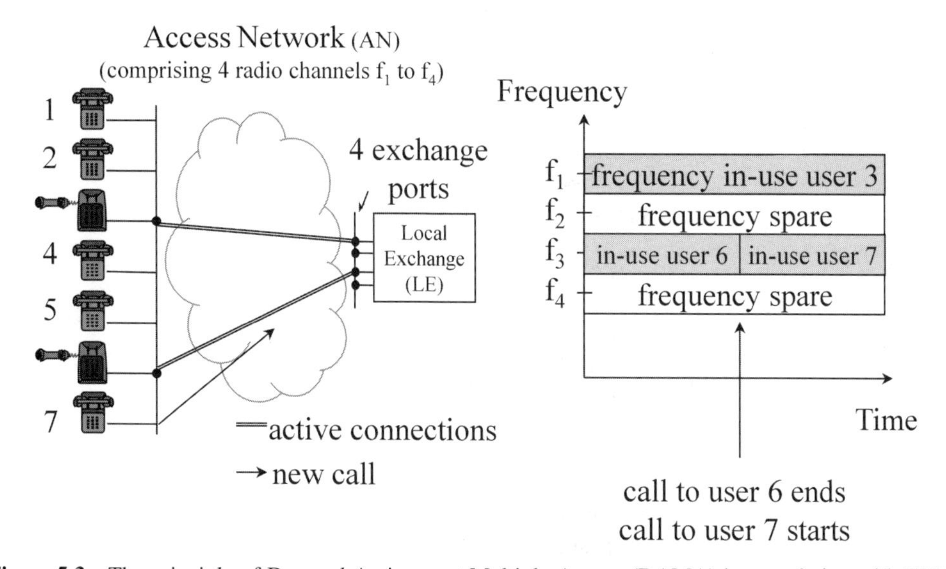

Figure 5.3 The principle of Demand Assignment Multiple Access (DAMA) in association with FDMA

Figure 5.3 shows how seven end users (the telephones numbered 1 to 7 on the left-hand side of the diagram) are sharing four radio channels (f_1 to f_4) in the access network and four exchange ports. At the beginning of the time shown (right-hand chart), only two users are active (users 3 and 6). These users are connected via radio channels f_1 and f_3, respectively — to ports 1 and 3. After a period of time, user 6 hangs up and user 7 makes a call. By chance, user 7 gets allocated channel f_3 and port 3. This is a *demand assignment* of frequency f_3. Equally, one of the spare channels f_2 or f_4 could have been demand-assigned to user 7; this is only a matter of chance.

5.3 TDMA (Time Division Multiple Access)

An alternative means of subdividing the available radio bandwidth in a PMP system for *multiple access* is by means of *TDMA* (*Time Division Multiple Access*). TDMA is based upon *TDM* (*Time Division Multiplexing*). In TDMA the entire radio bandwidth (in our example, of 100 kHz) is used by each of the end stations in turn, each using the signal for a quarter of the available time. A burst of information is sent by the end station in a pre-determined *timeslot* (say of 125 μs duration) allocated by the base station. The base station transmits continously on its transmit frequency, each station selecting the information only from the timeslot which is relevant to itself (Figure 5.4).

In 'classical' TDMA systems, as in 'classical' Time Division Multiplexing, the allocation of the timeslots in both *downstream* (i.e. from the base station to the remote end stations) and *upstream* (i.e. from the remote end stations to the base station) directions is on a *synchronous* basis (Figure 5.5). The timeslots are shared on an equal basis in strict

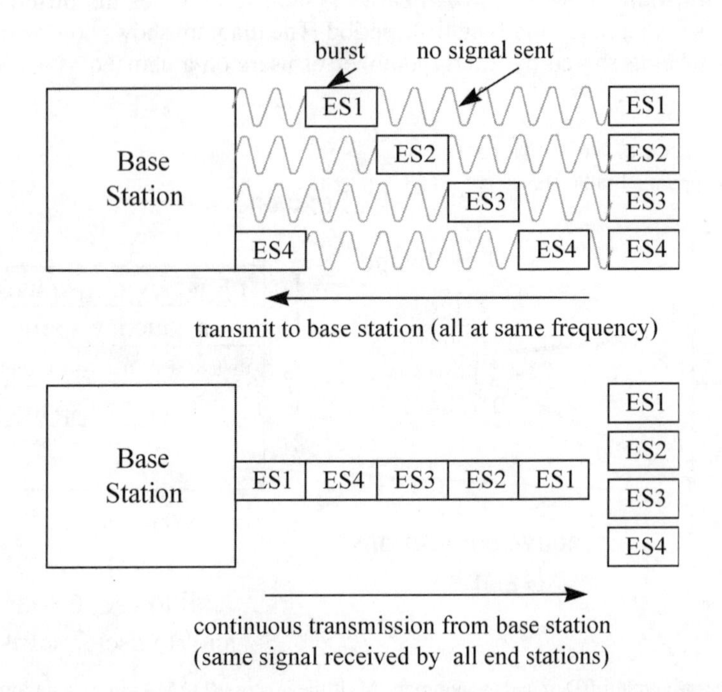

Figure 5.4 The principle of *TDMA* (*Time Division Multiple Access*) ES=End Station

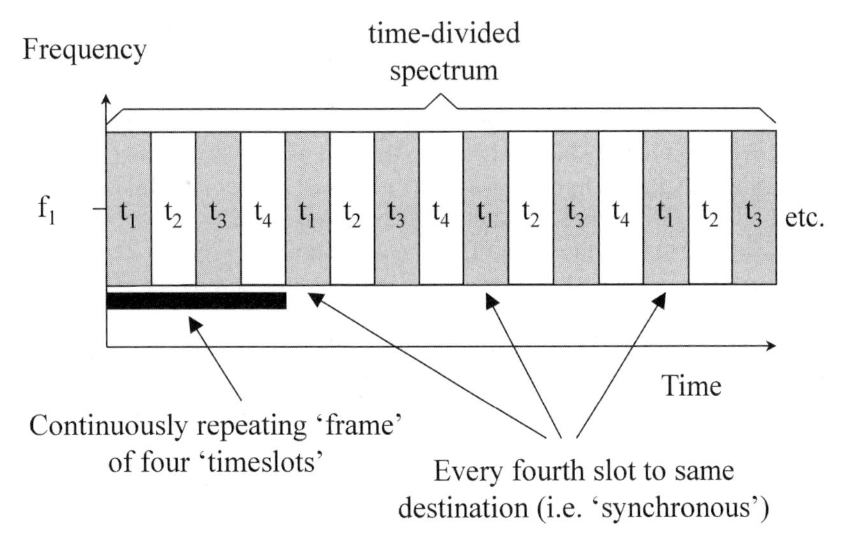

Figure 5.5 The synchronous allocation of timeslots in 'classical' TDMA

rotation: 'one for me, one for you, one for him, one for her, two for me, two for you, two for him, two for her, three for me, etc.'. In this way, the total bitrate of the radio channel (e.g. of 8 Mbit/s) is split into equal and discrete chunks. The division of 8 Mbit/s between four users would give each a separate 2 Mbit/s communications channel (Figure 5.5).

Like FDMA systems, TDMA systems can be realised in either PAMA or DAMA versions. Figure 5.6 illustrates a DAMA-version of TDMA in which individual timeslots are being allocated to individual telephone channels. The example is exactly analagous to the FDMA example we discussed in Figure 5.3. Seven telephone users (numbered 1 to 7) are sharing four radio channels (separate timelots of TDMA) and four exchange ports. During the time shown, caller 6 clears his call and caller 7 takes over his channel. Notice

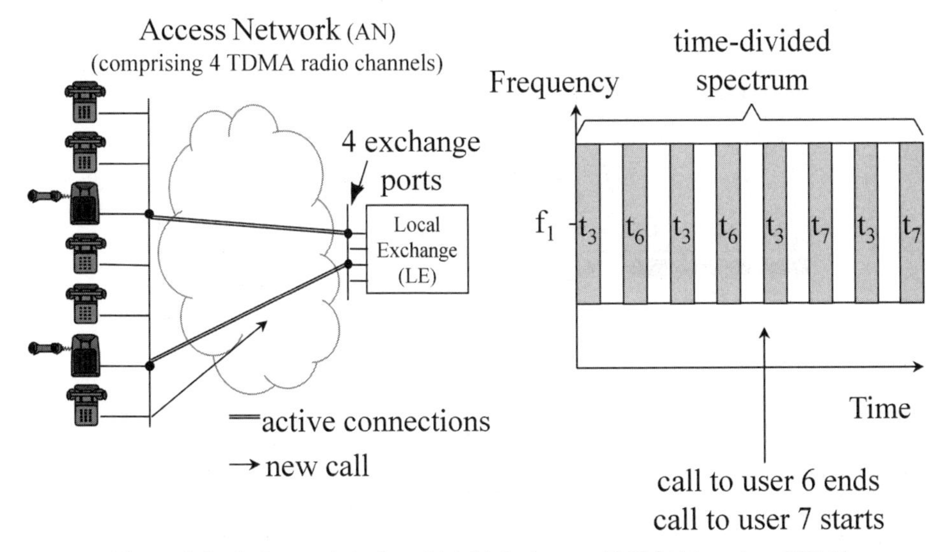

Figure 5.6 A Demand Assigned Multiple Access (DAMA) based on TDMA

how in TDMA the entire spectrum is used, but some of the timeslots are idle or empty, whereas with FDMA, some of the spectrum (channels 2 and 4 in our example) is not always in use.

Figure 5.7 illustrates a TDMA system used for *Pre-Assigned Multiple Access*. In our example, the distribution of timeslots between individual users is not an an equal basis. The user using timeslots labelled t_1 has the use of 50% of the bandwidth, while the users using timeslots t_2 and t_3 have 25% each. Such unequal distribution of the available bandwidth could also have been achieved in the equivalent FDMA system shown in Figure 5.2 by allocating the entire spectrum equivalent to both frequencies f_1 and f_2 to the same user. The allocation is shown in Figure 5.8.

The most recent DAMA techniques used in association with TDMA systems use more sophisticated means of allocating the individual slots of transmission capacity, allocating them on an *asynchronous* rather than *synchronous* basis. This type of allocation gives the potential for even more flexible use of the transmission capacity. Figure 5.9 illustrates a case in which the slots have been allocated in an *asynchronous* manner (so-called *Asynchronous Transfer Mode*, or *ATM*). In the example shown, the device using timeslots labelled t_1 has received (during the period shown in Figure 5.9) 5/15 of the bandwidth (if the total bitrate is 8 Mbit/s as in Figures 5.4 and 5.5, then this is equivalent to 2.7 Mbit/s). Meanwhile, the device using slots t_2 will receive 6 of the 15 timeslots (40% or 3.2 Mbit/s); the device using t_3 will receive 3 slots (20% or 1.6 Mbit/s) and the device using t_4 1 slot or 0.5 Mbit/s.

By allocating the slots on an asynchronous basis, a dynamic adjustment of the use of the slots can be made, allowing individual end users (equivalent to the users of slots t_1, t_2, t_3 and t_4 of Figure 5.9) to receive continuously varying bitrates (high bitrate one second, low or no communication in the next second). A need for such continuously varying bitrate is typical of modern data communications. Take, for example, an Internet user surfing the world wide web. Maybe only once every two minutes he sends an upstream command to request the 'download' of a given 'webpage'. Then for a few seconds he would like the

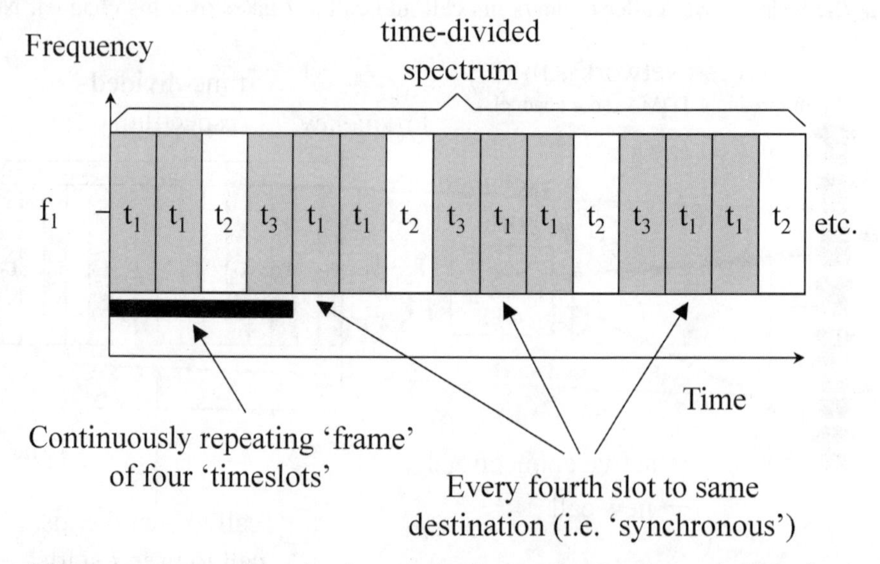

Figure 5.7 Communication channels of unequal bitrate using Pre-*Assigned Multiple Access* in association with TDMA

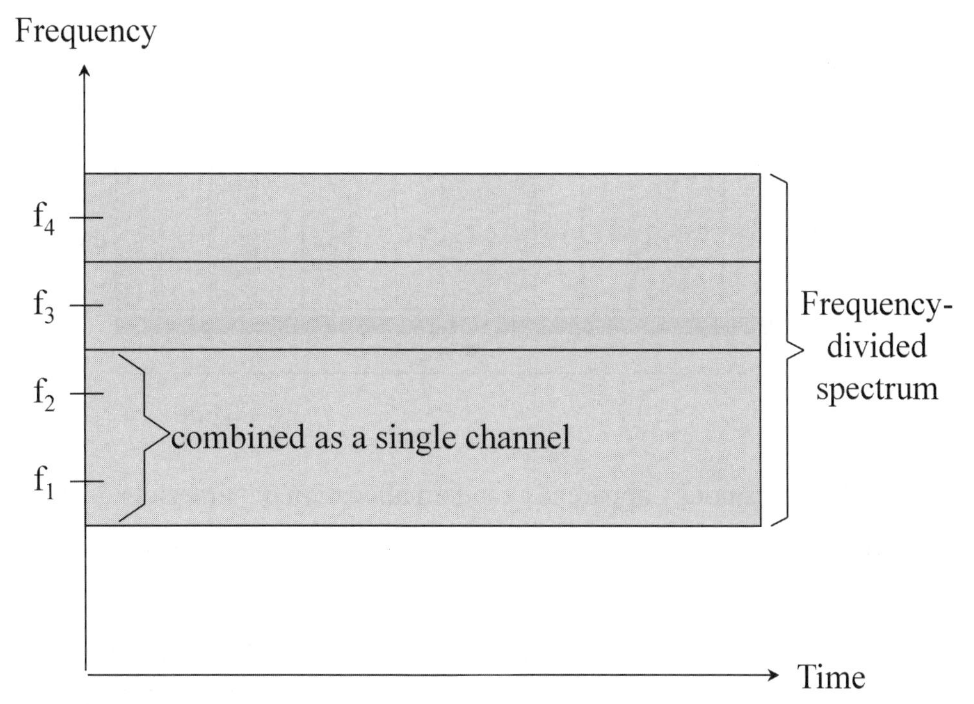

Figure 5.8 Communication channels of unequal bitrate using Pre-*Assigned Multiple Access* in association with FDMA

maximum available bitrate in order that the transmission takes place quickly. Thereafter (while reading the webpage) no transmission capacity is needed — and can therefore be made available to other users.

5.4 Frequency Hopping — A Hybrid Form of FDMA and TDMA

Frequency hopping could be thought of to be a mix of the FDMA and TDMA techniques. A number of bearer frequencies are used by each of the end stations in turn, but each only transmits on one of the frequencies for a short burst or timeslot. Immediately after the timeslot, the station moves to another bearer frequency. Each station thus hops in a given pattern between a set of pre-defined frequencies (*frequency hopping*). Figure 5.10 illustrates the technique. Thus, End Station 1 (ES1) transmits cyclically on frequencies $f1$, $f3$, $f4$, $f2$ and then $f1$ again, etc.

The advantage of frequency hopping, one of a number of *spread spectrum techniques*, is the relative immunity to radio *fading*. Since it is unlikely that each radio channel will be simultaneously degraded by fading, the effect of a single channel fading can be virtually eliminated by 'diluting' the degradation over a large number of active user connections. Another benefit of is the much greater complexity of equipment needed by criminals desiring to 'tap' (i.e. overhear) the radio channel.

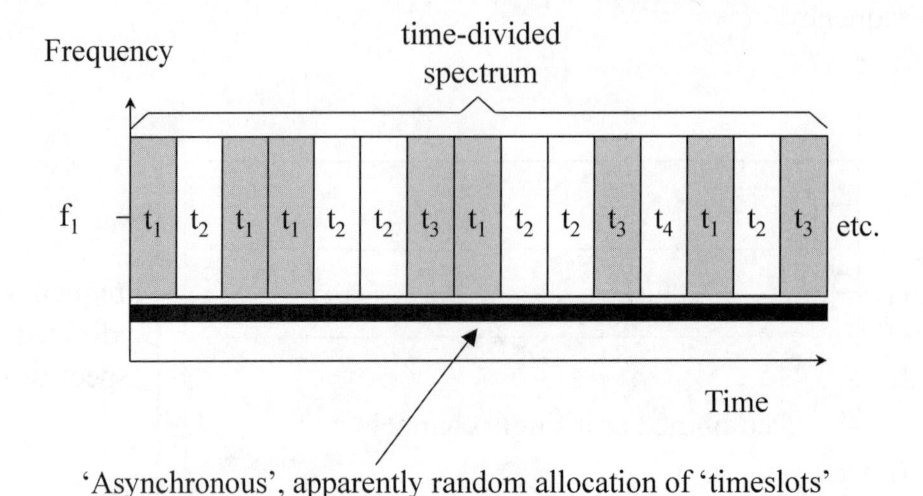

Figure 5.9 Asynchronous allocation of TDMA slots (*Asynchronous Transfer Mode, ATM*)

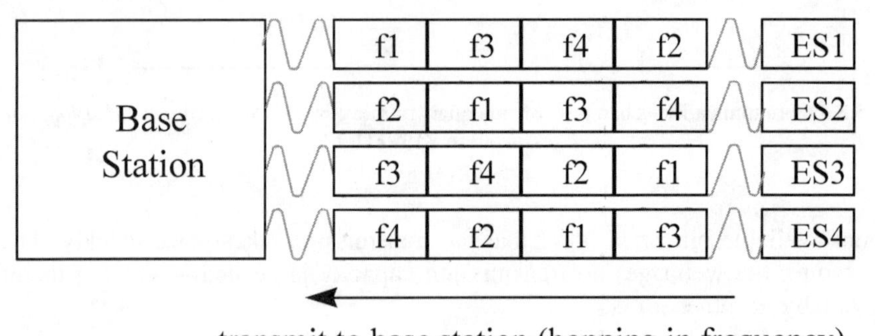

Figure 5.10 Frequency hopping

5.5 CDMA (Code Division Multiple Access)

CDMA (*Code Division Multiple Access*) is also a form of *spread spectrum* modulation. In CDMA, as in TDMA, all the remote users share the same radio channel for transmission, but *code* their signals differently, so that all the signals can be carried simultaneously.

In CDMA the carrier frequency is first modulated by the digital signal to be carried, and is then multiplied by a digital signal of a much higher bit rate (called the *chip rate*). The chip rate signal is actually a specific binary number (or code). It has the effect of *spreading* the modulated carrier over a very large frequency band, equal to the spreading factor (the ratio of the chip rate to the carried bit rate).

At the receiver, the signal is decoded using the same binary *code*, by division of the received signal. This regenerates the original modulated carrier.

Many signals may be modulated onto the same carrier frequency, but for each modulated carrier, a different code must be used for spreading, to ensure that all the 'spread' signals can be distinguished from one another. The 'spread signals' are recovered from the 'mixture' of modulated signals by division by their respective codes. During the division process, all the signals are of course divided by the code, but only one of the modulated signals is reduced to a signal closely grouped around the original carrier frequency. The other coded signals can thus be removed by *filtering* before the carrier is demodulated to recover the original digital bit stream.

5.6 Relative Strengths of the Different 'Multiple Access' Methodologies

FDMA, TDMA and CDMA can all be used as the basis of digital radio systems. It is erroneous to consider FDMA to be only an *analogue* technique, since each of the individual channels can be coded digitally. However, it is true to say that FDMA codes individual separate bitstreams onto separate radio channels — first dividing the spectrum, and then coding the individual channels. Meanwhile, TDMA codes the entire available spectrum into a single bitstream and then subsequently divides the resulting bitstream between individual users.

The prime advantage of TDMA when connecting the radio system to a digital fixed network is that the output of the base station towards the fixed network (for example, telephone exchange) is usually a single, high bitrate multiplexed digital bitstream (e.g. 2 Mbit/s). In the FDMA case, multiple 64 kbit/s telephone channels would be presented. Another advantage of TDMA over FDMA is that only a single, high bitrate radio modem is required in the base station (as opposed to one modem per channel in the case of FDMA). The disadvantage is that the control system required for TDMA is more complex than for FDMA.

The prime advantage of FDMA is the relative simplicity of the system. In addition, because at any point in time each radio channel is used only between one remote user and the base station (in effect point-to-point), the modulation scheme can be optimised to the prevailing propagation conditions.

The disadvantages of FDMA are the multiple modems needed at the base station, the multiple connections needed between the base station and the fixed network and the relatively fixed subdivisions of the available capacity (i.e. discrete, fixed bitrate channels). In comparison, TDMA can be relatively easily adapted to provide almost any subdivision of the capacity (channels of almost any bitrate), simply by adjusting the duration of the *timeslot* allocated to each end station.

The advantages of CDMA, like frequency hopping, is the resistance to multipath fading and interference resulting from the spread spectrum technique. In addition, CDMA can be realised relatively cheaply using digital signal processing, requires little coordination between the stations, and requires only a lower transmit power for low bit rate signals. The disadvantage is the relatively high bandwidth required for efficient operation and the close coordination required to ensure good isolation of the individual channels from one another.

Actual point-to-multipoint (PMP) radio systems may use a combination of modulation and multiple access techniques to achieve the desired combination of bitrates, pre-assignment or dynamically assigned *bandwidth-on-demand* or *statistical multiplexing*. Thus, for example, some of the most advanced modern PMP systems use a TDMA multiple access scheme which allocates slots based on ATM cells. This enables the use of ATM transmission on the radio interface. ATM is the most modern form of telecommunications transport, optimised for simulaneous transport of voice, data and video signals over a single transmission medium.

In cases where *bandwidth-on-demand* or *statistical multiplexing* schemes are applied to PMP radio systems, there needs to be a mechanism for resolving contention or congestion of the radio bandwidth (i.e. for determining what happens when there is not enough bandwidth to carry all the connections or communications demanded at a particular moment in time). In the case of 'classical' TDM-based systems, the channels are allocated on a first-come-first-served basis. When all the channels are in use, no further channels can be established until one of the channels is first cleared. In ATM-based systems, two different mechanisms are available for allocating the available bandwidth. A first-come-first-served basis at the time of establishment of a connection can be instigated by means of *Connection Admission Control* (*CAC*), or alternatively (particularly relevant where the capacity is to be *oversubscribed* by means of *statistical multipleying*, a priority mechanism can be used to determine the order of transmission of the ATM cells corresponding to the various connections (i.e. who gets first go). The priority scheme is usually based upon the ATM Forum *Weighted Fair Queueing* (*WFQ*) algorithm. We discuss the subject of ATM radio in more detail in Chapters 13 and 14.

5.7 Broadcast Systems and PMP

Point-to-multipoint systems, both satellite and terestrial-based, are well-suited to broadcasting, since a single radio channel (and thus single signal) can be received by multiple parties simultaneously. In most historical broadcast systems, the transmission was only in the downstream direction, from the central or base station to the remote end-users, either by terrestrial broadcast, or more recently, by means of wide coverage broadcast by means of satellite. More recently, though, systems with two-way communication have appeared. The base station may have the ability to selectively *multicast* the same message to a particular *Closed User Group* (*CUG*) of end-users. In addition, modern digital telephone channels also allow end-users to use an *upstream* communication channel to control the signals they receive, so making for *interactive* television. What is possible depends not only upon the radio modulation used for transmission and the multiple access scheme (e.g. FDMA, TDMA, CDMA), but also upon the *protocols* used within the radio system to control the connection. We discuss some of the various schemes in the later chapters of this book.

Part II

System and Network Design

6

Basic Radio System Design and Functionality

In this chapter we discuss the basic components which make up a radio system designed for operation in the millimetric or microwave bands (i.e. frequencies above 1 GHz). In particular, we consider the communications radio systems classified by ITU-R as *Digital Radio Relay Systems* (*DRRS*). Such systems typically comprise three main components; a so-called *Indoor Unit* (*IDU*), which comprises the termination point for the end-users equipment and the radio 'modem'; an *Outdoor Unit* (*ODU*) comprising the radio frequency transmitter and receiver and the antenna. The IDU is usually connected via coaxial cable to the ODU, which in turn is connected by means of 'waveguide' to the antenna. We shall discuss the detailed functions and the limitations of each of these components in turn.

6.1 Basic System Components

Figure 6.1 illustrates the basic components of a millimetre or microwave radio *terminal* (i.e. the equipment at one end of a radio link). The three basic components (*Indoor Unit*, *IDU*; *Outdoor Unit*, *ODU* and antenna) are present in all types of radio terminal, independent of whether the system is a point-to-point (PTP) or point-to-multipoint (PMP) system, or whether the terminal is a *base station* or a remote end-station.

The IDU usually comprises the radio modem. This device provides the termination point for the end-user equipment which is supplying the digital signal which is to be carried by the radio link. The prime function of the modem (and thus of the indoor unit), as we discussed in Chapter 4, is the conversion of the digital signal into a form suitable for *modulation* onto a radio *carrier signal*. In the transmit direction, the user signal is *modulated* and transmitted to the outdoor unit (radio) for transmission as a high frequency radio signal. In the receive direction, signals arriving from the outdoor radio unit are *demodulated* and presented to the end-user device.

A secondary function of the indoor unit is to provide electrical power for the radio terminal. Thus, it is usual to supply power to the indoor unit. The indoor unit, in turn, provides power to the outdoor unit. Furthermore, because the indoor unit is usually situated in a more accessible and less 'harsh' environment than the outdoor unit, it is also customary to include in it the digital electronics and software associated with the configuration and network management of the radio. The indoor unit usually either

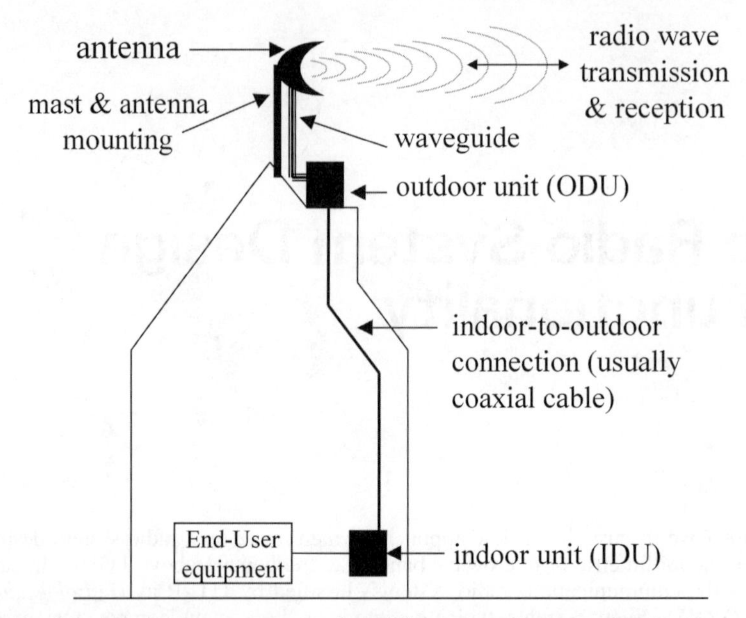

Figure 6.1 The basic components of a millimetre or microwave band radio system

includes a control panel or a port for connection of a PC or network management system for configuration and diagnosis purposes.

A coaxial cable is the usual form of indoor-unit-to-outdoor-unit connection. The cable carries the modulated radio signals between the two units, as well as supplying power to the outdoor unit.

The outdoor unit (radio unit) converts the modulated signal from its natural low frequency form into a high frequency radio signal in the correct radio band and channel for radio transmission. This signal is then fed to the antenna by means of a *waveguide*. Finally, the antenna serves to transmit the signal into the air in a direction aimed at the partner radio terminal at the other end of the *link*.

In the remainder of the chapter, we shall discuss the detailed functions of each of these components in turn, describing in particular the main system specification factors and parameters which distiguish one radio system from another.

6.2 Antenna

The purpose of the antenna, as we discussed in detail in Chapter 3, is the transmission of the high frequency radio signal into the air and the reception of the arriving signal. Usually, the antenna in a fixed wireless system (i.e. one in which the antenna is mounted permanently) is designed to have a relatively high *gain*. Antennas with a high gain are *directional* rather than *omnidirectional*. In other words, they tend to focus the outgoing radio signal in a single direction (i.e. towards the partner terminal at the other end of the intended link).

Omnidirectional antennas are used to 'broadcast' the signal in all directions. Such type of broadcast is needed at the central or *base station* of a point-to-multipoint (PMP) system or at remote stations when the radio is to be mobile. (For example, omnidirectional antennas are incorporated in the handsets of mobile cellular telephony handsets, since the device must transmit in all directions to maintain continuous contact with the nearest base station.)

There are various different antenna forms which are offered by different manufacturers for different frequency bands:

- Parabolic (dish antenna), typically 30 cm, 60 cm antennas (or even larger).
- Horn antennas.
- Lens-corrected horn antennas.
- Planar or flat-plate antennas.
- Yagi antennas (mainly used for frequencies in the VHF and UHF radio bands, for example, for television reception).
- Array antennas.
- Coil antennas (mainly used in low frequency bands, e.g. public broadcast radio receivers).

Frankly, the form of the antenna (given it meets the required specification) is rather unimportant in itself, other than the fact that its aesthetic appearance to the landlord, occupant or local authority may determine whether it is allowed to be installed on a particular building or not.

Certain local authorities, and even regional standards bodies, are considering, or have already imposed, maximum dimensions upon the size of antennas which may be installed (e.g. 45 cm largest single dimension).

People sometimes feel threatened by antennas. It seems that parabolic antennas make people more aware of the directional nature of the antenna, and make them concerned about how much dangerous radiation they may be being exposed to. For this reason, less obvious antennas, including planar antennas and flat plate antennas, have become more popular, as have 'integrated' antennas (in which the antenna is enclosed in a non-descript casing).

The technical parameters of most importance in selecting an antenna are the following:

- the radio band for which the antenna is designed;
- the antenna *gain*, the *front/back ratio* and the *antenna pattern*;
- the antenna *polarisation* and the *cross-polar discrimination*;
- the type of waveguide attachment necessary for connecting the antenna to the radio unit;
- the mounting arrangement for the antenna;
- the weight of the antenna;
- the wind loading forces inflicted on the antenna mounting pole and the stiffness of the pole required;
- the weather protection of the antenna.

Antenna Band of Operation

An antenna designed for use in a given radio band may not be suitable for use in another band. Use of the antenna in the wrong band will mean greatly degraded performance.

Antenna Gain, Front/Back Ratio and Antenna Pattern

A high antenna gain (usually quoted in units of dBi — decibels relative to an *isotropic* [or *omnidirectional*]) antenna is good for transmission along a single direction and powerful reception from the same direction.

The higher the gain of the antenna, the greater is the potential *range* of the radio link or the potential *availability* (the availability of a radio link is the percentage of the time that the link is functional — not disturbed by atmospheric disturbance to the radio signal or other radio interference. We shall discuss both the range and availability of radio systems in more detail in Chapter 7. High gain antennas are usually used at both ends of point-to-point (PTP) radio links and at the remote station sites of point-to-multipoint (PMP) systems (omnidirectional antennas or sector antennas, as we discussed in Chapter 3, are usually used at the central or base station sites of PMP systems).

A high antenna gain is a good thing for radio system performance, but high gain antennas are not without problems. The higher the gain of an antenna, the more directional it is, and the narrower is the path of the radio signal. This means that the antenna has to be much more accurately *aligned* during installation. (The *aperture* of the antenna [its *beamwidth*] can be much less than 1°.) Get a high gain antenna slightly off-centre and the performance of the radio is worse than had you used a lower gain antenna. Furthermore, higher gain antennas tend to be larger and heavier than lower gain antennas. They are therefore less acceptable to landlords and local authorities, and much more prone to being pushed-off their correct alignment during heavy wind.

Generally, it can be said that lower radio frequencies require larger antennas. A typical antenna dimension has to be around one quarter of the wavelength. So, for example a *longwave* broadcast antenna for public broadcast radio at 1500 m needs to be a mast several hundred metres high. However, the internal antenna for a mobile telephone on the GSM-1800 (1800 MHz) system need only be around 4 cm in length [wavelength $= 3 \times 10^8 / (1.8 \times 10^9) = 16$ cm].

An extra 3 dBi of antenna gain will double the signal strength (i.e. signal power) at a given distance from the antenna along the *pole* direction. However, doubling the power is achieved by halving the *beamwidth*. Thus, for example, a 30 cm diameter parabolic antenna designed for use in the 26 GHz radio band has a gain around 35 dBi and a beamwidth around the pole of about 2.5° (±1.3°). A 60 cm antenna for the same band, meanwhile, has a gain of around 41 dBi and a beamwidth under 1.4° (±0.7°). Similarly, a 45° *sector* antenna intended for use at a PMP base station will have a gain 3 dBi better than an equivalent 90° sector antenna (provided, of course, that the vertical beamwidth of both antennas is the same).

As we discussed in Chapter 3, it is unfortunately not possible to produce an antenna with 'perfect' radiation characteristics which transmit only in the desired direction. Other *sidelobes* (directions of relatively strong unwanted transmission) are always present. One of the most problematic directions of unwanted transmission can be in a direction effectively directly out of the 'back' of the antenna. For this reason, it is normal in an antenna datasheet or specification to state the *front-to-back-ratio* of the antenna. This value is normally quoted as a ratio in dB. Typical values of the front-to-back ratio are 35 dB to 50 dB. In general terms, the higher the gain of the antenna, the higher is the front-to-back ratio.

For radio frequency planning purposes (as we shall discuss in Chapter 9), it is normal to request the antenna manufacturer to provide accurate antenna diagrams, showing the antenna radiation patterns and sidelobes in *azimuth*, *elevation* and *cross-polar discrimination*. These are charts similar to Figure 3.7 of Chapter 3.

Antenna Polarisation

It is possible to *polarise* radio waves, by transmitting only one of the two electromagnetic *vector* components of the wave (either the *horizontal* component or the *vertical component*). Indeed, the construction of an antenna is usually so designed as to favour one of the two wave components, and the antenna is said either to be *vertically polarised* or *horizontally polarised*. The use of the two different polarisations enables the same radio frequency to be *re-used* by different radio systems operating near to one another.

As we shall discuss in Chapter 7, the vertical signal polarisation propagates slightly better than the horizontal polarisation. It is slightly more immune to radio fading caused by rainfall and other weather disturbances, as well as being less sensitive to multipath fading caused by signal reflections from roof surfaces near the antennas at either end of the link.

Some types of antenna are also *dual-polarised* (or *Co-Channel Dual Polarized, CCDP*). This means that they are designed to transmit equally in both signal polarisations. Such a type of transmission can have benefits for the overall link performance, since the link is less affected by interference, obstructions or rain fading which has affected one of the polarisations more than the other.

Antenna Waveguide Attachment

The type of connector used between the antenna and the radio unit, or between the antenna and the waveguide leading to the radio unit must be compatible with the waveguide attachment provided on the radio unit otherwise a convertor arrangement will be necessary. It is normal for radio manaufacturers either to provide the antenna or to recommend specific antennas and specialised waveguide connectors from other specialised antenna manufacturers. This ensures the compatibility of the waveguide connection between antenna and radio unit. A waveguide convertor is unsightly, cumbersome, can lead to additional operational problems, as well as cause unnecessary attenuation (i.e. weakening) of the radio signals, and therefore degrade overall system performance.

Alternatively, some manufacturers use a proprietary attachment interface for the waveguide attachment from the antenna to the waveguide, or for a direct attachment of the radio unit to the antenna.

The part of the antenna which is coupled to the radio terminal and serves to provide the radio signal to the antenna or to *illuminate* a reflector antenna (such as a parabolic antenna) is termed the *feed* of an antenna.

Antenna Mounting

It is usual for the antenna manufacturer to supply an antenna mounting designed specifically for the antenna. The mounting serves to secure the antenna firmly to either a steel pole or directly to a wall. However, during the installation phase of the antenna, it is also important that the mounting allows for easy alignment of the antenna. There needs to be a fine adjustment for both bearing (i.e. compass orientation) and azimuth (i.e. angle of

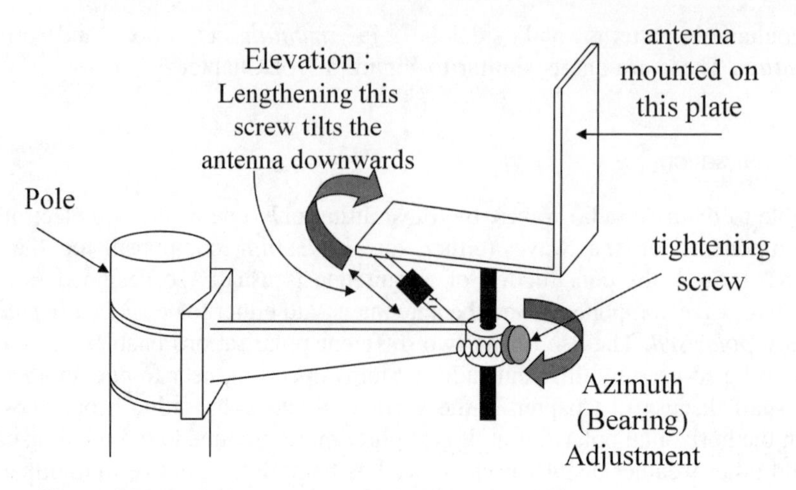

Figure 6.2 Typical antenna-mounting with adjustable bearing and azimuth

the antenna relative to the ground and horizon — pointing upwards, downwards or horizontally).

Figure 6.2 illustrates a typical antenna mounting arrangement, showing a mounting designed for installation on a pole. This design allows very easy adjustment of the *alignment* of the antenna in both *bearing* and *azimuth* directions, having already attached the antenna to the mounting.

Once the antenna has been aligned (we discuss how later in this section), locking screws or bolts are used to secure and maintain the correct orientation of the mounting.

Good antenna mountings are made of cast material (not pressed steel, which tends to deform during a long lifetime). The mounting should be rigid, and should be capable of withstanding continuous weather and temperature cyles over the entire lifetime of the radio system (probably many years). The mounting should not become loose in hot or cold conditions. There should be no surfaces for undue accumulation of water or ice. The metalwork should be either of stainless steel or anodised (including the bolts), so that it does not rust.

Weight of the Antenna

The weight of an antenna is important for two main reasons. First, the heavier the antenna, the stronger the pole or other support steelwork has to be to support it. Secondly, in cases where heavy and cumbersome antennas are to be mounted towards the top of high poles, the more complicated is the installation. Heavy devices may require a crane or hydraulically-raised working platform to be used by the installers, rather than simply climbing up rungs on the mast.

Antenna Wind Loading

For the purpose of designing the steelwork support structure or pole for mounting the antenna, the wind loading calculations are even more important than the weight of the antenna. Strong winds which are incident on large antennas will ‘load’ the steelwork with

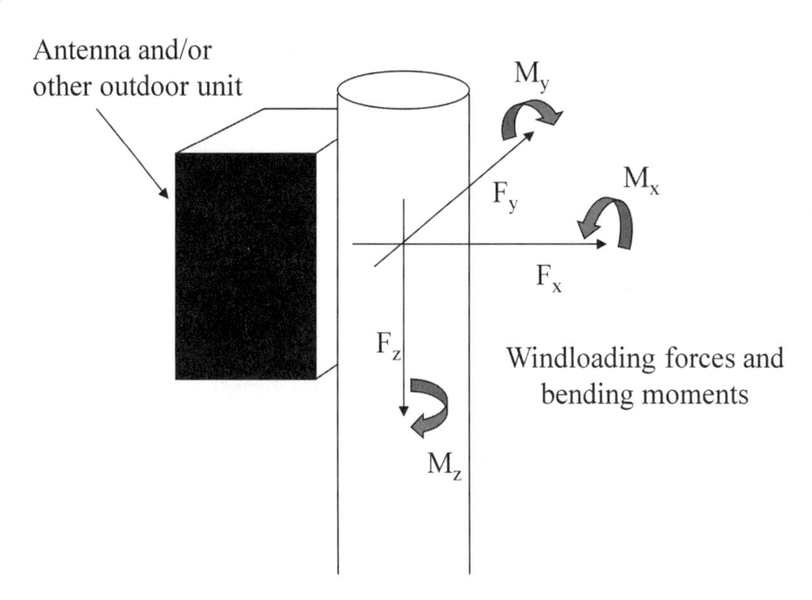

Figure 6.3 Windload forces on a mounting assembly resulting from the incidence of wind on an antenna

considerable torsional and bending forces. Since it is critical that the antenna does not either fall down or even lose its alignment, it is important that the steelwork be designed for adequate structural rigidity.

For the purpose of designing the steelwork, it is normal for the antenna manufacturer to provide a wind-loading chart or table, listing the resultant forces and bending moments, as illustrated in Figure 6.3, which arise from winds of differing strengths.

Figure 6.3 illustrates the various maximum forces (F_x, F_y and F_z) which may be inflicted on the mounting, as well as the bending moments (M_x, M_y and M_z).

It is usual for the manufacturer to provide as a minimum a table of the maximum values of each of the forces and bending moments, shown in Figure 6.3, for at least two different wind strengths (some manufacturers provide more detailed information, either technical formulae for calculating the forces, dependent upon wind direction and strength, or even software programmes for the same purpose). The maximum forces and bending moments are needed to calculate the necessary strength and stiffnes of supporting steelwork.

It is important to know the maximum forces on the mounting at the maximum *operational wind speed.* At this wind speed (regardless of the direction of incidence of the oncoming wind), the mounting structure should bend or deflect only in a minor way, thereby maintaining the *alignment* of the antenna within its *beamwidth* so that normal operation is possible. It is usual for the radio system operator to define his maximum *operational wind speed* requirements according to his local climatic conditions (e.g. 99.99% or 99.999% of the time windspeed does not exceed *X* km/h). The mounting and antenna should remain undamaged (other than damage caused by possible flying debris) at this windspeed.

It is also important to know the maximum forces on the mounting at the *survival wind speed.* This is the windspeed at which the system is designed to remain in tact (i.e. without breakage). At this windspeed, normal operation of the radio system may no longer be possible due to deflection of the mounting structure, causing misalignment of the antenna.

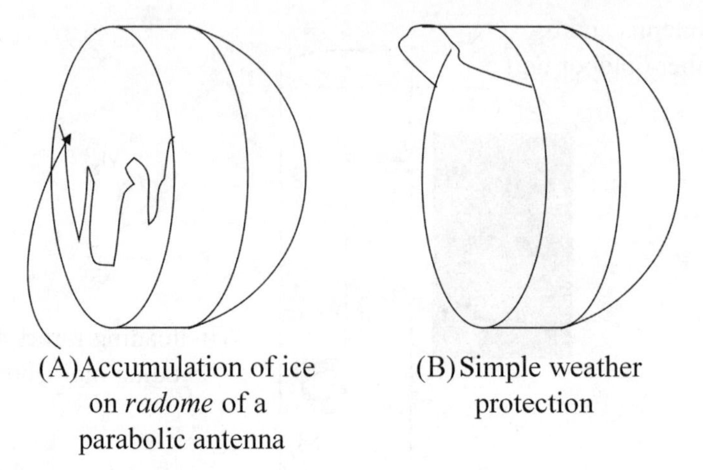

(A)Accumulation of ice
on radome of a
parabolic antenna
(B) Simple weather
protection

Figure 6.4 Accumulation of ice on the radome of a parabolic antenna, and simple means to avoid it

This speed is specified by the radio system to equal or exceed maximum expected windspeed in the region of operation.

Tall masts carrying numerous or large antennas may need to be designed to be able to carry considerable forces. In some cases, it may additionally be necessary to calculate in detail the mast anchoring arrangements, and even the strength of the building on which the mast installation is to be made (engineers refer to these as the *building statics* calculations).

Antenna Weather Protection

The antenna should be designed so that its performance is unaffected by the weather. For most antennas this simply means that the antenna should not present unduly large areas where water can accumulate or where direct sunlight might cause a problem. For antennas used in cold climates, it is also important that there should not be an accumulation of ice in the radio signal path. A thick accumulation of ice on the *radome* covering a parabolic antenna (Figure 6.4) can cause serious radio system degradation, since the water molecules in the ice tend to absorb, and therefore attenuate (weaken) the radio signal.

Ice can accumulate even on vertical surfaces (as illustrated in Figure 6.4a). To minimise the accumulation, it is usual to specify the use of water-repelling, Teflon-like material for critical surfaces, such as the *radome*, or alternatively to fit a small cover (Figure 6.4b). In very harsh climates, it may be necessary to specify antenna heating during very cold weather.

6.3 Waveguide

For the connection of the radio transmitter/receiver (*transceiver*) to the antenna, it is desirable to use a connection means which causes only very low resistance to radio wave propagation, and thus minimises the signal losses within the radio terminal. For frequencies up to about 2 GHz, coaxial cable is commonly used for this connection. The reason is that the cable is relatively cheap and relatively low loss at such frequencies.

At frequencies above 2 GHz, and even for long low frequency connections (i.e. below 2 GHz), it is normal to use *waveguide* rather than coaxial cable for the transport of radio signals from the radio unit (*transceiver*) to the antenna. Waveguide has the appearance of a pipe (usually rectangular in cross section). Waveguide is very accurately produced metal tubing, with smooth inside surfaces of critically defined dimensions, suited to the carriage of a particular radio frequency or small range of frequencies (i.e. radio band). The per metre attenuation of waveguide is much lower than that of coaxial cable. For waveguide runs much longer than about 20 m it is also usual to take precautions to ensure that only dry air is present within the waveguide, in order further to limit the signal attenuation.

The waveguide and, just as important, its flange are usually defined either in terms of *IEC* (*International Electrotechnical Commission*) or *EIA* (*Electronics Industries Assocation*, of the USA) standards:

- *waveguides* conform either to IEC specifications 153 IEC-R-xx or to EIA WR-xx (xx represents a series of different inidividually numbered specifications);
- *flanges* conform either to IEC specifications 154 IEC-PAR xx, 154 IEC-PBR xx, 154 IEC-PCR xx, 154 IEC-PDR xx, 154 IEC-PFR xx, 154 IEC-UAR xx, 154 IEC-UBR xx or 154 IEC-UER xx or to EIA CMR xx or EIA CPR xx.

Figure 6.5 illustrates a typical waveguide. Table 6.1 lists the specifications of standardised waveguides and their dimensions.

During installation, particular care should be taken to prevent damage to waveguide. The waveguide should be treated as a finely tuned mechanical component. It is crucial that the waveguide remains clean inside and free of even small obstacles. In addition, when bolting different lengths of waveguide together, it should be ensured that the waveguide is correctly orientated and accurately aligned. Small ridges and misalignments can lead to significant signal attenuation, and thus to radio performance degradation.

Waveguide connections between the radio transceiver (outdoor unit) and the antenna should be kept to the minimum possible length in order to minimise the (unnecessary) radio signal attenuation, which detracts only from the radio link *range* and *availability*. It also minimises cost, since waveguide is a relatively expensive component.

From Table 6.1 it is clear that the signal loss per metre of waveguide increases rapidly for frequencies above about 10 GHz. For this reason, it is normal at these higher

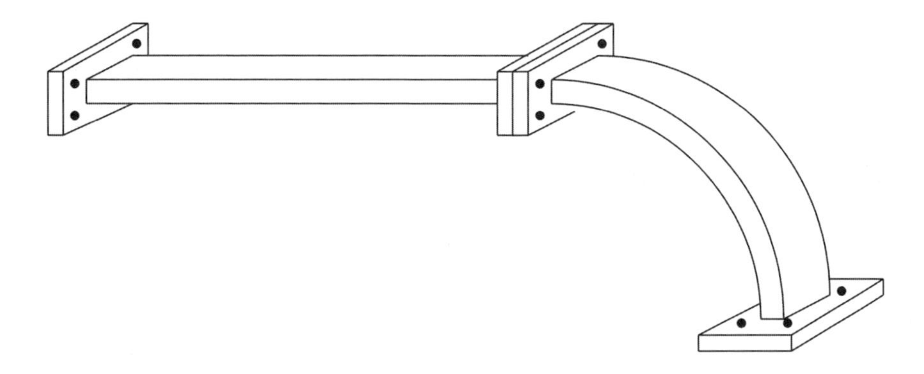

Figure 6.5 Typical waveguide

Table 6.1 Waveguide specifications

Waveguide designation	Radio band name	Frequency range	Attenuation	Waveguide dimensions (inside)
IEC R 3 (EIA WR 2300)	400 MHz	320–490 MHz	0.001 dB/m	580 × 290 mm
IEC R 14 (EIA WR 650)	L-Band	1.14–1.70 GHz	0.010 dB/m	165 × 83 mm
IEC R 22 (EIA WR 430)	W-Band	1.72–2.60 GHz	0.019 dB/m	109 × 55 mm
IEC R 32 (EIA WR 284)	S-Band	2.60–3.95 GHz	0.037 dB/m	72 × 34 mm
IEC R 48 (EIA WR 187)	C-Band	3.95–5.85 GHz	0.070 dB/m	48 × 22 mm
IEC R 70 (EIA WR 137)	6 GHz, 7 GHz	5.85–8.17 GHz	0.114 dB/m	35 × 16 mm
IEC R 100 (EIA WR 90)	10 GHz, 11 GHz	8.20–12.4 GHz	0.217 dB/m	22.9 × 10.2 mm
IEC R 140 (EIA WR 62)	Ku-Band (13 GHz, 15 GHz)	12.4–18.0 GHz	0.351 dB/m	15.8 × 7.9 mm
IEC R 220 (EIA WR 42)	K-Band (18 GHz)	18.0–26.5 GHz	0.723 dB/m	10.7 × 4.3 mm
IEC R 260 (EIA WR 34)	23 GHz, 26 GHz	22.0–33.0 GHz	0.868 dB/m	8.6 × 4.3 mm
IEC R 320 (EIA WR 28)	Ka-Band (38 GHz)	26.5–40.0 GHz	1.162 dB/m	7.1 × 3.6 mm

frequencies for manufacturers to design their systems for direct attachment of the radio unit to the antenna, since only a few metres of waveguide can detract significantly from the possible link *range*.

However, despite the system performance degradation which results from long waveguide runs, some radio network operators choose to use a relatively long waveguide connection to the antenna, in order to allow the remote placement of the outdoor unit (radio unit). Such operators, for example, may choose to place the radio unit at the bottom of the mast, so that field technicians can carry out maintenance work or equipment repair or replacement without having to climb the mast.

As well as rigid waveguide (as illustrated in Figure 6.5) there is also flexible waveguide. Flexible waveguide is typically more expensive than rigid waveguide and more prone to being damaged. It does, however, have advantages for certain types of antenna installations: for example, for the installation of *back-to-back antennas* connected only by waveguide, which are sometimes used as *passive radio repeaters* (we return to repeaters later in this chapter).

6.4 Outdoor Unit (ODU) or Radio Transceiver

The *Outdoor Unit* (*ODU*) comprises the *Radio Frequency* (*RF*) (also called *High Frequency* (*HF*)) transmitter and receiver. In simple terms, the ODU function is simply to 'translate' the (low frequency) digital signal received from the indoor unit into a radio signal in the transmit direction, and to convert the received radio signal into a digital signal to pass to the indoor unit. The content of the digital signal, its bitrate and synchronisation, as received from or passed to the modem is usually unaltered by the outdoor unit (radio).

Figure 6.6 illustrates a functional block diagram of a modern digital radio (outdoor unit). The incoming radio signal from the partner radio *terminal* at the other end of the link is picked up by the antenna and passed to the outdoor unit (radio unit) by means of the waveguide. A diplexor in the outdoor unit splits the incoming signal from the outgoing signal, which is being carried to the antenna for transmission. Optionally, a Low Noise Amplifier (LNA) is sometimes also used in the antenna or front end of the outdoor unit to amplify weak incoming signals. A low noise amplifier strengthens very weak signals without adding extraneous 'noise' — this helps to preserve the clarity and quality of reception. Low noise amplifiers are relatively expensive components.

The *Automatic Gain Control* (*AGC*) is a signal amplifier of variable strength which is designed to ensure that the signal strength on entry to the radio receiver is constant, independent of the signal strength received by the antenna. In other words, the AGC is designed to compensate for unpredictable radio signal losses during transmission caused by radio path attenuation (e.g. caused by bad weather).

The received radio frequency signal is demodulated to either an *Intermediate Frequency* (*IF*) or a *baseband* (*BB*) *signal* for onward transmission to the indoor unit over the *indoor-to-outdoor* connection cable. In the case of Figure 6.6, this is achieved in two stages, by mixing with demodulation signals. The first stage of demodulation is carried out by mixing a frequency to shift the RF signal to a given channel frequency within the radio band (e.g. channel 1 of the 26 GHz band). The second stage of mixing then demodulates the signal from this standard frequency to the standard IF frequency. By using two stages of demodulation like this, it is possible to adjust the frequency of the first demodulation stage by using software to control the synthesiser frequency used for the demodulation. This allows the receiver to be tuned to various different channel frequencies within the band. The second stage of demodulation can then be a 'standard' frequency shift corresponding

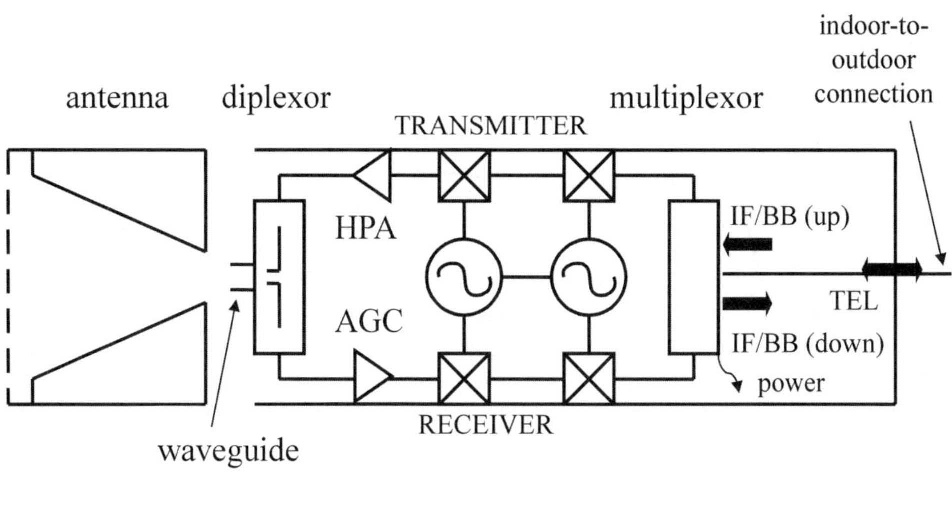

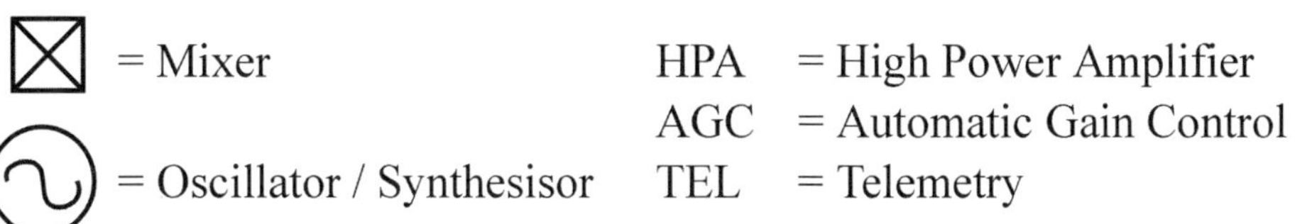

Figure 6.6 Functional block diagram of typical *outdoor unit*

to the radio band frequency — so allowing the use of a fixed frequency oscillator. Such a receiver design allows both accuracy and flexibility. The frequency 'shifting' is achieved simply by *mixing* and subsequent *filtering* of the resultant signal, as we discussed in Chapter 2.

In the case of Figure 6.6, the indoor-to-outdoor connection cable is by means of a single coaxial cable. This requires the use of two separate intermediate frequencies for the transfer of the incoming (*down* from 'outdoor' to 'indoor') and outgoing (*up* from 'indoor' to 'outdoor') signals. An intermediate frequency, as the name suggests, is a frequency intermediate between the original frequency of the user signal (comparatively low frequency) and the high frequency radio signal. A *multiplexor* is required to combine (in the downward direction) and separate (in the upward direction) the two intermediate frequency signals signals (see Figure 6.6). Alternatively, where *baseband* signals (i.e. the original low frequency user signals) are used on the indoor-to-outdoor connection, two separate cables may be required.

In the design of the outdoor unit shown in Figure 6.6, a telemetry signal is also added by the multiplexor, so that the outdoor unit can communicate with its respective indoor unit. This allows the control and monitoring of the outdoor unit from the indoor unit (or by means of a remote connection to the indoor unit) from a remote network management centre location.

Not shown in the receiver part of the outdoor unit of Figure 6.6, but sometimes necessary, is an *equaliser*. In cases where the incoming radio signal is carrying a relatively large amount of information (e.g. is a *high fidelity* signal, or a high bitrate signal coded with higher order modulation (as we discussed in Chapter 4), it may be necessary to add an equaliser. During transmission of a radio channel with a wide frequency bandwidth, it often happens that the frequencies at the lower end of the radio channel are attenuated relatively

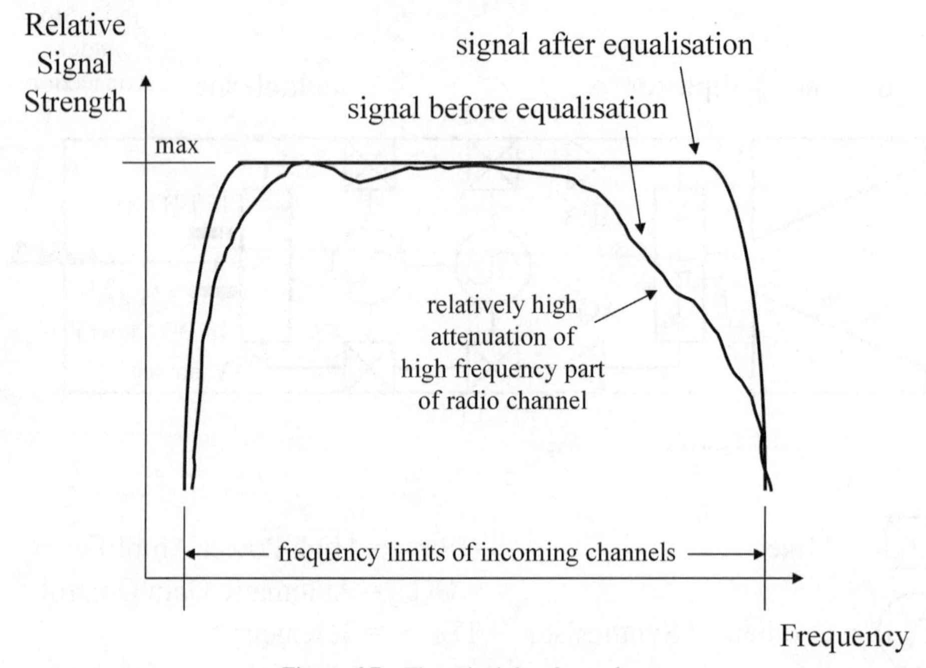

Figure 6.7 The principle of *equalisation*

more or less than the frequencies at the highest extreme range of the channel. An equaliser is designed to compensate this effect.

An equaliser has a function similar to the AGC already discussed. In particular, the equaliser is designed to compensate with different amplification the different frequency ranges within the overall incoming signal (Figure 6.7).

The transmitter part of the outdoor unit works like a receiver in reverse. The incoming signal from the indoor unit (received by means of the indoor-to-outdoor connection cable) is recived by the multiplexor (Figure 6.6). The direct current (power component) of this signal is removed to power the outdoor unit electronics. The remaining (user) signal is shifted from the baseband or intermediate frequency by means of the two stages of mixing. The first stage mixes an RF carrier signal to shift the signal into the appropriate radio band (e.g. 18 GHz, 23 GHz or 26 GHz, etc.). The second stage of modulation mixes a secondary RF signal component (controlled by a synthesisor) to shift the resultant signal to the correct radio channel within the band for transmission. The resultant signal is then applied to a *High Power* (radio frequency) *Amplifier* (*HPA*), and then conveyed by means of the diplexor and waveguide to the antenna for transmission.

The inside of an outdoor unit (high frequency radio unit) is not like the inside of a PC or other domestic electronic device. Rather than rows of digital electronic components, resistors and integrated circuits, there are bulky components made by high precision machining, for the diplexor, high power amplifier and mixers are all waveguide-like in appearance — accurately machined chambers and channels within a series of precision-manufactured metal blocks. The electronic components (if present) are most likely devices for controlling and monitoring the outdoor unit.

Once enclosed in its casing, the outdoor (radio) unit is a plain-looking article (Figure 6.8). However, its design is not without considerable challenge. The most stringent international specifications define 'extreme' ambient temperature range of operation (i.e the temperature of the atmosphere outside the casing) from $-33°C$ to $+55°C$. Inside the casing the temperature could be even higher (i.e. $+70°C$ or more). These temperatures must be withstood without any form of cooling or other ventilation. The casing itself must be water and weather-proof.

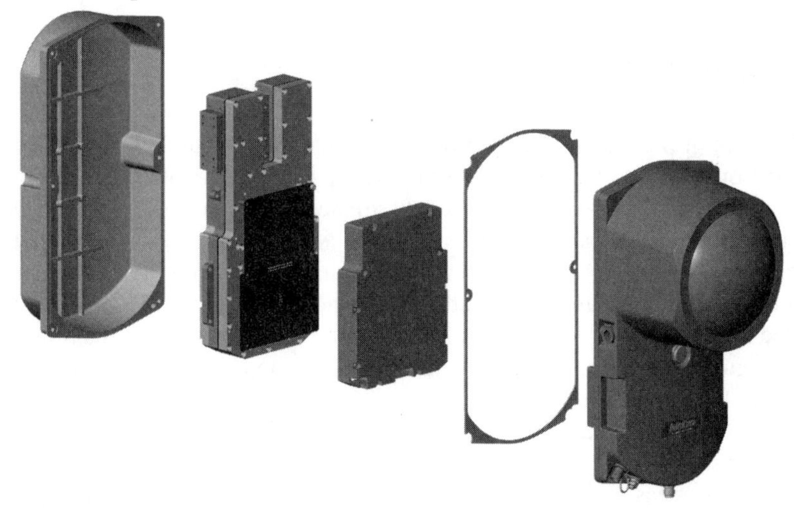

Figure 6.8 Typical radio *outdoor unit* — inside and outside views (Reproduced by permission of Netro Corporation)

If you look closely at Figure 6.8, you will notice that there are three external connectors to the outdoor unit. Two of these are for the waveguide (to the antenna) and for the coaxial cable connection (to the indoor unit). The third connector is for the attachment of a measurement device. This port allows a technician, during installation of the antenna and outdoor unit (ODU), to monitor the *Automatic Gain Control* (*AGC*) or the *Received Signal Level* (*RSL*). This value is used for the correct and exact *alignment* of the antenna.

The automatic gain control, as discussed earlier in the chapter, is an amplifier applied to the incoming radio signal to adjust its strength to a set given value. The more the signal has to be amplified to reach the set level, so the weaker the received signal must have been. Conversely, the less gain we have to apply, so the stronger the signal must have been. The amount of gain which needs to be applied by the AGC is thus a direct measure of the incoming signal (correctly, received signal level) strength.

The received signal strength is dependent upon the transmitted signal strength, the *path loss* (i.e. the signal attenuation incurred during radio transmission through the atmosphere across the link) *and* upon the *alignment* of the transmitting and receiving antennas. Since the transmitted signal strength can be set to a known value, and since it is normal also to install the antennas and radios during good weather (for which the path loss can be accurately calculated according to the length of the link, as we discuss in Chapter 7), then the AGC or RSL measurement can be used as a direct measure of the accuracy of the alignment of the antennas at either end of the radio link.

The third connector (or 'port') visible in Figure 6.8 is typically designed for the connection of a simple voltmeter. The voltage value presented at this connector is either a direct measure of the AGC or a calibrated value directly proportional to the RSL. Knowing the transmitter output power of the radio at the other end of the radio link; the length of the link and the gain of the two antennas (at each end of the link), it is possible to calculate (using the method presented in Chapter 7) the exact voltage to be expected as AGC or RSL. Only when both antennas are exactly aligned will it be possible to measure this value. If either or both antennas are misaligned, then the received signal strength will be considerably degraded.

6.5 Antenna Alignment

The alignment of high gain (highly directional) antennas is usually carried out at the same time as the installation of the outdoor unit (radio). First, both antennas are firmly mounted on their mountings (Figure 6.3) and crudely aligned on the correct bearing, pointing at the remote radio *terminal* at the other end of the link. Next the outdoor units (radio transceivers) are installed and powered on. Next, each antenna in turn is 'swept' (i.e. slowly rotated), first in a horizontal arc, then in an azimuth arc (vertical arc), until the AGC or RSL voltage (described in the previous section) achieves a value equivalent to the maximum strength of received signal (measured using a voltmeter connected to the AGC or RSL port — as described in the previous section).

Having achieved the maximum value, the first antenna is temporarily secured, while the second antenna is 'swept'. The antennas are swept in turn until the maximum received signal level is measured in both sweep directions by means of the AGC or RSL measurement port. The maximum measured value of AGC or RSL should be close to or identical with the predicted theoretical value. If there is a significant difference between the

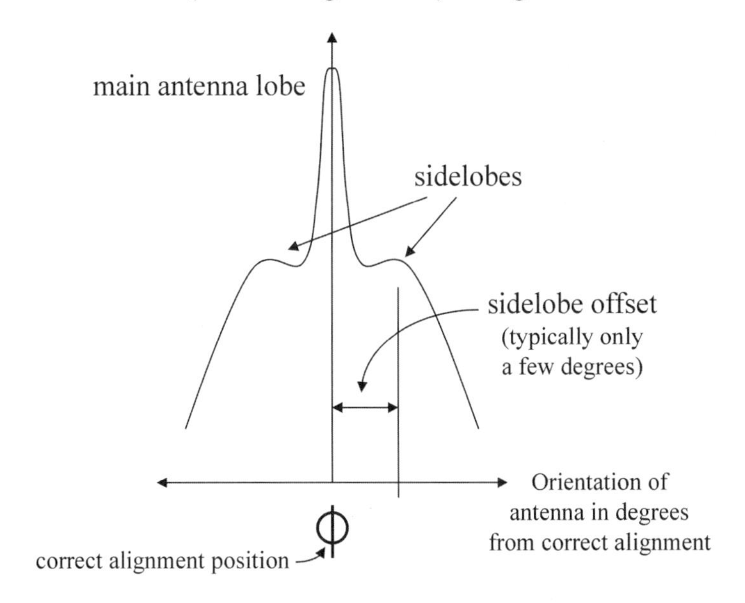

Figure 6.9 Take care for sidelobes when aligning high gain antennas!

measured and theoretical values it should be possible to explain the difference in terms of the path obstructions.

Figure 6.9 illustrates the typical RSL value measured when 'sweeping' the antenna across the bearing and/or azimuth orientation. It is no coincidence that the shape of the curve is the same as the 'lobe diagrams' of antennas which we discussed in Chapter 3 (the curve *is* the antenna diagram). Particular care has to be taken during the alignment process to ensure that the 'maximum' value measured on the RSL or AGC measurement port is the exact expected value at the peak of the main antenna lobe, and not merely the 'maximum' value corresponding to one of the sidelobes. It is easy to be lazy during installation and give up seeking the expected 'maximum' value of the main lobe in favour of a lower 'maximum' value corresponding to a sidelobe. However, this will greatly degrade the performance of the radio link during bad weather or heavy radio *fading* (as we shall discuss in Chapter 7). The goal is to install a high quality radio link — not a rain detector!

6.6 Qualities of a Good Outdoor Unit (Radio Transceiver)

The following characteristics differentiate the qualities of a radio transceiver (outdoor unit):

- the *maximum transmitter power* and the control of this power (e.g. *Automatic Transmit Power Control, ATPC*);
- the *receiver sensitivity*;

- the *receiver overload*;
- the *accuracy* and *stability* of the oscillator;
- the radio band and channel raster (or channel rasters — including duplex channel spacings) which are supported;
- the *channel size*, *modulation* techniques and bitrates which are supported (must conform to *spectrum mask*, as discussed in Chapter 2);
- the *tuning range* of the radio;
- the sensitivity to *adjacent channel interference*;
- the sensitivity to *co-channel interference*;
- the sensitivity to *cross-polar interference*;
- the generation of minimal *spurious emissions*;
- the means and ease with which the *AGC* (*Automatic Gain Control*) or *RSL* (*Received Signal Level*) voltage can be measured for purpose of *antenna alignment*;
- the remote and/or software configurability of the unit;
- the operating temperature range and relative humidity range;
- the surge current and lightning protection;
- the *Mean-Time-Between-Failure* (*MTBF*).

We discuss each of these briefly in turn.

Radio Transmitter Power and Power Control

The strength of the radio signal emitted by the radio transmitter is a key determinator of the overall system range. The stronger the signal, the more likely it is that the signal can still be correctly received and decoded at the remote end of the link even during periods of heavy *interference* or *radio fading* (e.g. caused by bad weather).

Typically, the output power of a millimetre wave or microwave radio is quoted in dBm (decibels relative to 1 milliWatt). The typical specified transmitter output power is in the range 15 dBm to 25 dBm. (At higher frequencies, e.g. 38 GHz, it is difficult to produce more than about 15 dBm, whereas an output power of 21 dBm for a 10 GHz radio unit is not uncommon.)

The formulae for conversion of the output power between dBm and Watts are as follows: (1 mW = 1 milliWatt)

$$\textit{Output power in dBm} = 10 \times \log_{10} \text{(output power in milliwatts divided by 1 mW)}$$

$$\textit{Output power in Watts} = 10^{\text{((output power in dBm} - 30)/10)}$$

However, while a high maximum transmitter power is good for ensuring maximum radio link range during inclement weather, it can also create a problem during good weather, since the strong signal tends to interfere more seriously with other radio systems using the same radio band.

Figure 6.10 illustrates two radio links operating in the same band, located at some distance from one another but with a similar orientation. The transmitter power of 'Link A' has been set to a maximum in order to overcome the attenuation caused by bad weather and still have the power to reach the remote terminal (Figure 6.10a). The problem is that during good weather this very strong signal has sufficient power not only to reach the remote terminal of Link A but also to overshoot into Link B (Figure 6.10b), causing unwanted interference. For this reason, it us usual during the *frequency planning* (see Chapters 7 and 9) of a new link to limit the transmit power used on each link.

(A) System link range is determined by the case of bad weather

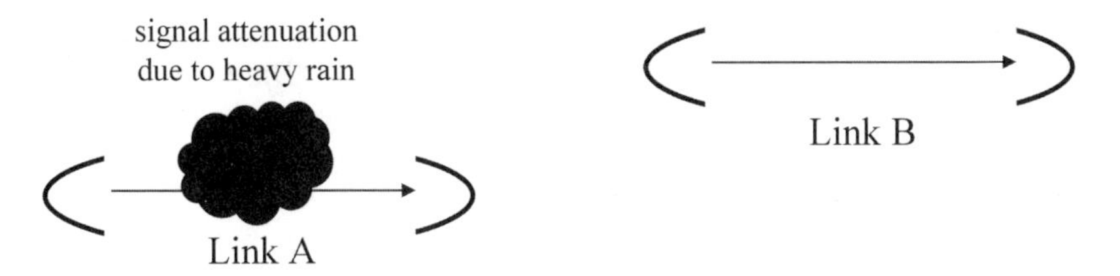

(B) Interference is greatest during good weather

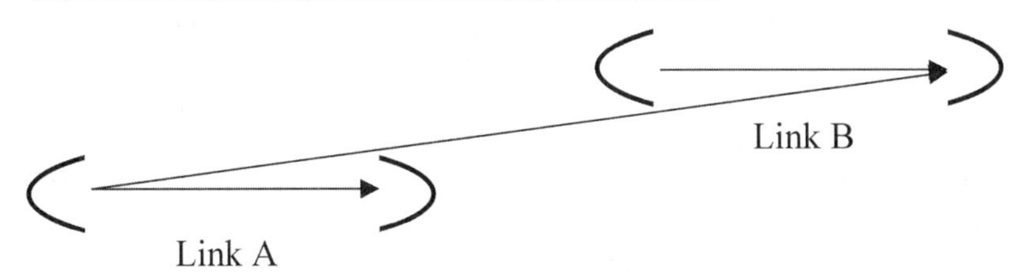

Figure 6.10 The problem of maintaining high transmitter power during good weather

Some radio system manufacturers offer a feature called *Automatic Transmit Power Control* (*ATPC*). This is designed to reduce the transmitter power of the radio during good weather, in order to minimise possible interference to neighbouring radio links. ATPC works by continuously monitoring the received signal strength at the remote terminal. Should the received signal strength drop (due to bad weather), then a signal is sent to the remote terminal to increase its transmitter power. Vice versa, if the received signal level increases above the level necessary for reliable reception, a signal is sent to the remote terminal to reduce output transmitter power.

Receiver Sensitivity

Like the maximum transmitter output power, the *receiver sensitivity* limits the maximum range of a radio system. The receiver sensitivity *threshold* is the minimum strength of received signal which the receiver is able to decode accurately. The definition of 'accurate reception' may vary. ETSI (European Telecommunications Standards Institute) nowadays defines the receiver thresholds as the values at which reception of a signal of a *maximum BER* (*Bit Error Ratio*) of 10^{-3} and 10^{-6} is achieved. However, manufacturers may quote the receiver sensitivity thresholds for different BER values (e.g. 10^{-3}, 10^{-6}, 10^{-9}) and recommend a different BER target for radio network planning purposes (so that their equipment meets its design operating performance).

Typically, the receiver sensitivity, like the transmitter power, is defined in units of dBm (decibels relative to 1 milliWatt). Typical values of receiver sensitivity in the Gigahertz (GHz) radio bands are around −70 dBm to −95 dBm, depending upon the data rate, the frequency of operation, the BER target and the modulation scheme. The larger the negative value of the receiver sensitivity, the more sensitive is the radio receiver. Thus, a radio with a sensitivity of −90 dBm will continue accurately to receive signals which are 20 dB weaker than the minimum strength signals which a receiver of sensitivity -70 dBm can receive.

When comparing the receiver sensitivity thresholds of different radio systems, you must make sure to compare comparable values. A threshold of -85 dBm for BER 10^{-6} may be equivalent to a threshold of -88 dBm for BER 10^{-3} (the size of the difference depends upon the radio design). In other words, you can get away with a weaker signal if the quality of the decoded signal does not have to be so good!

The simplest modulation schemes (as discussed in Chapter 4) are the most robust, and continue to be distinguishable even at very weak signal strengths. Thus, simple radio modulation schemes tend to offer the best receiver sensitivity but at the cost of relatively inefficient spectrum usage (the derived bits/s per Hertz). As the modulation scheme gets more complicated, and it becomes necessary to distinguish similar signals from one another during decoding, then it becomes necessary to maintain somewhat higher receiver signal levels, in order that the quality of reception remains acceptable. Radios using higher modulation schemes thus tend not to have such sensitive receiver thresholds.

Taking some concrete examples, a manufacturer of a given point-to-point radio system may use the same radio for different types of transmission. According to the particular modulation scheme used and the bitrate being carried, the receiver sensitivity threshold may vary. Typical values for different bitrates and modulation schemes are as follows:

- 26 GHz radio used as a 4-QAM (*Quadrature Amplitude Modulation*) PTP link for 2×2 Mbit/s links has a receiver sensitivity threshold of -88 dBm.
- The same radio, when used for 4×2 Mbit/s has a receiver sensitivity threshold of -85 dBm
- The same radio, when used with 16-QAM for carriage of 34 Mbit/s has a receiver threshold of -70 dBm.

Receiver Overload

The *receiver overload* is the maximum allowable signal strength at the receiver for good quality reception. As with the receiver sensitivity threshold, the value is quoted for a given signal bit error ratio. Signal strengths higher than this value will cause further bit errors. Much stronger signals still may even damage the receiver.

Frequency Stability

A radio system datasheet or specification will usually define the *frequency stability* of the transmitter and the methodology used to control the frequency. A typical frequency stability value is $\pm 0.0015\%$, achieved using *Phase Locked Loop* (*PLL*) circuitry. This means that the actual channel centre frequency of the chosen radio channel will deviate from the defined value by less than 0.0015% (also sometimes quoted as 15 ppm [parts per million]).

Radio Band of Operation, Raster and Duplex Spacing

Due to the complex nature of the components and radio circuitry necessary in the radio bands above about 1 GHz, radio systems designed for these ranges can usually only be operated in a restricted number of configurations. Thus, the individual radio units (i.e. individual hardware units) are restricted to a given radio band (e.g. 10 GHz ETSI or 24 GHz FCC, etc.), to a given channel raster (FCC or CEPT) and for a given *duplex spacing* (frequency separation of transmit and receive radio channels in duplex operation – - as discussed in Chapter 2). When choosing the correct radio unit, all of the values must conform with the individual project requirements. Usually, all of these factors are prescribed by local regulations.

Radio Channel Size, Modulation Scheme and Bitrate

Given the regulations which apply to a given radio band, it may be possible for radio network operators to exercise their own choice with respect to the radio *channel size* (e.g. 3.5 MHz bandwidth channel, 7 MHz, 14 MHz, 28 MHz, 56 MHz or 112 MHz, etc.), or in regard of the modulation technique to be used (e.g. 4-QAM or 16-QAM). These choices determine what bitrate may be carried across the radio link, and how much radio spectrum is consumed for the purpose. Spectrum may be saved by using higher modulation (e.g. 16-QAM instead of 4-QAM or 64-QAM instead of 16-QAM, etc.). This preserves the spectrum for use by other users and may save the operator spectrum charges. The 'downside' is the poorer receiver sensitivity threshold of the radio (this limits the link range, as we discussed earlier).

Only the most modern radios can easily be re-configured to accommodate different modulation schemes and bitrates. The ability to reconfigure the radio in this manner, however, may be an important differentiator for a 'fixed network operator' who wishes to be able to 'upgrade' his customers network capacity at short notice. In addition, by offering a smaller number of more flexible different hardware types, modern manufacturers are able to limit the range and number of different components which must be held in stock by operators as spare parts.

Radio Tuning Range

The *tuning range* of a radio has a significant impact upon the number of different individual hardware units a radio network operator has to use in his network. This therefore affects the number and type of different items which must be stocked as spare parts. The wider the tuning range of a radio, the more radio channels for which it is suited, and the wider is its possible range of use.

Figure 6.11 illustrates the structure of the 38 GHz CEPT band. This band is allocated by most European national regulators for PTP (point-to-point) usage. The total band (of almost 2.5 GHz bandwidth) is sub-divided into upper and lower sub-bands, each of approximately 1.25 GHz. The lower band is used for one direction of transmission and the upper band (always with a duplex spacing of exactly 1260 MHz) is used for the corresponding duplex channel (transmission in the opposite direction). For each PTP link, two radio frequencies are allocated for each duplex channel. (e.g channel 1 lower band starts at 37.058 GHz, its duplex partner starts at 38.318 GHz). Two different radio terminals are then used to provide the radio link; one terminal at each end of the link. At one end the transmit frequency is 37.058 GHz and the receive frequency is 38.318 GHz and at the other end terminal the transmit frequency is 38.318 GHz and the receive frequency is 37.058 GHz.

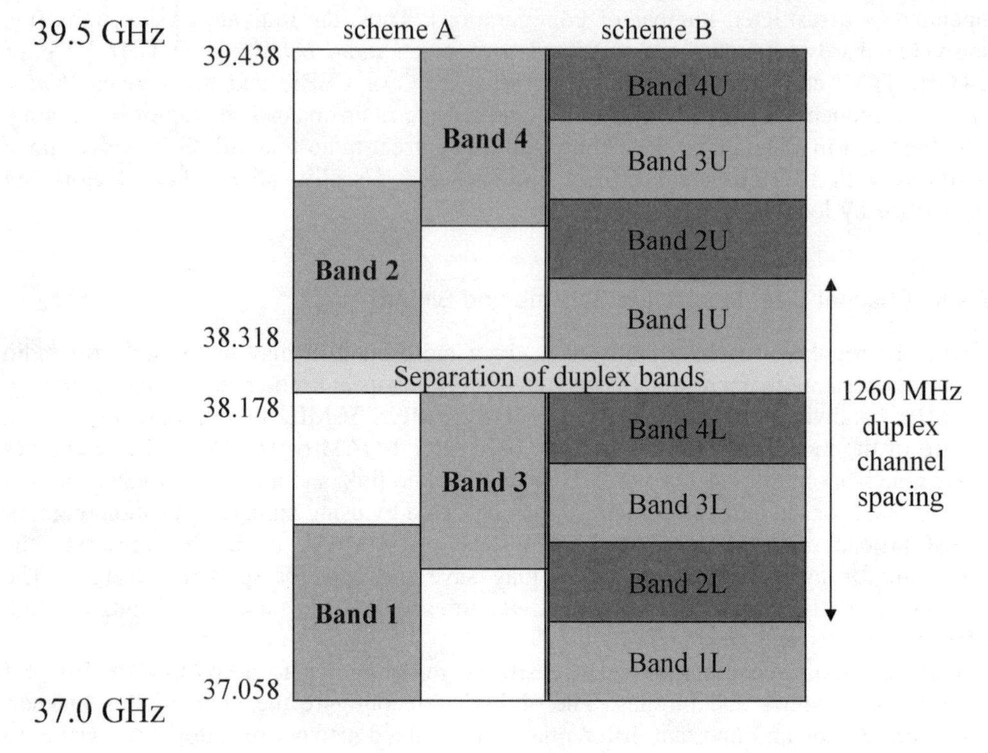

Figure 6.11 The effect of the tuning range of a radio — example in the 38 GHz CEPT band

Figure 6.11 shows how two different manufacturers (scheme A and scheme B in the diagram) have offered different types of outdoor unit radio hardware to cover the 38 GHz band. In scheme A, there are four different types of radio hardware (called 'Band 1', 'Band 2', 'Band 3' and 'Band 4'). Band 1 transmits in the lower frequency band (from about 37.0–37.6 GHz) and receives in the upper band frequency (from about 38.3–38.9 GHz). Meanwhile, the 'partner' radio (Band 2) transmits and receives in opposite frequency ranges (transmit 38.3–38.9 GHz and recieve 37.0–37.6 GHz). Because this manufacturer offers radios with a relatively wide tuning range (of about 600 MHz), only two pairs of radios (Band 1 + Band 2 plus Band 3 + Band 4) are required to cover the entire 38 GHz band. This equates to four different types of outdoor unit (hardware types).

In contrast, the second manufacturer (scheme B) only offers radio hardware with a tuning range of approximately 300 MHz. In this case, four separate pairs of radios are required to cover the entire band (Band 1L + Band 1U, Band 2L + Band 2U, etc.) (eight hardware types). Not only do the extra hardware types increase the number of spare parts which must be stocked, they also complicate the logistics of installation and service. You have to know which radio channel the link is to be operated in, before you can order the part for installation. Before swapping a defective unit during the operational phase, the field technician may have to return to the warehouse to pick up the correct spare part. In scheme A, a single pair of spare parts are good for nearly two-thirds of the band! If the operator decides to operate only in these lower channel frequencies, then the same pair of units can be used for any of the links!

Adjacent Channel Interference

A well-designed radio is relatively immune to interference from radio signal power in the *adjacent* radio channel. The relative immunity of the system to such *adjacent channel interference* is usually defined in terms of the maximum relative strength of signal in the adjacent channel which can be tolerated by the radio receiver while maintaining reception according to a given BER target. Maybe surprisingly, it is normal to be able to continue to receive correctly, even when the power of the interfering signal is many times stronger than the desired signal.

The value is usually quoted in terms of the required minimum ratio of the *carrier* signal strength (C) to the *interferer* signal strength (I) and given in dB. The parameter is often quoted as C/I (adjacent channel) = x dB.

A typical system value is C/I (adjacent channel, minimum) = − 10 dB to − 12 dB (for BER 10^{-6}). (the minus sign denotes that the *interfering* signal may be up to 10 dB to 12 dB stronger in power than the desired *carrier* signal).

Co-channel Interference

Co-channel interference is that caused primarily by distant radio systems using the same radio channel (i.e. interference like the overshoot of Figure 6.10).

As with the adjacent channel interference, co-channel interference value is usually quoted in terms of the required minimum ratio of the carrier signal strength (C) to the interferer signal strength (I) and given in dB. The parameter is often quoted as C/I (adjacent channel) = x dB. Unlike the adjacent channel interferer, a co-channel interferer (with the same polarization) must be considerably weaker in strength than the desired carrier signal if good reception is to be maintained (i.e. the C/I value is positive).

A typical system value is C/I (co-channel, minimum) of about + 13 dB (for BER 10^{-6}).

It is clear that a signal in the same radio channel is much more likely to cause interference than an adjacent channel. Hence, the much greater ratio (25 dB) of carrier signal to interferer required.

(Note: manufacturers conventionally quote C/I values for co-channel and other types of interferers according to relevant standards (e.g. ETSI standards). While it is clearly correct to do so, it is not always clear that the value quoted is only achievable in conjunction with a relatively strong received carrier strength. In other words, the network planner has either to reckon with a receiver sensitivity degradation in environments with strong interferers, or has to increase the C/I values he uses in his planning (i.e. he plans to reduce the interference).)

Cross-Polar Interference

As we shall discuss in more detail in Chapter 9, the re-use of radio frequency is sometimes crucial to the ability to obtain sufficient network capacity from the available radio spectrum. Sometimes, even the two different *polarisations* (*vertical* and *horizontal*) of the same radio channel are used as if they were two separate radio channels. For such cases, it is also important for the radio system to be relatively immune to interference from signals in the opposite signal polarisation (the so-called *cross-polarisation*). The separate polarisations are achieved by specially designed antennas. As with the adjacent channel interference and the co-channel interference, the *cross-polar interference* (the ‘polarisation interference') is usually quoted in terms of the required minimum ratio of the carrier signal strength (C) to the interferer signal strength (I) and given in dB. The parameter is often quoted as C/I (adjacent channel) $= x\,\mathrm{dB_c}$. A typical value for cross-polar discrimination is 35 $\mathrm{dB_c}$.

Since a signal-to-noise ratio needs to be maintained for good receiver decoding, and since discrimination is lost in practice due to antenna manufacturing and the tolerances of normal (simple) installation, as well as radio field effects, the typical value used for radio planning is around $-25\,\mathrm{dB_c}$ (for BER 10^{-6}) (in other words, as with the adjacent channel, the signal strength in front of the receiving antenna of the *interferer* of opposite polarisation may actually be considerably stronger than the desired *carrier*).

Spurious Emissions

Spurious emissions are unwanted signals transmitted by the radio transmitter outside of the frequency band corresponding to the spectrum mask of the intended radio channel. Regional standards and equipment approvals bodies specify and measure conformance to given acceptable levels of spurious emissions. Typically, it is usual to check the spurious emissions to a frequency of at least three times the operating frequency of the radio (i.e. to the *third harmonic* frequency). It is typically at the *harmonic* frequencies (multiples of the desired frequency) where problems may be encountered.

Automatic Gain Control Voltage or Received Signal Level Voltage Measurement and Calibration

As discussed earlier in the chapter, it is common to present a measurement port on the outdoor unit (radio unit) for purpose of assisting in the alignment of the antenna(s). This port usually presents a measurement voltage, either related to the AGC (Automatic Gain Control) voltage or RSL (Received Signal Level), as we discussed. The availability of such a port on the outdoor unit greatly eases the operational logistics associated with antenna alignment. However, in addition, the ease with which the presented voltage value can be measured varies greatly from one different type of equipment to another.

In some (generally older) types of equipment, one of the voltages within the AGC circuitry is presented directly. When this AGC voltage is used, it often has to be individually calibrated for each individual hardware unit. In other words, a pre-installation laboratory calibration of each link is required. During this calibration exercise, the equipment is teporarily set up as a 'link' in the laboratory, and various calibrated signal strengths are applied to the receiver. For each received signal strength, the actual AGC voltage is measured and plotted. During the actual installation and alignment, the calibration chart is used to determine the correct antenna alignment.

In more recent equipment, some manufacturers have sought to provide a voltage at the measurement port which is directly and linearly proportional to the RSL, thereby eliminating the requirement for pre-installation calibration of individual hardware pieces.

Outdoor Unit Configuration and Monitoring by Software

As with other types of telecommunications equipment, the most modern fixed wireless access radio systems allow for remote reconfiguration and monitoring in software. The range of the parameters which can be measured and controlled (the so-called *Management Information Base* (*MIB*)) determines how much can be done remotely. Ideally, the MIB is an open, standardised MIB, based upon a standard management protocol such as *SNMP* (*Simple Network Management Protocol*) or *CMIP* (*Common Management Information Protocol*).

Operating Temperature Range and Allowed Humidity of Outdoor Units

Since the outdoor unit usually has to be fully enclosed and weather-proof, it is difficult for its *environment* to be 'climate-controlled' (e.g. air conditioned or fanned). Radio units differ in their extent to withstand extreme temperature and humidity conditions. The 'extreme' climate ranges specified by ETSI for European countries (e.g. in the Alps) are:

- Ambient temperature from $-35°C$ to $+55°C$;
- Ambient relative humidity from 0–95%.

Well-designed units are also capable of withstanding prolonged periods of vibration, as may be experienced when mounted on single pole masts which may tend to vibrate somewhat in strong winds.

Surge Current and Lightning Protection

Since the outdoor unit is exposed to the weather, and in particular is attached electrically to a highly directional antenna in a position which is likely to be prone to lightning, it is usual for the radio to be designed in such a manner that it can withstand surges of current appearing from either the antenna or the waveguide.

Outdoor Unit Mean Time Between Failures

Since high frequency radio components are difficult to manufacture, it is the reliability of the outdoor unit (radio unit) which largely determines the reliability of a radio system. This reliability is usually quoted in terms of the *Mean-Time-Between-Failures* (*MTBF*). A typical MTBF (per radio link) achievable in practice is in the range 10–12 years. In other words, a typical link will fail on average every 10–12 years. Or said in another way, for each 10–12 years of operation within a network there will be one failure on average (i.e. about 1 of each 10–12 links fail in any given year, i.e. 8–10% fault rate of links per year).

When comparing the MTBFs of different systems, you need to make sure that like-for-like are compared. It is not correct to compare the MTBF of a *terminal* with the MTBF quoted by a different manufacturer for an entire *link*. The reciprocal values of the MTBFs must be added. Thus if the link comprises two terminals, of terminal MTBF M_1 and M_2, then the link MTBF M_L will be given by the formula

Adding MTBFs (mean-time-between-failure)

$$\frac{1}{M_L} = \frac{1}{M_1} + \frac{1}{M_2}$$

so that if both terminals have an MTBF of 10 years, then the 'link MTBF' will be only 5 years!

Obviously, practical MTBF values can only be measured reliably when thousands of radio units have been installed and have been in operation for several years. It is thus necessary for new equipment to use theoretical calculation means for estimating MTBFs. There are several recognised calculation methodologies which have been standardised by different technical standards bodies. These methods calculate the theoretical MTBF based upon the total number of components and upon the reliability of these individual components. Generally, such theoretical calculations underestimate the MTBF achieved in practice (assuming good quality manufacturing and proper care during installation and operation).

Factors which tend to limit the lifetime of a radio unit are component *ageing*, long-term *temperature cycling* and prolonged operation of the system at full power. Ageing is the 'wearing out' of the components over time. The process is speeded up in cases where the system is used continuously at its limits (e.g. at full power), or is subject to continuously varying extreme environmental conditions (e.g. daily temperature cycles from direct sunlight to very cold freezing nights). The more 'adventurous' a manufacturer is in specifying the maximum transmitter power of his radio, the shorter the *equipment lifetime* (and thus MTBF) is likely to be.

6.7 Indoor-to-Outdoor Cabling

Nowadays it is common to use a single coaxial cable for attachment of the indoor unit to the outdoor unit, and to provide power to the outdoor unit over this cable, thereby eliminating the need for a power supply on the roof or mast-top.

A cable specified for the carriage of an Intermediate Frequency (IF) signal from indoor unit to outdoor unit, and vice versa, is fully specified in terms of the following parameters:

- Maximum loss of the highest IF frequency (e.g. 30 dB);
- Nominal cable impedance (typically 50 Ohms, 50 Ω);
- Maximum loop resistance (of the inner and outer conductor, in order that there is not undue loss of the power feed) around 5 Ohms, 5 Ω;
- Cable type (e.g. screened or double screened);
- Sometimes also the frequency response of the cable is important. (All frequencies within the signal should be attenuated and carried approximately equally.)

A series of different types of coaxial cables are standardised by EIA using the commonly used nomenclature for such cables, RG/U xxx, where xxx represents a series of integer numbers for different cable specifications. These cables differ in their diameter, performance and price. A summary of such cables is provided in Appendix 5.

In general terms, it can be said that smaller diameter cable tends to be cheaper and easier to install, but suffers from greater signal attenuation per metre of cable. For longer cabling runs between indoor and outdoor units, wider diameter cabling may be necessary. While there may only be a few storeys of a building between the equipment room in the basement where the indoor unit is installed to the roof (where the outdoor unit is installled), the cable run length may be much greater than you initially expect, due to the layout of conduits and cabling ducts within and between the different floors (storeys) of the building.

When installing cables between floors or rooms of a building, there are usually local building and fire regulations which must be observed. It is common practice, particularly between the different floors of a building and different 'fire cells' within a given floor, to 'seal' the conduits at the floor or 'cell' boundaries with mortar (cement) as a protection against the spread of fire along cable conduits (since cable sheathings are typically highly flammable).

When installing cable, you should take care not to bend the cable to a corner which is tighter than the specified *minimum bending radius* of the cable (typically 20 mm or more). Bending the cable too tightly can lead to unexplainable serious radio performance degradation during operation, due to excessive signal attenuation or distortion in the cable, or (according to some modern national regulations), after each change of direction from vertical to horizontal.

The indoor-to-outdoor cable should be earthed at the point of entry from the roof into the building, to prevent current surges and lightning bolts being transmitted into the building or to the indoor unit. In addition, the cable sheath should also be earthed at a point near to is connection with the outdoor unit. Where there is also a significant length of cable between the outdoor unit and the building entrance point, it is normal to provide further cable earthing points, approximately every 20 metres.

6.8 Indoor Unit (IDU) or Radio Modem

The *Indoor Unit* (*IDU*) of a radio terminal comprises the radio *modem*, the user port interfaces for connecting external user communications equipment, the control circuitry and the power supply circuitry. In simple terms, the IDU function is to convert the digital input signal received from the end users communication equipment into the modulated form necessary for radio transmission. First the signal is converted to an intermediate frequency format for transmission to the *Outdoor Unit* (*ODU*). The content of the digital signal carried by the radio (outdoor unit), its bitrate and synchronisation, are all controlled by the indoor unit.

Figure 6.12 illustrates the functional block diagram of a typical indoor unit of a digital radio terminal. The unit comprises seven basic functions, which we shall discuss in turn:

- User data multiplexing, conversion and/or protocol preparation for radio transmission;
- *Forward Error Correction* (*FEC*) function;
- Modulation/demodulation (modem function);
- Terminal control unit;
- Telemetry signalling for intercommunication of indoor and outdoor units;
- Power supply unit;
- Cable multiplexor.

User Data Multiplexing and Transmission Protocol Preparation

The indoor unit receives data from end user communications equipment (e.g. a telephone, a computer, a data network, a *Local Area Network* (*LAN*)) in one of a number of possible standardised interface formats. The number of ports and the type of ports provided on the indoor unit varies from one radio equipment to another. Typical point-to-point (PTP)

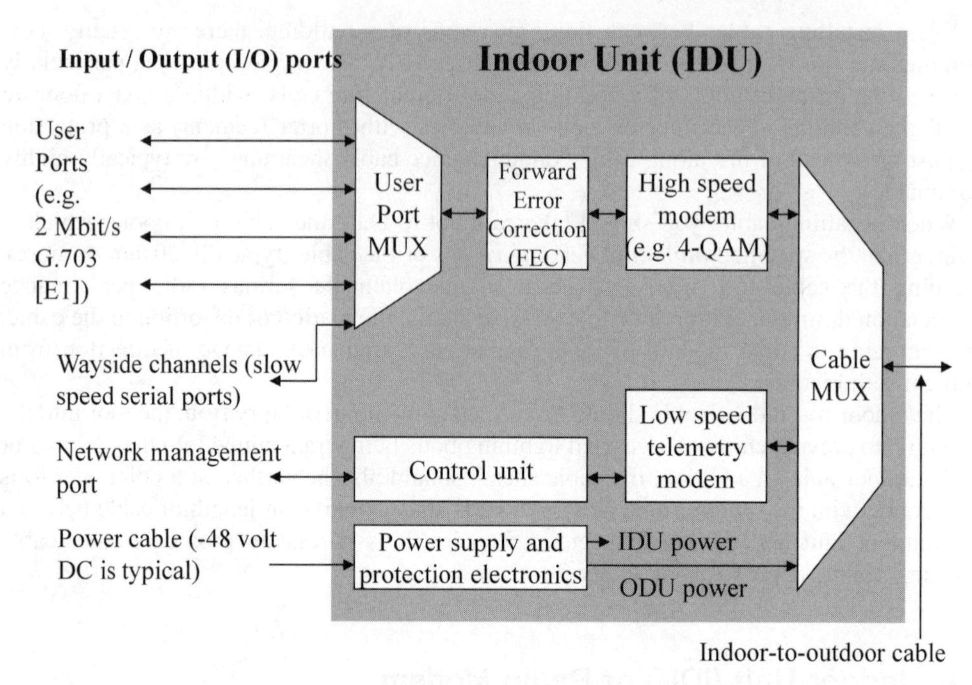

Figure 6.12 Functional block diagram of typical radio terminal indoor unit

microwave radio systems offer multiple 2 Mbit/s (so-called *E1-interface*) or 1.5 Mbit/s (so-called *T1-interface*), for example 2 × E1, 4 × E1, 8 × E1, etc. The PTP radio link thereby provides the equivalent of a direct wire connection between the two corresponding E1 ports at either end of the *link*. Point-to-multipoint (PMP) systems, meanwhile, seek to consolidate and switch the connections within the radio network. Thus in PMP networks, the interface type and the configuration of the individual end-user connections may be changed from one end of the link to the other, as a result of the switching undertaken within the network.

The user input/output (I/O) ports are presented to a multiplexing and/or switching device within the indoor unit. This device may perform any number or combination of three basic functions:

- Multiplexing of the individual connection data streams into a single highspeed digital bitstream;
- Switching and/or concentration of the individual data streams into a single, formatted digital bitstream;
- Signalling of control information (i.e. *protocol information*) to enable the two radio terminals at either end of the link to optimise the use of the radio channels and other resources within the particular PTP or PMP radio system.

A simple multiplexing function is included in most PTP radio systems in cases where multiple connections are carried simultaneously from one end of the link to the other (e.g. 2 × E1 PTP system or 4 × E1 PTP system). This is shown as the 'User port MUX' in Figure 6.12.

In contrast to PTP, the 'User port MUX' of Figure 6.12 is replaced in most PMP systems by a switching and/or concentration function and a communication function for sending control and coordination messages (so-called *protocol control information)* between base stations and remote end-user terminals. Typically the base station indoor unit of a PMP system comprises a real 'switch' function (so that individual remote end users can be connected to one another in any combination). Meanwhile the individual remote end-user terminals comprise a 'concentration' function which enables them to 'activate' or 'deactivate' their usage of the radio channel (i.e. the 'air interface'), restricting their usage to the times when they have a need to support an active connection. So that this switching process (and radio channel allocation process) runs smoothly and in an coordinated manner between base station and remote end-user terminals, it is usually necessary to add *Protocol Control Information* (*PCI*) (i.e. system control messages).

The end-user input/output (I/O) ports may take any number of different forms. PTP systems typically offer *serial port* interfaces conforming to one of the common leaseline interface types, either ITU-T G.703 (e.g. E1 [2 Mbit/s], T1 [1.5 Mbit/s], E3 [34 Mbit/s], T3 [45 Mbit/s], STM-1 or OC-3 [155 Mbit/s], X.21 [typically n × 64 kbit/s to 2 Mbit/s] or V.35 (typically $n \times 64$ kbit/s to 2 Mbit/s]. Such interfaces are carried by PTP systems 'transparently' (i.e. without alteration). The signals retain the same bitrate and *synchronisation*. The *jitter* of the signal (slight variation of the pulse lengths and timing over time) is also largely unaffected by the radio system.

Point-to-multipoint (PMP) systems also offer leaseline-like interfaces like PTP systems, but in addition, many systems also interpret both data *protocol* and voice *signalling* information to determine the desired destination of switched connections. PMP systems therefore typically also support one or more of the following interface types:

- *Plain Old Telephone Service* (*POTS*), also called *a/b-interface* (analog telephone signalling);
- *Basic Rate ISDN* (*BRI*);
- *Primary Rate ISDN* (*PRI*);
- *Channel Associated Signalling* (*CAS*) (Analogue telephone signalling via digital line plant);
- *Frame relay*;
- *Ethernet* (*10baseT or 10/100baseT*);
- *Internet Protocol* (*IP*);
- *ATM* (*Asynchronous Transfer Mode*).

We return in later chapters to the 'network integration' of these interface types (i.e. how to build a public switched data or telephone network incorporating PMP radio and offering these interfaces to end customers).

Before we leave the subject of the end-user input/output (I/O) ports, we should also briefly mention the *wayside channels*, *engineering order wire* and *external alarm relay* channels, which are typically also available (these are particularly common in the case of PTP systems). One or two *wayside channels* are typically made available from one end of a PTP system to another. A wayside channel is a low bitrate additional data connection between the two ends (thus a 2 × E1 or 4 × E1 PTP system might carry 2 × 9.6 kbit/s or 2 × 64 kbit/s *wayside channels* in addition to the main *payload*).

Wayside channels are intended for the interconnection or networking of 'peripheral' equipment in the remote operations location. Thus, for example, over the wayside channel of a PTP link interconnecting a main switch location to a remote operations room, it would be possible to connect directly to the control interface of remote switching equipment for network management purposes.

The ability to relay of *external alarms*, like the ability to carry of wayside channels is often built-in to radio systems to allow remote monitoring of remote operational locations which are connected to the main switch or network backbone site only by radio. Thus, for example, by relaying the 'door open' signal or the 'room temperature alarm', the network operations staff can be kept abreast of problems at the remote site.

An *engineering order wire* is another form of wayside channel, and one also intended for network operations staff usage. Specifically, the engineering order wire is a telephone connection between the two ends of the link which allows technicians during installation and maintenance work to talk with one another. With the increasing use by field technicians of mobile telephones, the need for engineering order wires has somewhat receded.

Forward Error Correction (FEC)

Because radio systems are susceptible to fading and, in particular, to *burst fading* (i.e. large runs of corrupted bits), it is usual in digital radio systems to employ *Forward Error Correction* (*FEC*) to detect and correct errors in the received bitstream at the receiver end.

The 'forward' nature of the error correction allows the detection and correction of the errors without the need for retransmission of the digital information which would otherwise result in unacceptable further delays to the signal. An advanced algorithm is used for the correction mechanism. An example of such an algorithm is a Hamming or a Reed–Solomon code.

To illustrate the principle of forward error correction we consider the form of FEC which is standardised as part of the *Digital Video Broadcasting* (*DVB*) standard.

A transmission *frame* is built around each *block* of so many consecutive bytes (say 187) of *user information*. To each block is added as a *header* a *synchronisation pattern* (typically of 1 byte in length). The synchronisation pattern identifies clearly the beginning of each frame. Following the synchronisation pattern and user information block, a *check sequence* or *Frame Check Sequence* (*FCS*) is added (e.g. of 16 bytes in length). The overall *frame length* in our example is thus 204 bytes, subdivided as we illustrate in Figure 6.13.

The frame check sequence may take any of a number of different forms, whereby the most commonly used codes in radio systems are *Hamming codes*, *Viterbi codes*, *BCH-codes* and *Reed–Solomon codes*. We detail in Appendix 6 the various codes, but to get at

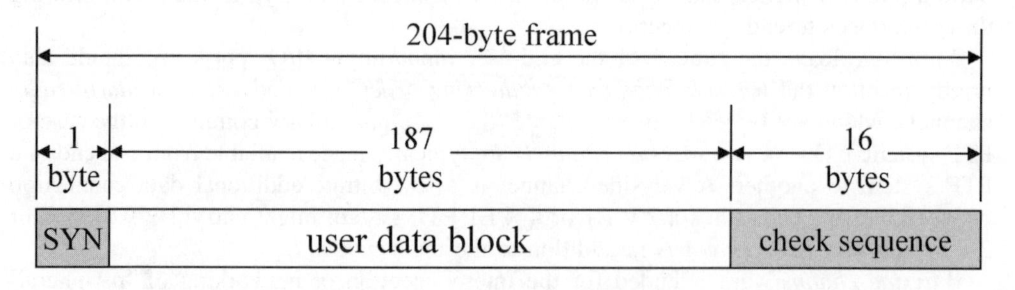

Figure 6.13 Typical framing structure for forward error correction

(A) correct [i.e. transmitted) code for user data '0111'

check bit 1	check bit 2	bit 3	check bit 3	bit 5	bit 6	bit 7
1	0	0	0	1	1	1

(B) received code with error at bit 6

check bit 1	check bit 2	bit 3	check bit 3	bit 5	bit 6	bit 7
1	0	0	0	1	0	1

(C) received code with error at bit 7

check bit 1	check bit 2	bit 3	check bit 3	bit 5	bit 6	bit 7
1	0	0	0	1	1	0

Figure 6.14 Simple *Hamming code* for error detection and correction

least some appreciation of how we can detect and correct *bit errors*, consider the simple Hamming code error detection and correction methodology illustrated in Figure 6.14.

Figure 6.14 illustrates the addition of a three bit code integrated into the 4-bit user data pattern, which occupies bit positions 3, 5, 6 and 7. The three *check bits* are being used to make up *even parity*, respectively for the bit combinations 5–6–7; 3–6-7 and 3–5–7 (Figure 6.14a). Thus, because the values of bits 5, 6 and 7 contain an *odd* number of values equal to '1', the first check bit is set to value '1'. This gives an *even* number of values '1' over the four bits 5-6–7 plus the first *check bit* or *even parity bit*. The second *check bit* provides for similar even parity over the four bits 3–6–7 plus the second check bit. The third check bit provides for *even parity* over bits 3–5–7 plus the third check bit.

The code has been designed to detect and correct a single bit error during transmission, without the need to retransmit the sequence. We consider the case of a single bit error affecting the user data during transmission (i.e. corrupting one of the bits 3, 5, 6 or 7). Figures 6.14b and 6.14c illustrate the received code in the case of bit errors either at position 6 (Figure 6.14b) or position 7 (Figure 6.14c).

During the detection process, the check bits are used to check the even parity again, and values are given to each of the check bits to indicate whether there is a parity error or not. In the case of incorrect parity, value '1' is assigned. Meanwhile '0' represents correct parity. The checks of parity on receipt (in the case of Figure 6.14b) yield the following results:

- Check bit 1 of the received pattern plus bits 5–6–7: received:1 calculated:0
  incorrect, set value 1
- Check bit 2 of the received pattern plus bits 3–6–7: received:0 calculated:1
  incorrect, set value 1
- Check bit 3 of the received pattern plus bits 3–5–7: received:0 calculated:0
  correct, set value 0

The resulting bits signal the binary value '110' — in decimal value '6'. In other words, we have detected an error in the bit at position 6, so we are also able to correct the error.

In Figure 6.14c, the values of the received parity checks are '100'. This compares with the calculated values '011'. So that the check bits 1, 2 and 3 are respectively 1, 1 and 1 or '111' (decimal value '7'). (As an aside, you might like to note that the comparison of the two values '100' and '011' is achieved easily by *modulo 2 addition*. This is an important assistance for electronic circuit designers.)

In the case of no errors, the receive end parity check will deliver the value '000'.

So, you might ask, what about an error occuring in the string of parity check bits? The answer is simple, we add another parity bit to check whether an error affected the other parity bits. If this parity bit is incorrect on receipt, then we ignore the other parity check bits and make no corrections to the user data bits. On the assumption that there will be a maximum of 1 bit in the sequence which is corrupted, we are able to *detect* and *correct* the error!

Since radio transmission is prone to *burst errors* (strings of consecutive bit errors) rather than single bit errors, it is normal to *randomise* the order of the user data bits and check bits within the user data block of Figure 6.14. This has the effect of distributing the errors across the different individual connections carried within the block. This has two advantages. First, it is easier to design correction codes to cope with individual error. Secondly, it is less likely that a single end user will have to shoulder the burden of any *unrecoverable* errors. We illustrate this technique in Figures 13.3 and 13.4 of Chapter 13.

More advanced forward error correction codes work in a similar manner to our example. The various codes have different relative strengths and are designed for different length user data blocks and different error rates.

Before we leave the subject, you might also have observed that we have 'used' 4 of the 8 transported bits for the error correction and detection, and thus reduced the maximum user bitrate we can carry. This is a correct observation. Typically, we have to use more powerful FEC to improve the reliability of transmission across radio links using the higher modulation schemes which we discussed in Chapter 4. It is thus true to some extent that what we can in higher bitrate with higher modulation schemes we must give up (at least in part) to carry the extra bits needed for the FEC. It is up to the radio system designer to ensure the optimum mix of modulation, and FEC to achieve the required user data bitrate and received signal quality.

Modulation and Demodulation (the Radio Modem)

The radio modem unit of Figure 6.12 performs the modulation of the data signal to be transmitter, converting it from digital format into intermediate frequency using the techniques we discussed in Chapter 4. It also performs demodulation of the incoming intermediate frequency signal, for presentation to the input/output ports.

Terminal Control Unit

The terminal control unit of Figure 6.12 comprises the electronics which monitor the radio terminal as a whole, and respond to software commands received from a network management system or other equipment control or configuration terminal. Modern equipment is typically designed to be largely software-based, and to conform with standard software and network management interfaces (including SNMP and CMIP) and is capable of being monitored and configured from a remote location.

The control terminal or network management terminal may take a number of different forms. An external control terminal (e.g. a personal computer) is designed to be attached to the indoor unit of Figure 6.12 for purposes of monitoring and configuring the indoor unit and the outdoor unit of the same terminal. In most systems, a communication channel via a network management wayside channel over the link also allows the same (local) control terminal to be used to monitor also the remote terminal (both indoor and outdoor units).

Telemetry Communication between Indoor and Outdoor Units

The telemetry modem is used to create a communications channel for network management, monitoring and configuration purposes, allowing the outdoor unit to be controlled from a control terminal connected to the indoor unit. This allows even mast-top equipment to be diagnosed and reconfigured without having to climb the mast. The telemetry modem need only use a relatively simple modulation technique. The signal is transmitted only as far as the outdoor unit. Network management communications with the remote terminal, by contrast, need to be incorporated in *wayside channels*, as we discussed earlier in the chapter, in conjunction with the user port multiplexor.

Power Supply Unit

Most radio equipment is designed to work from a direct current power source, typically 24 volt DC (in North America) or −48 volt (in Europe). Typically, the equipment will be rated at for the given voltage (e.g. −48 volt), but also permit the use of an 'unsteady' supply provided the supply remains within a given tolerance band (e.g. tolerance band typically ±20%, say −39 volts to −62 volts). The power supply should conform with the nominal required voltage. A normal supply of −60 volts (given the above tolerance range) will not do!

Typical consumption of a radio terminal is about 50 Watt per modem. Much of this power is 'wasted' in the electronics rather than transmitted as radio signal energy. The unit cooling or airconditioning therefore needs to compensate this heat dissipation. Particularly, the indoor unit may be equipped with fans for forced cooling or have a perforated casing to allow convective cooling. The heat dissipation can be a problem in restricted spaces packed with other equipment or modems. The ventilation holes in the casing can also be a problem if there is any possibility of water intrusion—for example, due to the collection of condensation.

Sometimes indoor units are placed in 'outdoor cabinets' (e.g. small weathertight air conditioned cabinets). Such installations need to be planned carefully in order to keep the indoor unit maintained within its specified operating temperature and humidity range (typically 0°C to +40°C and humidity 0–95%, but *non-condensing*). The elimination of condensation can be a challenging problem!

Cable Multiplexor

The cable multiplexor (Figure 6.12) serves only to share the use of the indoor-to-outdoor connection cable between the modulated user data signal, the control telemetry communication and for the conveyance of power to the outdoor unit.

6.9 Radio Repeaters

Figure 6.15 illustrates three different types of *radio repeaters* used in *radio relay systems*. These include two types of *passive repeater* and one type of *active repeater*. Passive repeaters serve to 'bend' radio signal transmission to reach areas which might otherwise receive no reception. A typical usage for a passive repeater is to provide for radio signal coverage behind an obstacle (e.g. perhaps the remote end of a PTP link is on the rooftop of a relatively low building which is hidden by a neighbouring tall building. As shown in

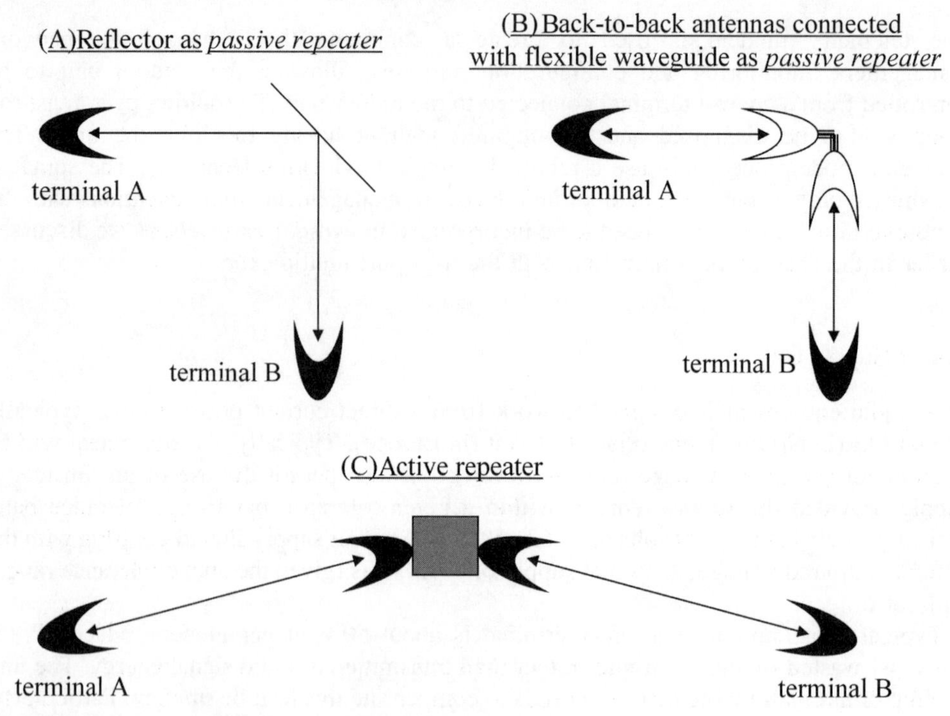

Figure 6.15 Repeater types used in radio relay systems

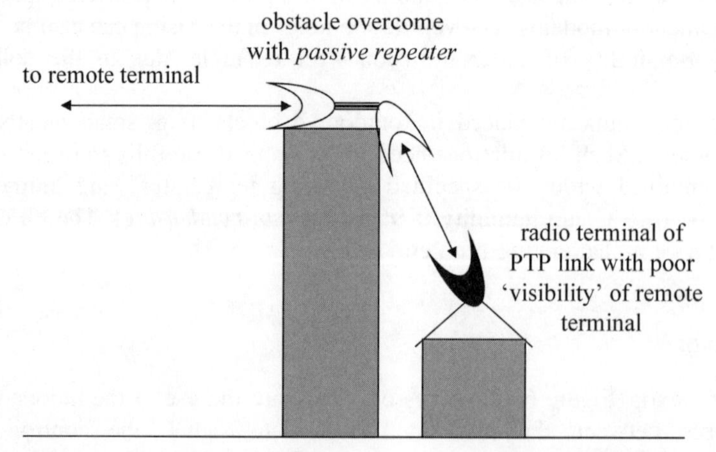

Figure 6.16 Typical use of a *passive repeater* to deflect radio signals around an obstacle

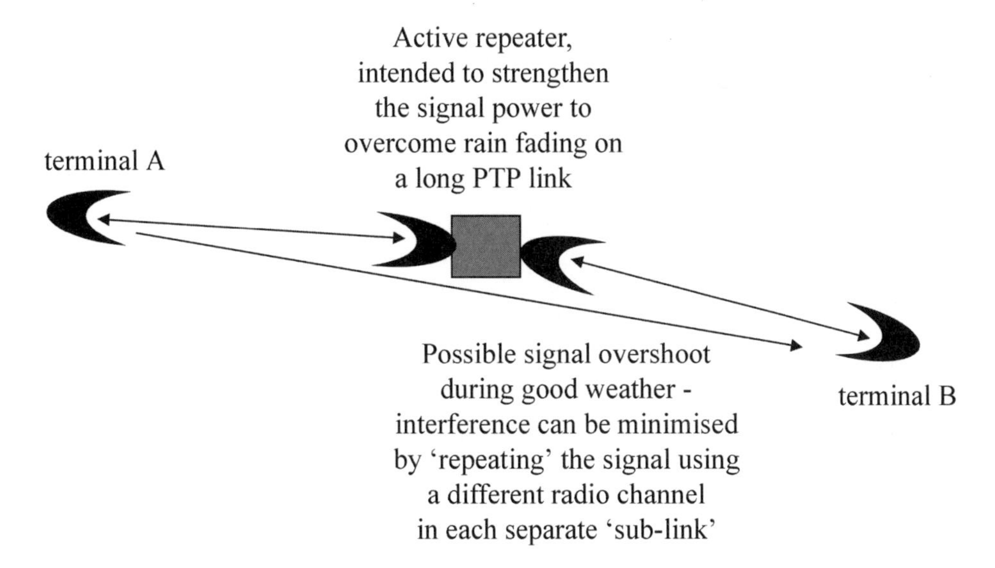
Active repeater,
intended to strengthen
the signal power to
overcome rain fading on
a long PTP link
terminal A
Possible signal overshoot
during good weather -
interference can be minimised
by 'repeating' the signal using
a different radio channel
in each separate 'sub-link'
terminal B

Figure 6.17 Using an active repeater to boost signal strength and change radio channel to prevent overshoot interference of the repeated sub-links

Figure 6.16, the signal coverage behind the obstacle can be improved by using a passive repeater.

Passive repeaters typically take the form of either *reflectors* (Figure 6.15a) or of back-to-back antenna configurations (Figure 6.15b). Reflectors are generally cheaper, but more difficult to install and *align* correctly. In contrast, *back-to-back antennas* are more expensive, but easier to align. In effect, the antennas of the two sub-links (either side of the repeater) are aligned separately as PTP links and then simply linked together using flexible waveguide. A passive antenna thus serves to deflect the available signal strength to the destination where it is needed. It does not strengthen the signal power, but does not need a power source at the repeater location and is relatively cheap.

In contrast to *passive repeaters*, *active repeaters* (Figure 6.15c) are relatively expensive. They also require a power source at the repeater location, but they offer the additional benefit of being able to boost the radio signal strength, thereby making longer links possible. In addition, some active repeaters also provide the option to change the radio channel frequency (i.e. shift the transmission to another radio channel frequency). This may be beneficial in preventing the two PTP sub-links from interfering with one another due to 'overshoot' during good weather (Figure 6.17).

7

Radio Propagation, System Range, Reliability and Availability

The operation of a radio system within its system 'range' is critical to ensuring good quality of transmission. In this chapter we discuss how to caculate the range of a radio system, and we learn about the parameters and other factors which define the range and affect the performance. We discuss how the weather and the gaseous constitution of the atmosphere affect the performance and reliability of a radio system. We also learn that the 'range' of a radio system is not a fixed value — radio waves do not conveniently cover the intended coverage area and then stop at a boundary line we call the 'range' of the system. Rather, the range is only a nominal value, defining the maximum distance the two endpoints of the radio link may be from one another if a given quality of link performance and reliability is to be achieved. For different link quality targets, the range will differ. We discuss the methodologies and detailed mathematical formulae for calculating the range of a radio link or system, and define on the way a critical parameter for defining our reliability needs for the link — the 'availability' (the percentage of time the link is to meet the desired quality standard).

7.1 The Reliability, 'Range' and 'Availability' of a Radio System

There is no standard means of defining the 'reliability' of a radio system. Instead the standard terms are the system *range* and the radio link *availability*. The range and availability of a radio system are limited by the *transmitter power* output, the *receiver sensitivity*, and by the atmospheric and climatic conditions of the region of operation. Before we can determine exact values, though, we must define exactly what we mean by range and availability.

The reliable propagation of radio waves across a radio link is achieved when a signal of adequate strength for good quality demodulation arrives at the receiver. So that the signal arrives with sufficient strength, there has to have been sufficient signal power generated by the transmitter, adequate sensitivity of the receiver and only limited signal attenuation or amplification (i.e. loss or gain in signal strength) caused by the atmosphere. The atmospheric or *path attenuation* is usually caused by interference from other signals, absorption of the signal (by atmospheric gases or by rainfall) or by path reflections or other effects.

Of all the factors influencing the range of the system, the technical specifications of the radio system are most easily definable:

- The transmitter power of microwave radio systems is usually quoted in Watts (W), milliWatts (mW) or dBm (decibels relative to 1 milliWatt). Typical maximum values are in the range 15 dBm (32 mW) to 25 dBm (320 mW).
- The receiver sensitivity is usually quoted in terms of the minimum threshold *Received Signal Level* (*RSL*) required to be received by the receiver for a defined quality (or 'accuracy') of reception. CEPT and ETSI recommendations for digital radio suggest quoting the receiver threshold as the minimum power required for the receiver to achieve a *Bit Error Ratio* (*BER*) reception of a digital signal better than 10^{-6}.

The receiver sensitivity of a digital microwave radio system is typically around $-80\,dB_m$. If the received signal level is less than than this value, then the desired *threshold BER* will not be achieved. In this case, the received signal will contain more errors than the target BER of 10^{-6}, and the radio link is assumed to be out of operations limits or *unavailable. But there will still be reception of a signal—received signal level lower than the threshold value does not necessarily mean we have no reception.*

Expressed in dBm, the higher the negative value of the *receiver sensitivity* (i.e. the weaker the receivable signal) the more sensitive is the receiver. Sometimes, multiple values of sensitivity are quoted for a radio receiver. Thus a receiver with a sensitivity threshold (or *receiver threshold*) of $-80\,dB_m$ for $BER = 10^{-6}$ will have a threshold of approximately $-83\,dB_m$ for $BER = 10^{-3}$. In other words, the receiver can still receive even weaker signals, but at a lower level of accuracy. Said another way, the link of high quality is *unavailable*, but a lower quality link is *available*!

More difficult to define and predict than the basic radio performance is the path attenuation or *path loss* of the signal. This is the loss in signal power caused by the atmosphere between the two ends of the link. This attenuation is variable, depending upon current weather and atmospheric conditions. The availability of the radio link is thus dependent upon the local climate.

Heavy rainfall, atmospheric disturbances and radio signal absorption effects tend to attenuate (i.e. weaken) the signal so reducing the RSL. During good weather the signal strength will allow good reception, while during very heavy thunderstorms the received signal may contain an unacceptably high number of errors. The principle is straightforward, but the prediction of the exact quality of performance is much harder to handle. The important question is, for what percentage of the time the *Received Signal Level* (*RSL*) exceeds the minimum value required to achieve a given BER target. This percentage of the time for which the target link transmission quality is achieved is termed the link *availability*.

For estimating the likely effects of the climate on the operation of a radio system, ITU-R (the *Radiocommunications sector* of the *International Telecommunications Union*) has published a large number of recommendations and reports on the propagation of radio. These present a standardised method for calculating the range of a radio system.

The ITU-R reports and recommendations classify the regions of the world into different climate zones, each zone characterised by a typical climate profile drawn from extensive weather measurements over many years and named with a letter of the alphabet from A to S (Figure 7.1). The climate zones are characterised in terms of their 'normal' (i.e. statistically averaged) weather patterns, from which the 'expected' radio path loss for different types of weather and at different times of year can be calculated.

The range of a radio system, when calculated according to ITU-R recommendations, is the maximum distance at which the radio terminals may be placed apart from one another, in order to ensure a minimum BER quality for a given target availability. The range of the

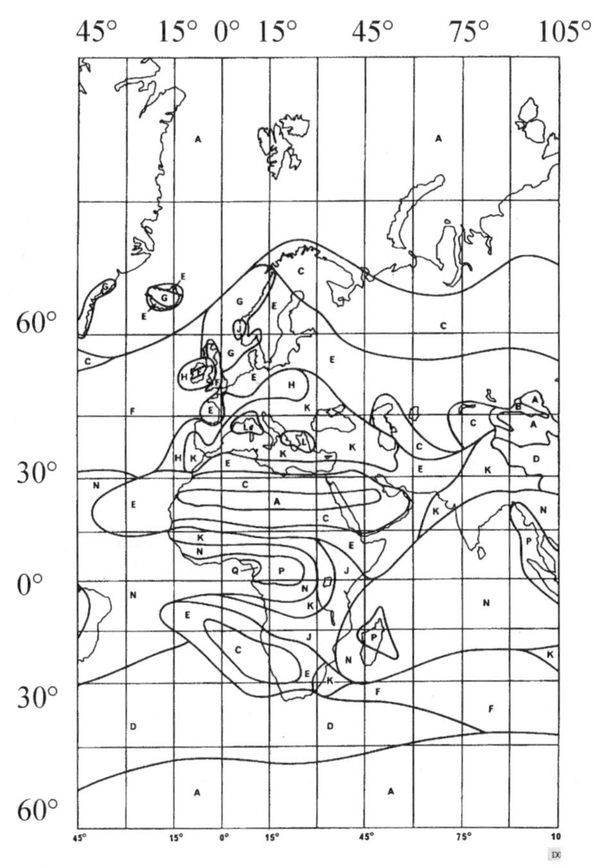

Figure 7.1 ITU-R climate zones (ITU-R Recommendation PN.837, Reproduced by permission of the ITU)

system is usually calculated in kilometres for a given ITU-R climate zone. When quoted, the defined availability target and the BER target should also be quoted.

Thus, a 26 GHz shorthaul radio system may be quoted as having a *range* of 15 km, given:

- a target BER (Bit Error Ratio) of 10^{-6} (i.e. a maximum of 1 error in every million bits sent) is achieved;
- a target availability of 99.99%;
- in the *worst month* (i.e. maximum of 4.5 minutes unavailable in the worst month—or alternatively over the year as a whole — outage 53 minutes per annum);
- in climate zone 'E';
- while carrying 2×2 Mbit/s bitrates (the threshold RSL may vary from one bitrate to another).

The range, when quoted on its own (without also stating the other five target parameters listed above), is meaningless. It is thus very dangerous to compare the system range quoted by different manufacturers of radio systems on their datasheets, without having checked that each of the five parameters above have been set the same for the two system range calculations.

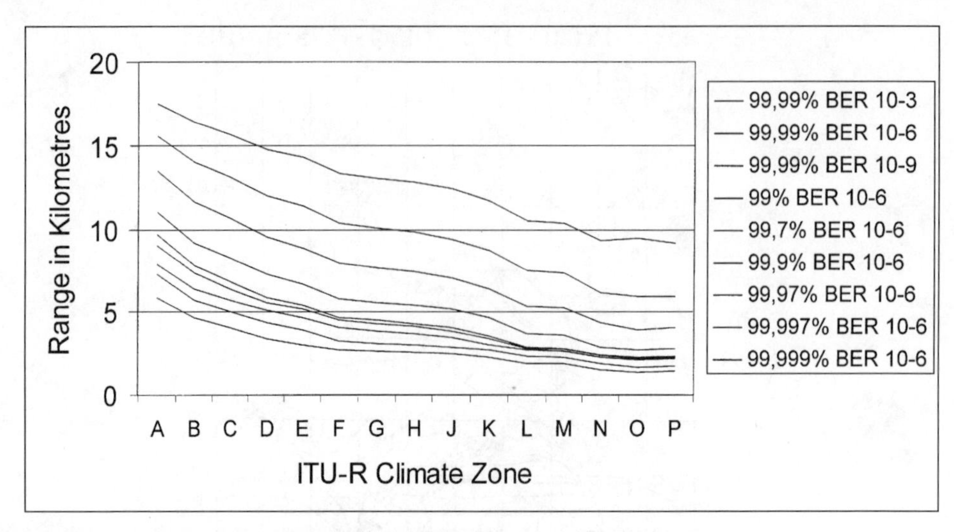

Figure 7.2 The *range* of a radio system depends upon the target values for link *quality* (*bit error ratio*) and *availability*, as well as being dependent upon the *climate zone*

Most fixed wireless access radio systems operate in either the millimetre wave or microwave radio bands. Attenuation in these bands is particularly great during heavy rainfall. It is the high density of water between the antennas of transmitter and receiver which leads to high signal attenuation, and thus to periods of *unavailability*. Unavailability generally thus results from the *critical rainfall rate* being exceeded for the given path length (typically during a very heavy thunderstorm).

It is customary to limit the unavailability and to design radio systems for a minumum availability of 99.99% (53 outage minutes per annum, based on annual statistics of ITU-R, or about 4.5 minutes per month if *worst month* radio planning is conducted). A calculation set out in ITU-R recommendations allows the radio planner, based upon the *climate zone*, the *transmitter power*, the *receiver sensitivity* and the *antenna gain* to calculate the maximum system range. The calculation method is based upon an ITU-R model, named the *ITU-R transmission loss concept*, which we discuss later in the chapter. However, before we move on, we summarise the *concepts* of radio system range and availability to underpin our understanding of the terms.

Neither of the terms *range* or *availability* are 'absolute' measures: both depend upon the targets we set for the 'performance' (quality) and 'reliability' of the radio link. Figure 7.2 illustrates an example of how the *range* of a 26 GHz radio system is affected by the target values set for climate zone, carried bitrate, BER target and availability. The values assumed are typical values for a point-to-multipoint (PMP) system operating with 90° base station sector antennas. From the diagram you will see how the range is reduced by:

- the ITU-R climate zone (the later the letter in the alphabet, generally the lower the range);
- increasing link availability target (the higher the availability needed, the lower the range);
- increasing link quality target (the lower the target bit error ratio required, the lower the range).

7.2 The Concept of 'Fade Margin'

Most of the time (i.e. during good weather), the received signal level is well in excess of the threshold receiver sensitivity, so a radio system operates without errors. It is only when the signal attenuation exceeds the *link budget* or *system budget* that problems occur (the problem is an unacceptably high level of errors, defined as link *unavailability*).

The *link budget* is the maximum signal loss which may occur between transmitting and receiving antennas. During good weather there is only the *free space loss*. The difference between this 'normal' level of attenuation and the 'threshold condition' is known as the *fade margin*. The fade margin is thus the extra level of signal attenuation which may be caused by weather *fading* before the system is considered unavailable. It is thus normal to plan radio links for a given minimum fade margin. This is the 'reserve' signal strength built in to overcome the attenuation caused by bad weather conditions. The fade margin is set by adjusting the transmitter output power accordingly.

Unfortunately, while a high fade margin is advantageous in protecting the link against weather fading it may cause interference problems during good weather (i.e. most of the time), since the transmitted signal will propagate much farther than it needs to (so-called *overshoot*). To avoid this problem, some radio systems are equipped with a transmitter power control function called *Automatic Transmit Power Control* (*ATPC*), as we discussed in Chapter 6. ATPC reduces the transmitted power during good weather, and automatically increases the output power at times of bad weather in order to limit the effects of fading.

7.3 ITU-R 'Transmission Loss Concept' (ITU-R Recommendation P.341)

ITU-R recommendation P.341 defines the standard terminology and notation used to characterize the signal losses which occur in a radio path. Figure 7.3, taken from this recommendation, sets out the basic 'model' for the *transmission loss concept*, and presents the standard nomenclature. A range of other ITU-R recommendations define how to calculate the values of the individual components of the signal loss. The recommendations are equally suited for calculating the losses associated with either PTP (point-to-point), PMP (point-to-multipoint) or broadcast radio systems.

It is normal to define the signal losses caused by the various attenuation effects in decibels (dB), as this simplifies the various calculations. We describe each of the losses in turn.

Total Loss (of a Radio Link), L_l or A_l

The *total loss* is the loss incurred between the output of the transmitter and the input to the receiver. This loss includes the losses in waveguides, radio signal filters, feeder connections and other components between the transmitter and the transmitting antenna, and between the receiver and the receiving antenna.

System Loss, L_s or A_s

The *system loss* is the loss incurred between the input to the transmitting antenna and the output of the receiving antenna. This includes all losses associated with the antennas and the radio path losses. In cases where the transmitter and receiver are connected directly to their respective antennas, the system loss is equal to the total loss.

The system loss should not be confused with the *system budget*, which includes the *free-space system loss* and operating *fade margin*. The system budget is an operational planning target for the radio power configuration of a link. At times when the actual system loss exceeds the system budget (caused by heavy rain) the link is said to be *unavailable*.

The system loss is the signal loss (expressed as a ratio in decibels) between the transmitter output and the receiver input signal power. It is defined as follows:

$$\textit{System Loss} \quad L_s = 10 \log_{10}(P_t/P_a) = P_t - P_a$$

Where p_t = RF (radiofrequency) power output of the transmitter, p_a = RF signal power at output of the receiving antenna/or at input to receiver, P_t = transmitter power output expressed in dBm or dBW, and P_a = receiver signal input power expressed in dBm or dBW.

Antenna Losses, L_{tc} (Transmitting Antenna) and L_{rc} (Receiving Antenna)

The antenna losses are those signal power losses occurring within the antenna. Typically, these losses are accountable to signal losses in the waveguide *feeder* to the antenna (i.e. the

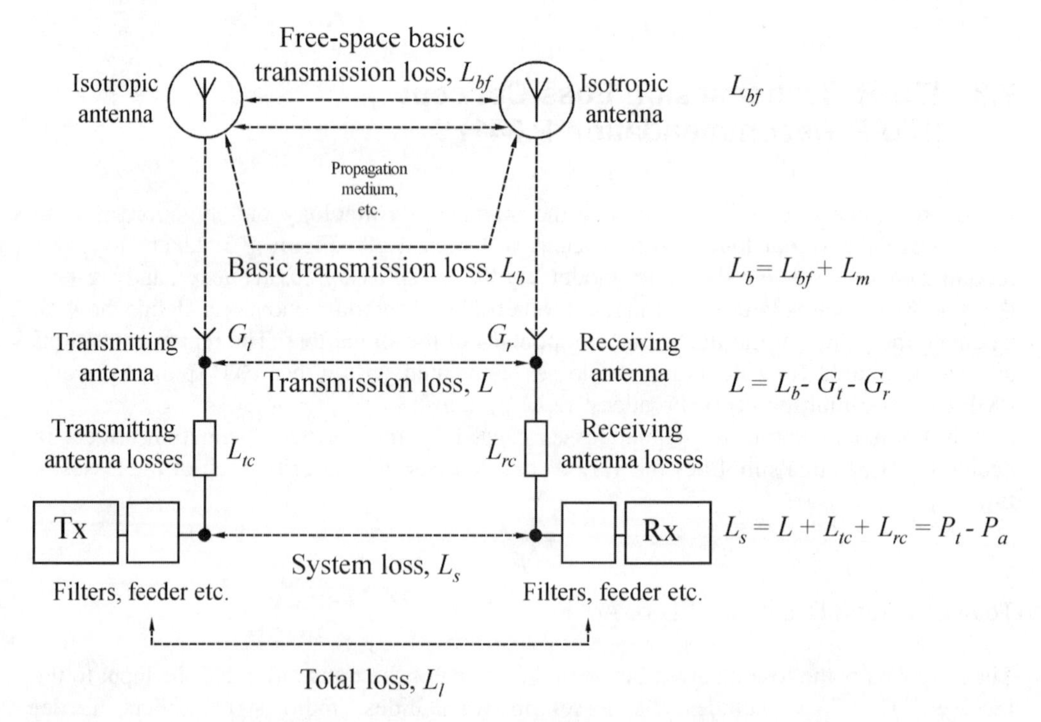

Figure 7.3 The transmission loss concept model of ITU-R recommendation P.341 (Reproduced by permission of the ITU)

length of waveguide which runs from the antenna attachment point (usually at the back of the antenna) to the feeder point of the antenna (e.g. at the 'focus' of a parabolic reflector antenna). Such antenna losses are usually designed to be small, and it may be acceptable to ignore these losses (unless the signal is very high frequency, and there is also a very long feeder arrangement). They are denoted mathematically in ITU-R recommendations as L_{tc} and L_{rc}

Transmission Loss or *Path Loss* (of a Radio Link), L, A or L_t

The *transmission loss* is the loss between the input to the transmitting antenna and the output from the receiving antenna, assuming the use of 'perfect' antennas (i.e. with no Radio Frequency (RF) losses in their feeders).

The transmission loss (also called the *path loss* or *ray path loss*) can be considered to be equal to the system loss in cases where the *antenna losses* are accounted for by 'effective' (or 'net') antenna gain values. (Such 'net-gain' values are often used, since it is relatively easy to measure the 'net' performance or gain of an antenna from its input point to the signal strength in the various transmission directions. It is much harder to try to measure separately the losses in the feeder and the actual antenna gain.)

$$\textit{Transmission Loss},\ L = L_s - L_{tc} - L_{rc}$$

where L_s = system loss, L_{tc} = antenna loss of transmitting antenna, and L_{rc} = antenna loss of receiving antenna.

Antenna Gain, G_t (Transmitting Antenna) and G_r (Receiving Antenna)

The antenna gains G_t and G_r used in radio loss calculations correspond to the maximum antenna *gain* along the main *lobe* of the antenna, i.e. along its *pole*. This we discussed in Chapter 3.

Basic Transmission Loss (of a Radio Link), L_b or A_i

The basic transmission loss is that loss which would be incurred between the output of the transmitter and the input to the receiver if both the transmitting and receiving antennas isotropic antennas. We ignore the practical problems (e.g. reflections) which might be newly introduced by such isotropic antennas:

$$\textit{Basic transmission loss},\ L_b = L + G_t + G_r$$

where L = transmission loss, G_t = transmitting antenna gain in $\mathrm{dB_i}$, and G_r = receiving antenna gain in $\mathrm{dB_i}$.

The basic transmission loss is considered to comprise two components, the *free space transmission loss* (L_{bf}) and the loss due to the propagation medium (e.g. due to the gaseous content of the air, due to weather, etc.). The loss due to the propagation *m*edium is correctly called the *loss relative to free space* (L_m), and is defined as follows:

$$\textit{Basic transmission loss},\ L_b = L_{bf} + L_m$$

where L_{bf} = free space transmission loss, and L_m = loss relative to free space (i.e. loss due to medium).

Radio, Link or System Budget

From the equations of the previous section, we are able to establish a relationship between the maximum system loss that a radio system can endure during normal operation. This is the *system budget* we already spoke of. It is derived easily as follows:

Basic transmission loss	$L_b = L + G_t + G_r = L_{bf} + L_m$
Therefore	$L = L_{bf} + L_m - G_t - G_r = L_s - L_{tc} - L_{rc}$
Therefore *system loss*	$L_s = L_{bf} + L_m - (G_t - L_{tc}) - (G_r - L_{rc})$
or	$L_s = L_{bf} + L_m - G_t' - G_r'$

where $G_t' = G_t - L_{tc}$ = measured gain of transmit antenna from input port, $G_r' = G_r - L_{rc}$ = measured gain of receive antenna from input port.

Since this is the maximum allowed loss, then *link or system budget must exceed the maximum expected value of* L_s:

System Budget $> L_{bf} + L_m - G_t' - G_t'$

The system budget is an important formula for radio link planning. The maximum value of the system budget is called upon during periods of maximum medium attenuation (L_m), caused by very heavy rainfall.

Free-space Transmission Loss, L_{bf} or A_0

The *free-space transmission loss* is the drop in signal strength which would occur between transmitter and receiver if an isotropic antenna were used for transmission in a *free-space* medium. A *free-space medium* is one perfect for radio transmission, with no obstacles or radio power absorbing qualities. The loss in signal strength in free-space is due only to the 'spreading out' of the signal over the ever-larger surface area of the spherical wavefront. The radio waves dissipate from the centre of a sphere, like ripples on a pond after dropping in a stone — but in three dimensions.

Free-space basic transmission loss, $L_{bf} = 20 \log_{10}(4\pi d/\lambda)$

where λ = radio signal wavelength, and d = distance between the antennas, assumed to be much greater than λ.

Loss Relative to Free-space (Loss in Medium)

The *loss relative to free-space* is the signal loss caused by attenuation due to the signal absorption or obstruction effects of real (i.e. non-perfect) transmission media. Such attenuation or *fading* of radio signals may occur for one of a number of different reasons:

- *fading* (absorption) due to *precipitation* (i.e. rain, snow, fog, clouds or other weather effects);
- signal *absorption* due to atmospheric gases or due to the *dielectric* state of the atmosphere (i.e. to what degree it is *ionised* or *non-ionised*);
- attenuation due to vegetation and sand or dust storms, etc.;
- *fading* (absorption) due to *multipath*, whereby different reflections of the original signal interfere harmfully with one another;
- losses due to signal diffraction, path obstruction or partial obstruction (this affects *groundwaves* which are propagated near the earth's surface);

- signal *polarisation* effects — the different propagation of differently polarised signals (e.g. *horizontal* and *vertical polarisation*), *polarisation coupling losses* which arise from a 'mismatch' of the antennas or their alignment given the physical nature of the local surroundings;
- magnetic and electrical effects of the earth's surface and geography;
- signal *reflection* or *scattering* of the signal (this particularly affects transmission through the earth's upper atmosphere — the *ionosphere*).

The different causes of fading have varying effects upon different radio frequency bands, so that, for example, rain fading is the most significant cause of attenuation for frequencies in the millimetre and micro-wave bands (particularly above 10 GHz. At lower frequencies (less than 10 GHz), rain fading is less limiting on the system performance and link range. At these frequencies, the effects of path refraction within the earth's atmosphere and other effects are significant in limiting link range.

Before we go on to discuss in detail the various types of fading and signal attenuation which go to make up the signal *loss relative to free space*, it will be useful to discuss the nature of radio wave propagation and the various types of different radio systems.

7.4 Radio Wave Propagation

When radio waves are transmitted from a point, they spread and propagate as spherical wavefronts. The wavefronts travel in a direction perpendicular to the wavefront, as shown in Figure 7.4.

Radio waves and light waves are both forms of electromagnetic radiation, and they display similar properties. Just as a beam of light may be *reflected*, *refracted* (i.e. slightly bent) and *diffracted* (slightly swayed around obstacles), so may a radio wave, and Figure 7.5 gives examples of various different wave paths between transmitter and receiver, resulting from these different wave phenomena.

Four particular modes of radio wave propagation are shown in Figure 7.5. A radio transmission system is normally designed to take advantage of one of these modes. The four modes are:

- *line-of-sight* propagation;
- *surface wave* (diffracted) propagation;
- *tropospheric scatter* (reflected and refracted) propagation;

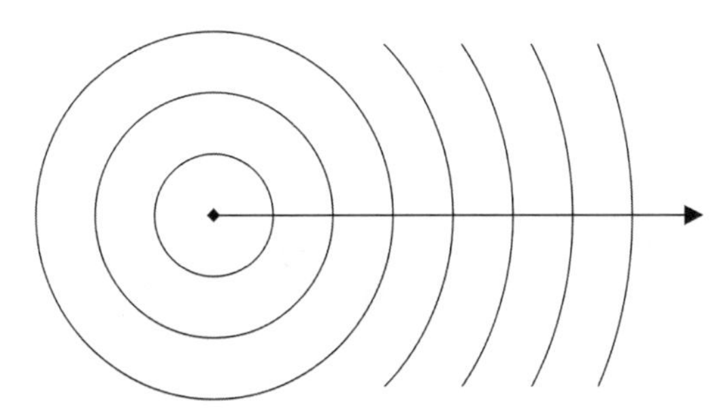

Figure 7.4 Radio wave propagation

- *skywave* (refracted) propagation.

A *Line-Of-Sight* (*LOS*) radio system relies on the fact that waves normally travel only in a straight line. The range of a line-of-site system is limited by the effect of the earth's curvature, as Figure 7.5 shows. Line-of-sight systems are therefore restricted to *shorthaul* applications (maximum 15–20 km range), and can reach beyond the horizon (up to 70 km) only when the terminals are installed on tall masts. Most fixed wireless access radio systems operate in the frequency bands above 1 GHz and are *line-of-sight* or *near-line-of-sight* systems (the latter terminology is sometimes applied to systems designed for bands at the lower end of the range, e.g. 3.5 GHz systems).

Radio systems can also be used beyond the horizon by utilising one of the other three radio propagation effects also shown in Figure 7.5 (surface-wave, tropospheric-scatter or skywave propagation). The radio frequencies which make these types of propagation possible are comparatively low (e.g. MHz rather than GHz). However, these forms of propagation can crop-up even in the GHz frequency range, and lead to propagation difficulties.

Surface waves have a good range, dependending on their frequency. They propagate by diffraction using the ground as a *waveguide*. Low frequency radio signals are the best suited to surface wave propagation, because the amount of bending (the effect properly called *diffraction*) is related to the radio wave length. The longer the wavelength, the greater the effect of diffraction. Therefore the lower the frequency, the greater the bending which is possible.

A second means of over-the-horizon radio transmission is by *tropospheric scatter*. This is a form of radio wave reflection. It occurs in a layer of the earth's atmosphere, called the *troposphere* and works best on Ultra High Frequency (UHF) radio waves.

The final example of over-the-horizon propagation given in Figure 7.5 is known as skywave propagation. This comes about through refraction (deflection) of radiowaves by the earth's atmosphere, and it occurs because the different layers of the earths upper atmosphere (the *ionosphere*) have different densities (Figure 7.6), with the result that radiowaves propagate more quickly in some of the layers than in others, giving rise to wave

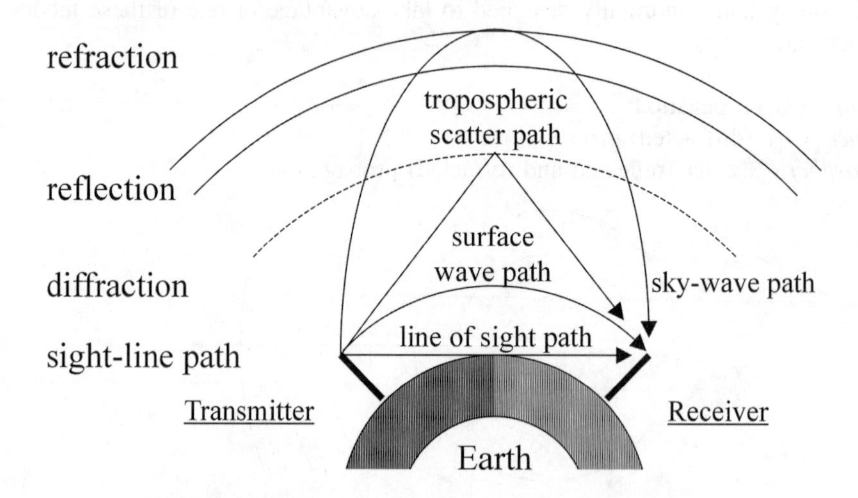

Figure 7.5 Different modes of radio wave propagation

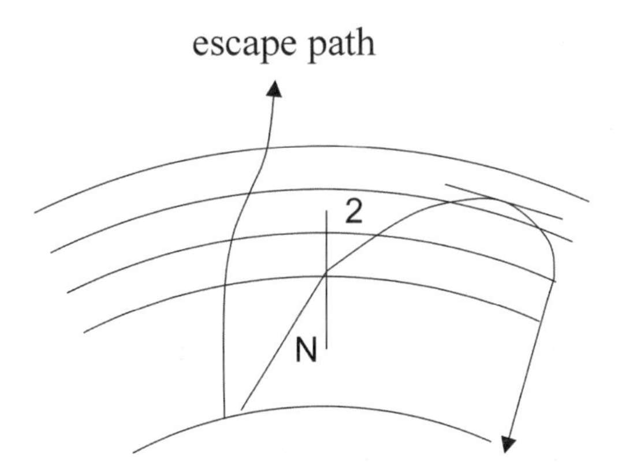

Figure 7.6 Refraction of radio waves: *skywaves* and *escape waves*

deflection between the layers. Not all waves are refracted back to the ground; some escape the atmosphere entirely, particularly if their initial direction of propagation relative to the vertical is too low (angle ϕ).

Even at microwave frequencies above 1 GHz, *surface-wave*, *scatter* and *refraction* path propagation is possible, but usually undesired. It results from signals straying from the direct line-of-sight path and propagating via alternative paths. Unfortunately, this inevitably leads to signal losses, *multipath* and degradation through interference of the main signal. The importance of the main signal losses due to these causes varies from band to band, as we shall see.

7.5 Free Space Transmission Loss— The Expected Received Signal Level in Good Weather

ITU-R recommendation PN.525 provides a simple formula for calculating the *free space transmission loss*, L_{bf}:

Free space transmission loss

$$L_{bf} \text{ (in dB)} = 20 \log (4\pi d/\lambda)$$

where λ = wavelength in m, and d = path length in m.

Alternatively:

$$L_{bf} \text{ (in db)} = 92.44 + 20 \log f + 20 \log d f = \text{frequency in GHz},$$

where, d = path length in km.

The *free space transmission loss* is the likely signal degeneration during good weather (i.e. at times when atmospheric disturbance effects and signal attenuation due to inclement weather are minimal). By working from the output transmitter power and the gain of both the receive and transmit antennas, the expected *Receive Signal Level* (*RSL*) can easily be calculated for good weather conditions.

Since it is normal to install radio links and align the antennas during good weather, the calculation of the expected RSL (free-space loss) is an important part of the installation process. This calculation leads us to the maximum Automatic Gain Control (AGC) or RSL voltage for antenna alignment as we discussed in Chapter 6:

Expected *RSL during good weather* (i.e. expected RSL during link installation and antenna alignment)

$$RSL \text{ (in dB}_\text{m} \text{ or dBW)} = P_t + G_t + G_r - 92.44 - 20 \log f - 20 \log d$$

where P_t = transmit power in dB_m or dBW (same units as RSL), G_t and G_r are the transmit and receive antenna gains in dB_i (i.e. gain relative to an *isotropic antenna*—the relative focusing of the signal in the desired direction), f = frequency in GHz and d = distance in km.

Taking a typical *shorthaul* PTP (point-to-point) radio system operating in the 38 GHz band, the typical range of the system is 6.5 km. The typical transmit power is 15 dBm (32 mWatt), and, with 30 cm antennas, the transmitter and receiver antennas will have gains around 39 dBi. From our formula, the RSL in good weather will thus be around − 47 dBm. Given that the *receiver sensitivity* is likely to be around − 80 dBm (for a target BER of 10^{-6}), this gives a margin for fading of the radio signal (at the maximum link range) of 33 dB.

7.6 Path Losses Relative to Free-Space (L_m) in 'Line-of-Sight' Radio Systems

ITU-R Recommendation P.530 defines the framework for predicting the propagation losses relative to free-space for *terrestrial line-of-sight systems*. Terrestrial line-of-sight systems in this sense are radio systems operating with both ends of the link on the earth's surface in the bands from about 1 GHz to 40 GHz. This description accounts for nearly all fixed wireless access systems.

At frequencies above about 1 GHz, radio signals do not easily diffract around obstacles or penetrate through buildings or vegetation. For the best reliability of the radio links, it is therefore normal during the planning and installation phase of a fixed wireless access system to ensure that there is a line-of-sight (i.e. a direct, unobstructed path) between the two antennas at either end of the link.

ITU-R recommendation P.530 presents the methodology for calculating the path signal losses of line-of-sight systems. We list the effects here in rough order of importance, before discussing each effect in turn:

- *fading* (absorption) due to *precipitation* (i.e. rain, snow, fog, clouds or other weather effects);
- signal *absorption* due to atmospheric gases or due to the *dielectric* state of the atmosphere (i.e. to what degree it is *ionised* or *non-ionised*);
- attenuation due to ground coverage—buildings, vegetation and sand or dust storms, etc.;
- *fading* (absorption) due to *multipath*, whereby different reflections of the original signal interfere harmfully with one another;
- losses due to signal diffraction, path obstruction or partial obstruction (this affects *groundwaves* which are propagated near the earth's surface);
- signal *polarisation* effects—the different propagation of differently polarised signals

(e.g. *horizontal* and *vertical polarisation*), *polarisation coupling losses* which arise from a 'mismatch' of the antennas or their alignment given the physical nature of the local surroundings;
- magnetic and electrical effects of the earth's surface and geography;
- signal *reflection* or *scattering* of the signal (this particularly affects transmission through the earth's upper atmosphere — the *ionosphere*).

Path Loss due to Rain Fading and 'Precipitation'

Signal attenuation due to precipitation, particularly due to heavy rainfall, is the main component of the *path loss relative to free space* (L_m — the 'medium' loss) in *Line-of-Sight* (*LOS*) systems operating above 10 GHz. (At lower freqeuncies there are additional problems, for example *multipath*, as we shall discuss later.)

It is the absorption of the signal by the water within the rain which causes signal attenuation. The absorption is greatest when the mass of water between the transmitting and receiving antennas is greatest, i.e. during very heavy thunderstorms.

Rain fading as affecting *Line-of-Sight* microwave radio systems can be estimated according to ITU-R recommendations 838 and PN.837. Recommendation 868 states that the attenuation, γ_R (in dB/km), caused by rain is as follows:

Attenuation due to rain (ITU-R recommendation 838: freqencies up to 40 GHz and path lengths up to 60 km)

$$\gamma_R \text{ in (dB/km)} = kR^{\alpha}$$

where k and α are values which depend upon local weather and climatic conditions, and R is the rainfall rate in mm/hour, where

$$k = [k_H + k_V + (k_H - k_V) \cos^2\theta \cos 2\tau]/2$$

$$\alpha = [k_H\alpha_H + k_V\alpha_V + (k_H\alpha_H - k_V\alpha_V) \cos^2\theta \cos 2\tau]/2k$$

θ = path elevation angle, and τ = polarisation angle relative to the horizontal. Values of k_H k_V, α_H and α_V are input from Table 1 of recommendation 838 according to the radio signal frequency. This table is reproduced in Appendix 8. Rainfall rate values R are input according to the climate zones and tables defined in recommendation PN.837, which is also reproduced in Appendix 8.

We explain how to calculate the path loss due to precipitation (rain) with an example.

We return to our example of the 38 GHz link of maximum range 6.5 km. Two sections previously, we calculated a fade margin of 33 dB at the maximum link range. The values of the constants for 38 GHz in climate zone E (American midwest and northern Europe) are $k = 0.279$ and $\alpha = 0.943$ (calculated according to ITU-R recommendation 838). Recommendation PN.837 tells us that for less than 0.01% of each *average* year (i.e. annual statistic) the rainfall exceeds 22 mm/hour. We can thus expect the attenuation per kilometre to exceed $\gamma_R = kR^{\alpha} = 5.1$ dB/km for (at most) 0.1% of the time. Therefore (for 99.99% of the time — the target *availability* of the link) we can expect the attenuation due to rain over a 6.5 km transmission link *not to exceed* $5.1 \times 6.5 = 33$ dB. The 33 dB loss at maximum range (6.5 km) during heavy weather (0.1% of the time) is exactly the *fade margin* we calculated in our previous example (the difference between the *system budget* and the *free space transmission loss*).

Effective path length

Actually, our example calculation of the *path attenuation exceeded for 0.01% of the time* (correctly written $A_{0.01}$) is not quite accurate, since a correction is normally applied to the path length *d*. The attenuation per kilometre, γ_R, should be multiplied by the *effective path length*, d_{eff}, given by the formulae listed below. The effective path length accounts for the fact that rainfall does not usually occur equally over the entire length of the path. Typically, particularly heavy downpours are 'squallish' in nature, tending to have a given 'rainfall cell diameter'.

Effective path length (ITU-R recommendation P.530)

$d_{eff} = d \times r$

where $d =$ actual path length in km, $r =$ correction factor $= 1/[1 + d/d_0]$,

$d_0 = 35_e^{-0.015\,R_{0.01}}$ ($R_{0.01} < 100$ mm/h) $d_0 = 7.81$ ($R_{0.01} > 100$ mm/h), and $R_{0.01} =$ rainfall rate (mm/h), which is exceeded 0.01% of the time.

In our example above, the rainfall rate exceeded 22 mm/h for 0.01% of the time in climate zone E (from recommendation PN.837). Thus, $R_{0.01}$ equals 22 mm/h. Therefore, $d_0 = 25.2$ and $r = 0.79$. The *effective path length*, d_{eff} equals 5.2 km. (A better estimate of the real range of the system for 99.99% availability at BER $= 10^{-6}$ is thus 6.5 km/0.79 = 8.2 km. We need to reiterate the process with path length = 8.2 km and re-calculate the new effective path length. By trial-and-error re-iteration we finally derive the actual value of the link range as 7.9 km.)

In the case of very long links subject to very heavy rainfall rates, the correction factor r will make the effective path length much shorter than the real path length. This is a reflection of the very high unlikelihood of very heavy rain along the entirety of a long path.

Different link availability targets

It is standard practice to design radio links for an *availability* target of 99.99% (i.e. using 0.01% rainfall rates) and for BER target of 10^{-6}. For determining the range or path losses of such systems, it is usual to calculate the 0.01% path attenuation, $A_{0.01}$ as we have done above ($A_{0.01} = \gamma_R \times d_{eff}$). However, there are times when other availability targets may be set (e.g. 1%, 0.1% or 0.001%). For the calculation of path losses for with such availability targets, other values of rainfall rates for each of the climate zones are also provided by recommendation ITU-R PN.837 (see Appendix 8). Alternatively, the scaling factors of Table 7.1 may be applied to the 0.01% path attenuation ($A_{0.01}$).

'Annual' and 'worst month' statistics for link planning

The methodology so far discussed for determining the signal attenuation caused by rainfall will predict the annual statistics of the path losses. Thus, in any given year the rainfall can be expected to exceed the 0.01% threshold value for about 53 minutes of the year. The 53 minutes will not occur all at once, but as the result of summation of smaller durations (thunderstorms do not generally maintain maximum rainfall for such long periods). In fact, the 53 minutes is the average likely *outage* time (i.e. unavailable time) of a radio system designed for 0.01% *annual availability*. Just because the annual availability is 99.99% (the

Table 7.1 Path attenuation due to rainfall scaling factors for percentages of time other than 0.01% (99.99% availability)

Availability target or percentage of time link suffers a particular loss	Attenuation scaling factor
Loss exceeded for 1% of time (99.000% availability)	$0.12 \times A_{0.01}$
Loss exceeded for 0.1% of time (99.900% availability)	$0.39 \times A_{0.01}$
Loss exceeded for 0.01% of time (99.990% availability)	$1 \times A_{0.01}$
Loss exceeded for 0.001% of time (99.999% availability)	$2.14 \times A_{0.01}$

threshold rainfall rate is exceeded 0.01% of the year) does not mean that the *monthly availability* can be expected to reach the same level.

Typically, rainfall is not spread evenly over the year, but varies according to the seasons, so that there are some *worst months* and some dryer months. To design a radio system for *monthly availability* of 99.99% (i.e. maximum *outage* due to rainfall in any one month) requires consideration of the *worst month* climate statistics of ITU-R. The methodology for this calculation is presented in ITU-R recommendation 841.

Path Loss Due to Other Types of 'Precipitation'—hail, snow, fog and cloud

Hail can cause significant signal degradation, especially when the hailstones are of a size roughly equal to the signal wavelength, but it is relatively uncommon and so has little effect on overall annual link availability.

In comparison with the signal loss which can be calculated as above using rainfall precipitation statistics, *terrestrial line-of-sight* radio systems operating in radio frequencies between about 1 GHz and 40 GHz are relatively unaffected by most other forms of precipitation—snow, fog and cloud. Such types of precipitation are quite 'dry' in comparison with heavy rainfall, and therefore are unlikely to limit the *range* or *availability* of a given radio *link*, and can be considered (in the case of terrestrial radio systems) to have been taken into account by the normal rainfall methodology. 'Wet' snow has a more degrading effect, but is generally considered to be covered by the general 'rain' statistics.

ITU-R recommendation P.840 presents a method for predicting signal attenuation due to clouds and fog, but this is meant for designing satellite radio systems, in which the signal has to penetrate through the earth's atmosphere into space. The *specific attenuation* caused by the water droplets present in fog and cloud typically lies around 1 dB per kilometre or less for frequencies up to about 40 GHz and temperatures above about 0°C (compare this with the 5 dB per kilometre in the example of the previous section). At higher altitudes, where the temperature is much cooler, the *specific attenuation* is greater.

Path Loss due to Signal Absorption by Atmospheric Gases

ITU-R recommendation P.676 presents the methodology for calculating the radio path signal attenuation caused by absorption of the signal power by atmospheric gases. The recommendation is valid for radio frequencies up to 1000 GHz. The methodology calculates the specific attenuation γ in decibels per kilometre (dB/km).

Figure 7.7 illustrates the specific attenuation caused by atmospheric gases across the full range of radio spectrum up to 1000 GHz.

Figure 7.7 illustrates two absorption curves, one for a *dry atmosphere* and one for a *standard atmosphere* (i.e. a typical real atmosphere as defined by ITU-R recommendations P.835, P.836 and P.453). In a dry atmosphere, the main attenuation is caused by oxygen (at around 60 GHz as well as *Debye spectrum absorption* below 10 GHz) and a pressure-induced nitrogen absorption above 100 GHz. In a standard atmosphere there is additional signal absorption caused by the water vapour in the air. Figure 7.8 illustrates the specific attenuation caused by the various different gaseous components of a normal atmosphere up to frequencies around 350 GHz.

Noteworthy is the particularly high attenuation (around 15 dB/km) of radio signals of frequency around 60 GHz. This is caused by signal absorption by oxygen molecules. This attenuation is one of the main causes of signal path loss at this frequency. However, for frequencies up to about 40 GHz, the attenuation caused by atmospheric gases is limited to about 0.1 dB/km, and is usually ignored in system range and availability calculations.

Attenuation due to Ground Coverage — Vegetation, Buildings, Dust and Sand Storms

The attenuation caused by ground coverage is the most difficult to accurately predict. An accurate methodology would require an extensive database of building heights and vegetation coverage. The standardised form of such databases has been defined by ITU-R (in recommendation P.1058), but the signal strength prediction methodologies are currently not very advanced. The best method of minimising such signal attenuation or fading is by thorough planning during the planning and installation phase of the link (i.e. by conducting a thorough line-of-sight (LOS)-check (as we shall discuss in Chapter 8) to make sure that the path is free of obstacles as far as possible.

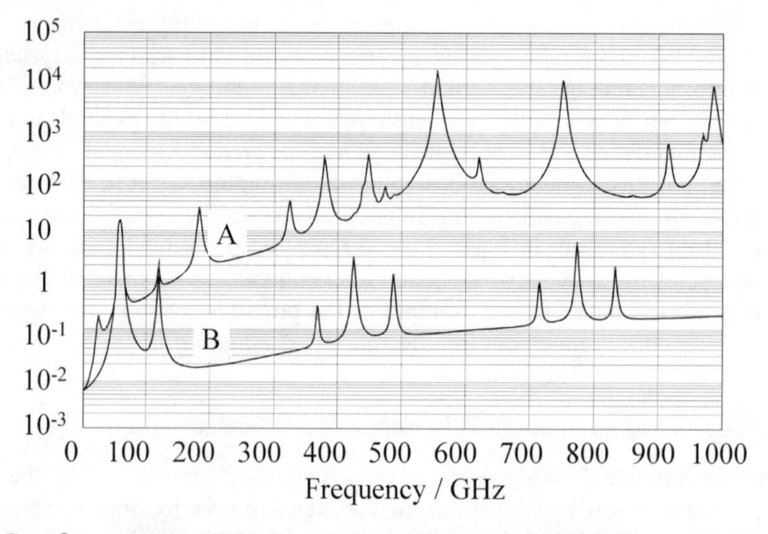

Figure 7.7 *Specific attenuation* (dB/km) caused by atmospheric gases (ITU-R recommendation P.676). Curves A: standard atmosphere (7.5 g/m); B: dry atmosphere. (Reproduced by permission of the ITU)

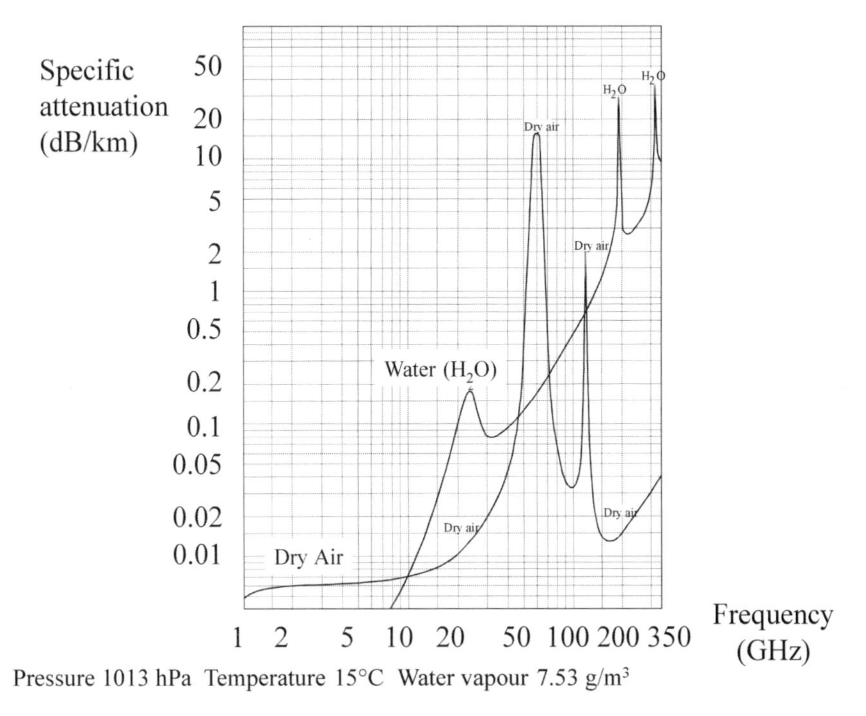

Figure 7.8 *Specific attenuation* (dB/km) caused by individual gaseous components of a *standard atmosphere* (ITU-R recommendation P.676). Curves A: standard atmosphere (75 g/m); B: dry atmosphere. (Reproduced by permission of the ITU)

ITU-R recommendation PN.833 provides a simple method for estimating attenuation caused by vegetation. The method is defined particularly for line-of-sight links which are only partly obstructed by jungle or woodland foliage.

Dust and sand storms in desert areas are also known to cause radio system degradation, but there are no reliable means of predicting likely signal strength as yet.

Multipath Fading

As we will discuss in more detail in Chapter 8, it is important when planning and installing line-of-sight (LOS) radio links to make sure that there is an unobstructed path between transmitting and receiving antennas. In particular, we shall discuss how it is important to keep the *first Fresnel zone* free of obstructions. The Fresnel zone is the area within which a glancing reflection from an obstacle may lead to a destructive alternative radio transmission path between the transmitter and receiver (Figure 7.9).

Where an alternative ray path (caused for example by a reflection as in Figure 7.9) has a length equivalent to to the direct path plus a distance equal to half the radio signal *wavelength* λ, then the two signals (the direct path and the alternative path signals) will arrive at the receiving antenna in *antiphase*. They will destructively *interfere* with one another, causing significant signal attenuation or *fading*. Such fading is called *multipath interference* or *multipath fading*.

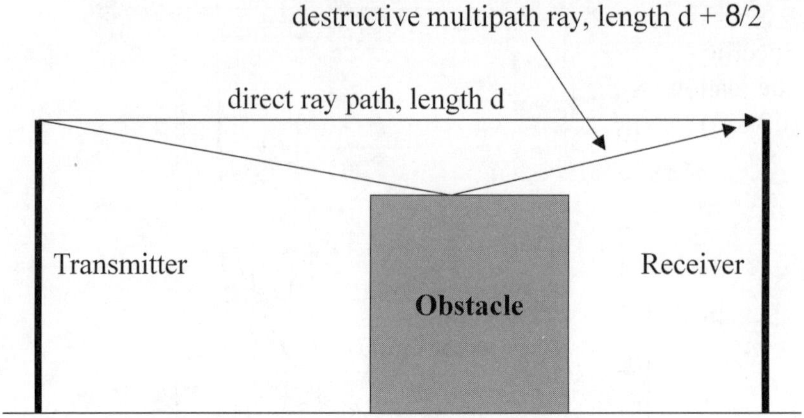

destructive multipath ray, length d + 8/2

direct ray path, length d

Transmitter

Receiver

Obstacle

Figure 7.9 Multipath interference caused by reflection

The simplest way to avoid multipath fading caused by reflections from obstacles (as shown in Figure 7.9) is to be thorough during link planning to ensure an obstacle-free first Fresnel zone (this we discuss in Chapter 8). Unfortunately, however, multipath fading can also be caused by alternative transmission paths of different lengths caused by refraction within the earth's atmosphere (Figures 7.5, 7.6 and 7.10).

For radios operating at frequencies below 10 GHz, the effects of refractive multipath fading (i.e. interference caused by the atmosphere) need to be taken into account, since relatively long link ranges are possible (according to rainfall precipitation calculations). Once again, ITU-R recommendation PN.530–5 offers a simple formula for calculating the probability of a multipath fade of a given *depth* as follows:

Radio fading due to multipath effects caused by refraction

Fade probability (% of year incidence) $\approx ^{-\Delta G/10} K\, d^{3.6} f^{0.89}\, 10^{-A/10}$

where $\Delta G = 11.05 - 2.8 \log d$ (for latitude 50°—Europe)

$A =$ depth of fade in dB

$f =$ frequency in GHz

$d =$ distance in km

$K \approx 10^{-6.5} P_L^{1.5}$

$P_L =$ factor depending upon atmospheric refractivity, for Europe and North America the value lies between 5 and 20.

For calculating fades in other regions, altitudes or latitudes, the different coefficients and formulae for the fade probability of ITU-R recommendation P.530 should be used.

Obstacles and Signal Losses due to Diffraction

As we have discussed, it is normal to plan to operate radio systems in the bands above 1 GHz in a line-of-sight (LOS) mode, i.e. with an unobstructed path between the transmitting and receiving antennas. This is because such high frequency radio signals do

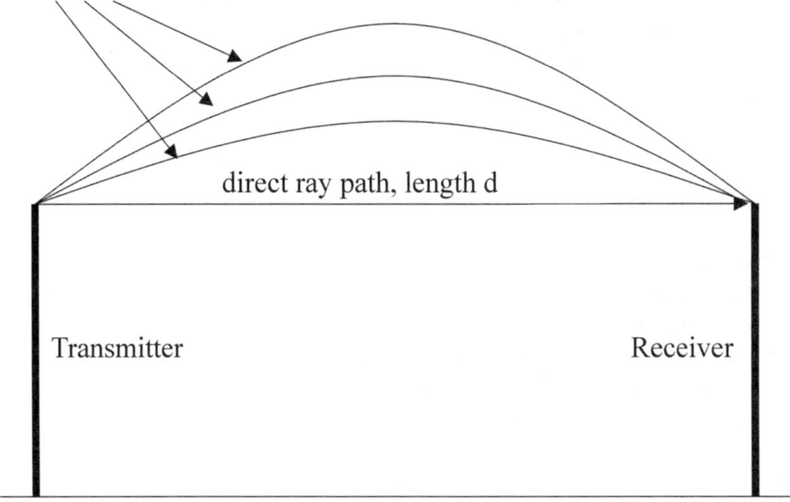

Figure 7.10 Multipath fading caused by atmospheric refraction

not *diffract* much around obstacles (i.e. the 'rays' are not able to 'bend' around obstacles). All very well, but in practice it is not always possible to ensure an entirely obstacle-free path. So we need to be able to calculate the *diffraction loss*. ITU-R recommendation P.526 provides detailed methodologies for calculating signal losses caused by path obstructions of different forms (from 'knife-edge' obstructions to rounded obstructions—including analysis of the obstruction caused by the sphere of the earth's surface on very long links).

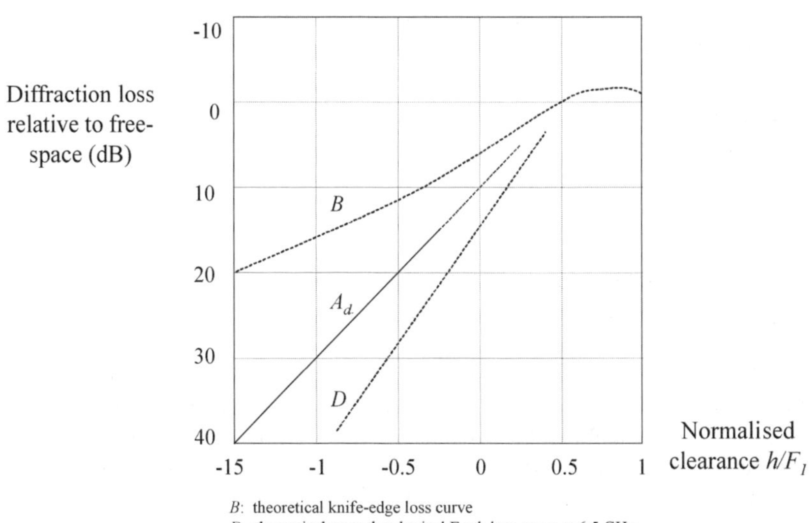

Figure 7.11 *Diffraction loss* for obstructed line-of-sight microwave radio paths (ITU-R recommendation P.530, Reproduced by permission of the ITU)

Figure 7.11 presents the upper and lower limits of the diffraction losses calculated by the methods of ITU-R recommendation P.526 and P.530. The *normalised clearance* is a measure of the distance of the direct ray path from the obstacle. For negative values of normalised clearance, the direct ray path is blocked. The radius of the first Fresnel zone is explained in Chapter 8.

The Effect of Radio Polarisation on the Signal Propagation and Strength

It is a fact that radio signals of the same frequency, but of different *polarisation* propagate with different effectiveness. Generally, *vertically polarised* signals have greater range than *horizontally polarised* signals. This is due to the geometry of precipitation and the earth's surface. Horizontally polarised signals are more prone to reflections from surfaces parallel to the earth's surface, and more prone to rainfall attenuation due to the elongated nature of falling raindrops. In the attenuation model for rainfall attenuation (ITU-R recommendation 838 — see Appendix 8) the value of the parameter α is greater for horizontal polarisation ($\tau = 0$) than for vertical polarisation ($\tau = 90°$).

Polarization scaling of path attenuation

The atmospheric attenuation of one polarisation (either vertical or horizontal) can be scaled according to the following simple formula (ITU-R recommendation P.530):

$A_V = 300A_H/(335 + A_H)$

$A_H = 335A_V/(330 - A_V)$

Another factor which has to be taken into account when using both the horizontal and vertical polarisations of a given radio channel (e.g. for purposes of frequency re-use as we shall discuss in more detail in Chapter 9) is the problem of the loss of *cross-polarisation* (or *cross-polar*) *discrimination*. The cross-polarisation discrimination is the ability of the antenna and radio receiver to pick up and decode only one of the polarisations without suffering interference from the oppositely polarised signal (which is being used by another radio system). The discrimination can be reduced by atmospheric effects or reflections which tend to convert signals of one polarisation into the other polarisation. ITU-R recommendation P.530 also provides a prediction method for estimating outage due to reduction of *cross-polarisation discrimination*.

The Effects on Radio Propagation of the Magnetic and Electrical Properties of the Earth's Geography

ITU-R recommendations P.527 and P.832 provide the methodology and geography-dependent parameter values for calculating the effects on radio propagation and signal attenuation due to the electrical conductivity of the earth's surface. The procedure helps in particular to determine the *permeability* (μ), *permittivity* (ε) and *conductivity* (σ).

In general, these recommendations are really aimed at lower frequency propagation effects (i.e. in the 1–30 MHz range) where the conductivity of the earth can signficantly affect the signal strength. Surface conductivity effects on the propagation of line-of-sight system frequencies are negligible, although the high relative conductivity of water may demand special planning effort to be applied to radio links which are to be operated across large expanses of water or snow-covered ground.

Signal Losses due to Reflections and 'Scattering' in the Atmosphere

Planning effort, as we shall discuss in Chapter 8, is required to ensure that line-of-sight radio system links are not subject to reflections from neighbouring buildings or obstacles. Otherwise the reflection and scattering effects of the earth's ionosphere and galactic meteorites which affect other forms of skywave radiocommunication (Figure 7.5) and satellite communication are not present in terrestrial radio systems.

7.7 Radio Noise

Stray radio signals in the earth's atmosphere affect the ability of radio receivers to accurately receive and decode the correct user signal. Good reception depends upon the strength of the desired signal being many times more powerful than the surrounding *noise*. In other words, a minimum *Signal-to-Noise* (S/N) or *Carrier-to-Noise* (C/N) has to be maintained to ensure good reception.

Radio noise can result from any number of different causes:

- atmospheric disturbances and 'static' (this can be caused by emissions from atmospheric gases or by lightning — it causes the 'crackling' reception on public broadcast radio);
- man-made noise, caused by unintended electromagnetic radiation from electrical machinery (including lift (escalator) motors, power lines, combustion engines, etc. This type of noise makes it essential, for example, not to install radio receiver equipment near escalator motors on roofs);
- noise from space (caused by celestial causes, meteorites, etc.);
- obstacles in the radio path.

It is important to take account of radio noise in the design of radio receivers. In particular, the radio receiver must should be designed to be able to operate in the intended radio frequency band despite the level of noise which can be expected in that given band and in the given world region of operation.

ITU-R recommendation PI.372 defines the terminology associated with radio noise, in particular defining the terms *noise factor* (f), *noise figure* (F), *antenna temperature* (or *brightness temperature*, t_a). The recommendation presents world maps and other charts which enable the radio designer and network planner to determine the appropriate *noise figure* of the radio receivers which he needs. It is inappropriate to use a radio receiver with a noise figure less than the value of the minimum external noise.

Noise Factor (f) and Noise Figure (F)

The noise factor (f) of a radio receiver is a defined parameter representing the summation of all external and internal noise sources affecting the correct reception of the desired signal. (External noise sources are, as we explained above, but we must also take into account the unintended electromagnetic disturbance caused by electronic components or inaccurate waveguides within the radio itself.)

The *external noise factor* is defined in terms of the noise power received by the antenna (this is the same as the noise power received from a perfect 'lossless' antenna), the temperature, the bandwidth of the receiver filter and *Boltzmann's constant*:

External noise factor, $f_a = p_n/kt_0b$

Where p_n = *noise power* received from an *equivalent lossless antenna* (in Watts)

k = *Boltzmann's constant* = 1.38×10^{-23} J/K

t_0 = reference temperature = 290 K

b = noise power *bandwidth* of receiver (i.e. filter) in Hz

Radio engineers tend to work with the *noise figure*. This is merely the noise factor expressed in decibels:

External noise figure $F_a = 10\log_{10}(f_a)$

$= P_n - B + 204$ (derived from above)

where $P_n = 10\ \log_{10}(p_n)$

$B = 10\ \log(b)$

$10\ \log_{10}(kt_0) = -204$ dBW/Hz (equal to -174 dBm/Hz)

The external noise factor is sometimes also expressed as a temperature, known as the *effective antenna temperature* or *brightness temperature* (t_a). (The brightness temperature is commonly used to define the noise sources from skywave and disturbances due to solar or celestial radiation, but is not so often used in terrestrial radio system noise calculations. We present the relationship of the *noise factor* to the brightness temperature for completeness.)

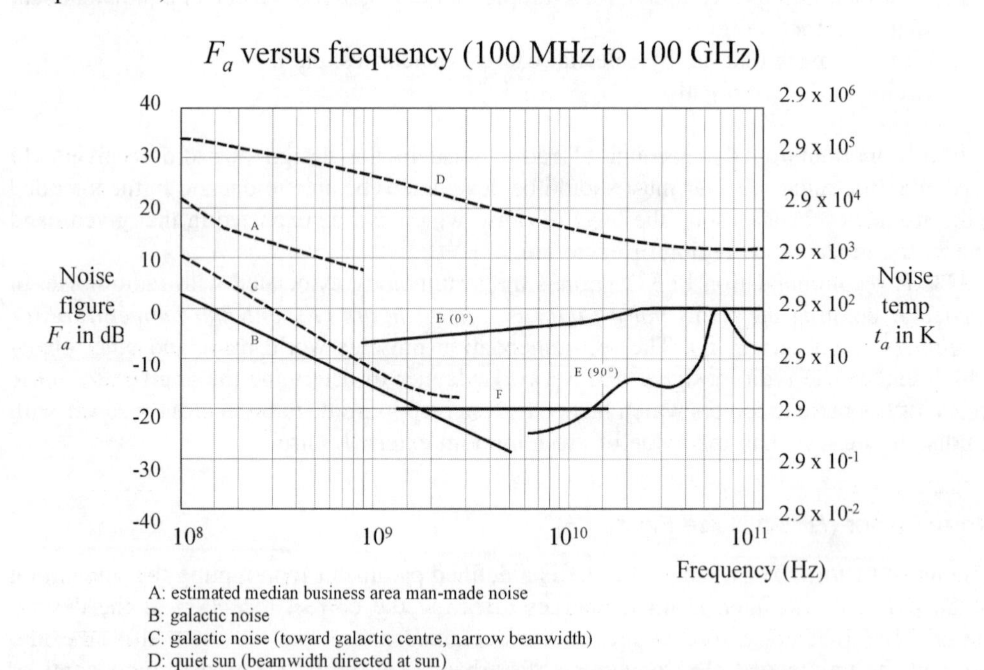

Figure 7.12 Expected external noise in the GHz radio frequency range (ITU-R recommendation PI.372, Reproduced by permission of ITU)

External noise factor, $f_a = t_a/t_0$

where t_a = *effective antenna* (or *brightness*) *temperature*

t_0 = reference temperature (290 K)

The above noise figure and temperature approximate to the system values of *low noise* radio receivers. In other cases, as we have said, we must also take account of the internal noise sources emanating from the transmission line, the antenna and the receiver itself. Assuming each of these devices is actually at the reference temperature (290 K = 17°C), then the total *system noise factor* and the *system noise figure* are given by the formulae:

system noise factor, $f = f_a - 1 + f_c f_t f_r$

system noise figure, $F = 10 \log_{10}(f)$

where f_a = external noise factor

f_c = antenna circuit noise factor

f_t = transmission line noise factor

f_r = receiver noise factor

Terrestrial radio systems are generally designed and planned according to the noise figure. Figure 7.12 reproduces the external noise sources and their strengths as predicted by ITU-R recommendation PI.372. Thus, a radio receiver in the GHz range (above 1 GHz) needs to be designed for a minimum noise figure of at least 0 dB. Taking into account the receiver noise and potential disturbance caused by the receiver pointing at the sun, a noise figure of 10 dB or more may be advisable.

7.8 Summary — Radio System Range and Availability and the Need to be able to Predict Radio Signal Strength

We have discussed at length in this chapter the prediction methods recommended by ITU-R for determining radio signal strength as well as the effect of losses and disturbances which can be expected when operating real radio systems. And why have we set out the methodologies and formulae in such detail? Because they are important to radio equipment and network designers in determining:

- the expected *availability* of a radio link of a given length meeting a given *BER* (*Bit Error Ratio*) quality target;
- the practical *range* of a radio system, given the specified quality targets of *availability* and *BER*;
- the *Received Signal Level* (*RSL*) strength which can be expected at the receive end of a link (this is important for example for aligning the antennas, as we discussed in this chapter and Chapter 6);
- the strength of the signal at any point, and thus the potential for causing interference to other radio systems operating in the same frequency band. Knowing the expected signal strength is the basis of *radio frequency planning* and *frequency re-use planning*, as we shall discuss in Chapter 9. In particular, frequency planning seeks to minimise the output

Table 7.2 Typical radio system range calculated according to the methodologies presented in this chapter

System type (PTP or PMP)	Radio band (GHz)	Transmitter power (max) (dBm)	Receiver sensitivity (BER 10^{-6}) (dBm)	Antenna size (diameter in cm)	Transmitter antenna gain (dBi)	Receiver Antenna gain (dBi)	Approx Range in Climate Zone 'E' (km)
PTP	7	27	− 91	240	43	43	*70
	13	27	− 90	240	42	42	*45
	15	27	− 90	120	43	43	*40
	18	24	− 88	120	45	45	33
	23	20	− 90	60	41	41	20
	26	18	− 89	60	42	42	16
	38	16	− 85	30	39	39	8
PMP	10	20	− 88	approx. 20	13	25	approx. 10 km
	26	19	− 85	approx. 15	15	28	approx. 4 km
	28	18	− 85	approx. 15	15	28	approx. 4 km
	39	17	− 84	approx. 12	15	32	approx. 2 km

Values are typical values of commercial systems.
(*Note that the range of systems operating below about 18 GHz is not primarily limited by rainfall attenuation)

power from the transmitter to the minimum value required to meet a given radio link performance and meanwhile cause a minimum of disturbance to other radio links.

Typical Radio System Range and Availability Charts

Table 7.2 presents without further discussion some typical values of radio system range for different types of point-to-point (PTP) and point-to-multipoint (PMP) systems.

8

Radio Path and Radio Network Planning Considerations

Apart from the design of the radio equipment, the most important factors determining the performance of radio systems are the good location of radio antenna sites, good radio path planning and the choice of an interference-free radio channel. Only with good path planning can an operator achieve freedom from interference and radio fading, and a high system availability. Radio path planning begins with determining the likely range of a given radio system (the length of a point-to-point link or the radius of a given point-to-multipoint cell) as we discussed in Chapter 7. This assumes prior knowledge of the radio band of operation (e.g. 26 GHz, etc.), the climate zone (Appendix 8) and the technical performance specifications of the radio system to be used (transmitter power, antenna gain, receiver sensitivity, BER and availability targets, etc.). After this comes the critical task of radio network and path planning. This begins with determining the planned locations of the radio stations. Site-survey visits or computer-based tools are then used to predict the likely radio propagation between the two end-points of a link, or over the area of a given base station's cell. Path problems, particularly the effect of obstacles should be assessed. Finally, a radio channel for operation of the system within the given band is allocated by the radio frequency re-use plammer (usually supported by a computer tool). The channel is chosen to be different from the radio channels used at neighbouring radio stations, in order to minimise the 'interference' between the stations.

8.1 Radio Planning Terminology and Quality Targets

The radio communication sector of the International Telecommunications Union (ITU-R) is the worldwide authority on radio communication. The terminology, quality targets and methodologies to be used in the planning of fixed wireless systems (ITU-R's so-called *fixed service radio-relay systems*) are defined in the *F-series* of ITU-R recommendations.

The terminology used for radio-relay systems is defined in recommendation 592. This recommendation sets out the names of the different types of radio communication (point-to-point, point-to-multipoint, etc.) as well as the terms for the different types of signal modulation and system configuration. The recommendation also sets out the precise definitions of quality parameters which are used to define the performance and range of radio systems. These parameters include:

- *Bit Error Ratio* (*BER*) (the ratio of the number of errored bits to the total number of bits received);

- *Residual Bit Error Ratio* (*RBER*) (the BER in the absence of weather-induced fading; in other words, the BER resulting from the equipment design and reliability);
- *Errored Second* (*ES*) (a time period of 1 second during which one or more errors have been received);
- *Severely Errored Second* (*SES*) (a time period of 1 second during which the BER is greater (worse than) a specified target).

These quality parameter definitions are central to the design of radio systems, as they determine the performance targets, which in turn define the design range and reliability of radio links.

A number of further ITU-R recommendations set out so-called *hypothetical reference circuits* for different types of radio connections and network applications. These hypothetical reference circuits are imaginary 'worst case' radio connections for a given combination of the type of network connection (e.g. international link, national trunk network part of a connection or WLL) and the type of application (e.g. 'ISDN', 'analogue telephone channels', etc.). The hypothetical reference connections define, for example, the maximum number of cascaded radio links which may be used in an international connection, and the specifications and quality targets which each sub-link must adhere to so that the overall end-to-end link quality is acceptable for the intended application.

Yet more ITU-R recommendations then lay out, for each of the different hypothetical reference circuits, the specific quality targets which should be met when designing and planning radio systems for the given network application. These recommendations set hard targets for the:

- The link *availability* objective (*availability* as we discussed in Chapter 7);
- The link *error performance objectives* (defined in terms of the *Bit Error Ratio* (*BER*), *Errored Second Radio* (*ESR*) and *Severely Errored Second Ratio* (*SESR*) as we defined above);
- Allowable *interference* (from neighbouring radio systems) and allowable *noise* power.

A complete list of the relevant ITU-R recommendations appears in Appendix 7. We shall not discuss all of the configurations in detail in this chapter. Instead, we present a more general and pragmatic approach suited to radio link and cell planning for practical fixed wireless access systems. As we do so, we shall refer to the relevant specific ITU-R F-series recommendations.

8.2 The Need for a 'Line-of-Sight'

As we learned in Chapter 7, the radio frequency bands most commonly used for fixed wireless access systems are at frequencies exceeding about 1 GHz, so that most fixed wireless access systems fall into the ITU-R classification *terrestrial line-of-sight systems*. Line-Of-Sight (LOS) systems are so-called because of the inability of the radio signals at high frequency to propagate 'around corners' or to *diffract* around obstacles. There must therefore be a direct and unobstructed radio signal transmission path (a so-called *line-of-sight*) between transmitting and receiving antennas. If you can see the opposite station, the transmission path will be unobstructed for a high frequency radio signal as well as for light rays!

Confirming that there is a line-of-sight between the two proposed end-points of a radio link is normally done prior to installation of equipment during the site survey phase. The confirmation process is usually referred to as a *line-of-sight-check* or *LOS-check*. We present a checklist for LOS-checks later in the chapter.

8.3 Choosing the Ideal Locations for Radio Stations

You can always generate a line-of-sight between any two proposed end-points of a *short haul* or *medium haul* radio system, it is only a question of high how the towers or masts will have to be for mounting the transmitting and receiving antennas! On the other hand, it might be easier to try to choose the end-points of the link to be in prominent geographical locations or on the top of tall buildings.

Tall buildings, hill tops and other prominent locations benefit from good visibility, and are ideal sites for radio base station sites for point-to-multipoint radio systems or 'hub' sites for concentrating multiple point-to-point radio links, in cases where long link range or coverage is desired using a minimal number of sites. ''But'', you protest, ''it is not always the case that the nearest public telecommunications switch site is an 'ideal' high building suited to being a point-to-multipoint base station'' and ''the two end-points of a desired PTP 'leaseline'-type link are rarely both in good radio sites''. So what can be done to get around these problems?

The easiest way to get around the problem of a lack of line-of-sight is to find an intermediate point which does have good visibility to both end-points. Figure 8.1 shows how a radio base station for *point-to-multipoint* coverage of a 'cell' area within a city can be remotely located from the switch site on a tall building, and be connected to the switch site either by terrestrial fibre or by means of a high capacity *point-to-point* link.

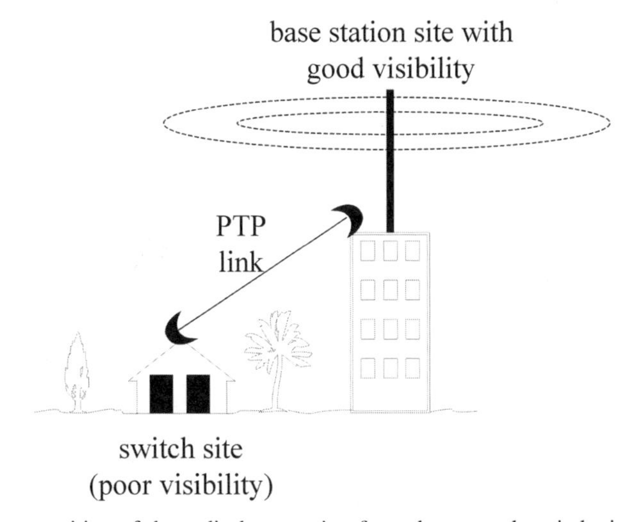

Figure 8.1 Remote siting of the radio base station from the network switch site to improve visibility

The desired end-points of point-to-point links are often on low buildings with poor line-of-sight visibility. A tall building near the poorly visible site can be used as a *passive repeater* (Figure 8.2, and as we discussed in Chapter 6). Alternatively, *active repeaters* could be used to enable PTP connections to reach beyond the normal range of a single link.

Such simple methods as illustrated in Figures 8.1 and 8.2 are used to overcome many of the problems of radio coverage (obstructions, shadows, multipath interference, etc.). If all else fails, you need a bigger mast at the end-point location!

The tallest building is not always the best one for a base station! The problem of very tall buildings is that they can be seen from miles around — which means that they can also cause radio interference for miles around too! Sometimes, it is the desire of the radio frequency planner (particularly in point-to-multipoint and other *cellular* coverage area radio networks) to restrict cell sizes on purpose in order to reduce the distance at which the same radio channel frequency can be *re-used* without causing unacceptable degradation due to. Taking advantage of natural obstacles and local geography can be a good way of reducing interference between *adjacent* base stations (Figure 8.3).

Another factor which should be considered when selecting very tall buildings or other very high points as base stations, is the possibility of having poor coverage in the area immediately surrounding the base station, due to a limited antenna elevation aperture (Figure 8.4).

8.4 The Fresnel Zone

Fresnel zones are elliptically shaped three-dimensional volumes, something like a rugby ball or american football, which surround the main direction of a line-of-sight radio path (as shown in Figure 8.5). The *first Fresnel zone* (often lazily referred to as '*the* Fresnel zone') should be kept clear of obstacles so that destructive radio reflections from objects within this zone do not lead to serious interference of different reflected versions of the radio signal arriving at the receiver.

The surfaces of the various Fresnel zones (i.e the surfaces of the elliptical 'footballs') correspond to points at which, if a reflected path into the receiving antenna were to be generated, it would have a very destructive interfering effect on the signal received via the direct-ray path. Figure 8.6a illustrates the reflection from an obstacle near the direct-ray (or

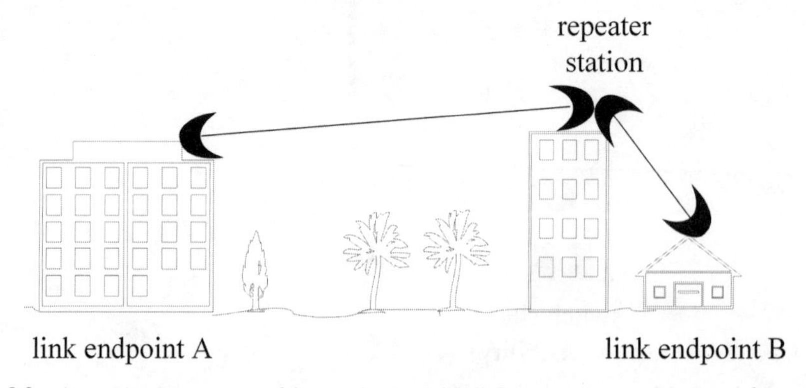

Figure 8.2 A *passive repeater* used in a point-to-point link to overcome obstructed *line-of-sight*

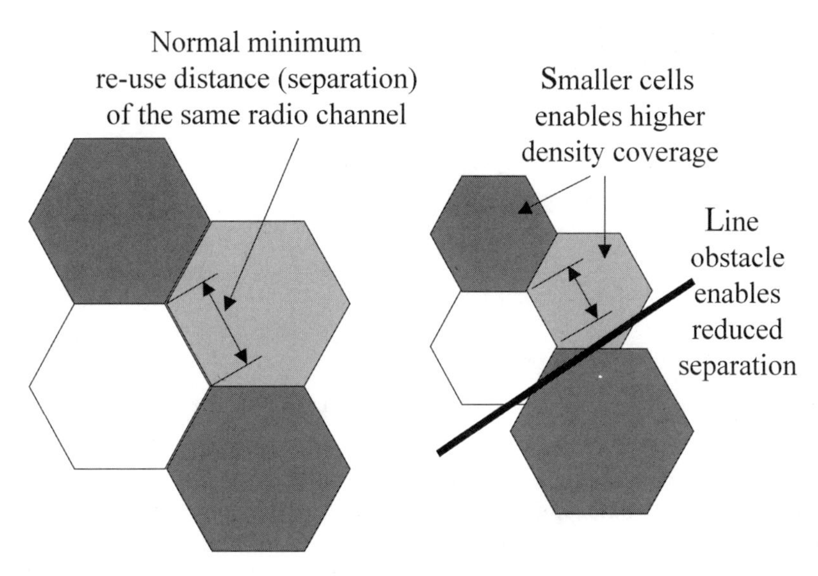

Figure 8.3 Taking advantage of *line-of-sight* obstacles to resolve interference and increase the potential for *frequency re-use*

line-of-sight path). Figure 8.6b shows the destructive effect of interference caused in the case that the reflected ray path is exactly half the wavelength (of the radio carrier signal) longer than the direct path. In this case, the two different signals (the direct one and the reflected one) arrive at the receiver in *antiphase*, summing to give a net signal of very little strength (i.e. causing high levels of *attenuation* or *interference fading*).

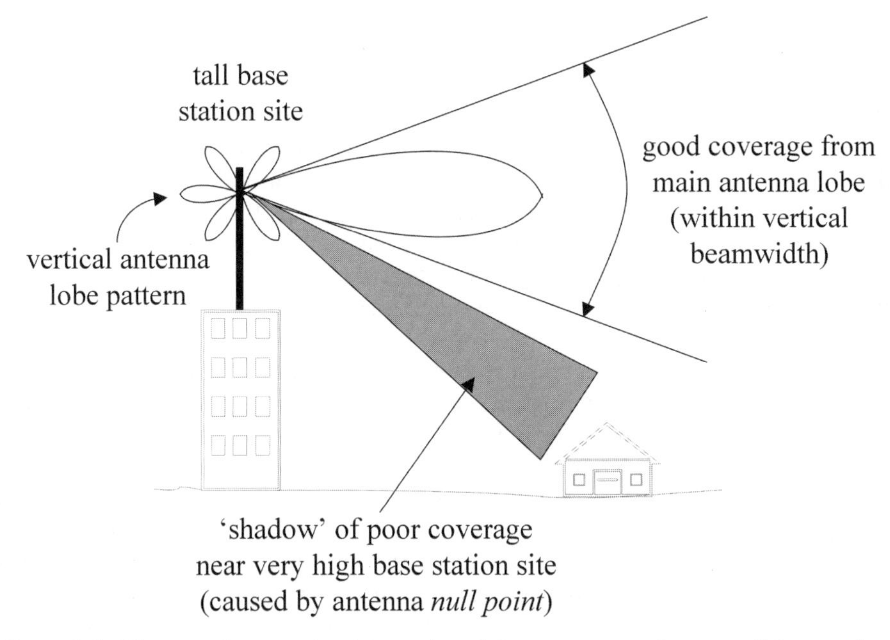

Figure 8.4 There can be poor signal strength and thus coverage under very tall base stations!

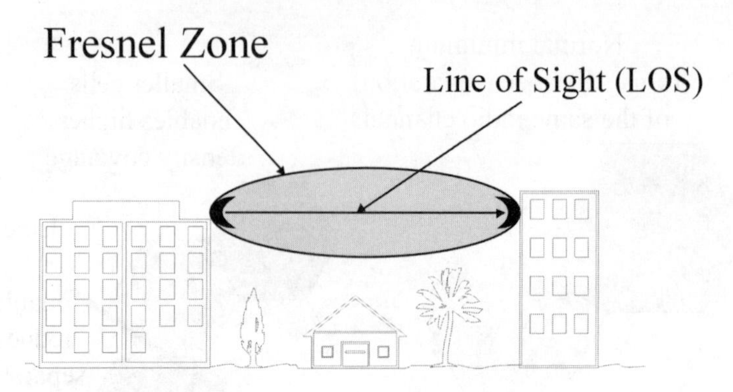

Figure 8.5 Obstacles should be avoided in the first *Fresnel zone*

The worst interference in Figure 8.6 results when the reflected signal arrives in exact antiphase from the direct path signal. This corresponds to a path difference of half the radio carrier signal wavelength (i.e. $\lambda/2$). In other words, the path lengths a and b, when added, are a constant $\lambda/2$ longer than the direct path, d. Anyone who remembers from his school days the trick of drawing an ellipse by using a length of string (of constant length, in this case equal to $d + \lambda/2$) looped around two posts (the positions of the transmitter and receiver) will recognise that there are a whole series of possible reflection points which could cause the destructive reflections, and that these lie on a series of ellipses.

Fresnel zones (Ellipsoids)
Eccentricity of ellipse $\varepsilon = 2d/(2d + n\lambda)$ where $n = 1,2,3,4,5...$
Length of ellipse $= d/\varepsilon$

$$\text{Height of ellipse} = \frac{d}{\varepsilon} \times \sqrt{(1 - \varepsilon^2)}$$

The first Fresnel zone corresponds to the volume within the ellipse with a path length difference of *one* half wavelength (i.e. $n = 1$ in the formulae above). Reflections near the boundary of the first Fresnel zone will cause serious signal interference. For this reason, it is usual to try to ensure (during link planning) that the first Fresnel zone (sometimes simply called *the* Fresnel zone is free of obstacles. An obstacle intruding into the zone can be circumvented by using a taller mast, for example, or perhaps by locating the antenna at a different corner of the building. (Actually, reflections near the 3rd, 5th, 7th and other odd order Fresnel zones can be as disruptive as first Fresnel zone reflections, but these are often ignored by planners in practice, due to the much greater complications in assessing them. Signal delays of three, five and seven half-wavelengths are less destructive than one half-wavelength delays due to the changing nature of the modulated signal — because the signal has changed slightly, the interfering signal is no longer a perfect negating match. In addition, reflections from even order Fresnel zones (2nd, 4th, 6th, etc.) tend to strengthen the signal by 'positive' interference.)

For practical purposes, it is easiest for the planner (during his site survey visit) to take with him a pair of binoculars and try to determine whether a constant diameter 'pipe' of clear space (free of obstacles) is available between the proposed transmitter and receiver antenna locations. The necessary diameter of the 'pipe' depends upon the length of the link (d above) and the frequency of the radio channel to be used. We can calculate the diameter from the formulae below:

Maximum diameter of (first) Fresnel zone (at midway point of link) $= 2\sqrt{(d\,n\lambda)}$ (all distances and wavelengths in metres)

Alternatively, maximum diameter $= 17.3 \times \sqrt{(d\,n/f)}$

Where d = length of link (transmitter to receiver) in km, n = whole number, characterizing the Fresnel zones, and f = radio channel frequency in GHz.

Thus, at a frequency of 38 GHz over a link of length 8 km, the first Fresnel zone has a diameter at the mid-point of the link of $17.3 \times \sqrt{8 \times 1/38)} = 8$ metres. Meanwhile, a 7 GHz link of 70 km length has a Fresnel zone of maximum diameter 55 metres. This 'free space pipe' should be checked for obstacles at the time of the link site survey (i.e. during the link planning stage prior to equipment installation).

The radius of the first Fresnel zone at any point on the radio link path is given by
$F_1 = 17.3 \times \sqrt{\{(d_1 d_2)/(f\,d)\}}$

Where d = length of link in km, d_1 and d_2 are distances in km from the end terminals to the point in question, f = radio channel frequency in GHz.

The formula presented above is useful for determining whether an obstacle is intruding into the Fresnel zone (as in Figure 8.7).

Obstacles in the Fresnel Zone

In the case of unavoidable intrusion of an obstacle into the Fresnel zone of a particular radio link, it is advisable to estimate the resultant loss of received signal strength. The methodologies for calculating the *diffraction losses* caused by different types of obstacles are covered by ITU-R recommendation P.526. In Figure 11 of chapter 7 we presented the graph used in this recommendation for calculating diffraction losses caused by obstacles.

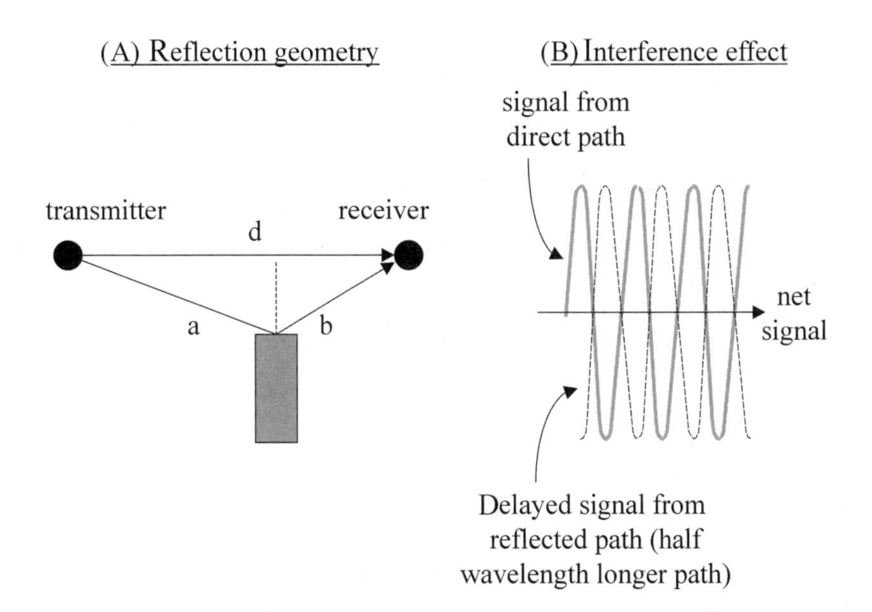

Figure 8.6 Interference caused by reflected ray paths arriving at the receiver out of phase

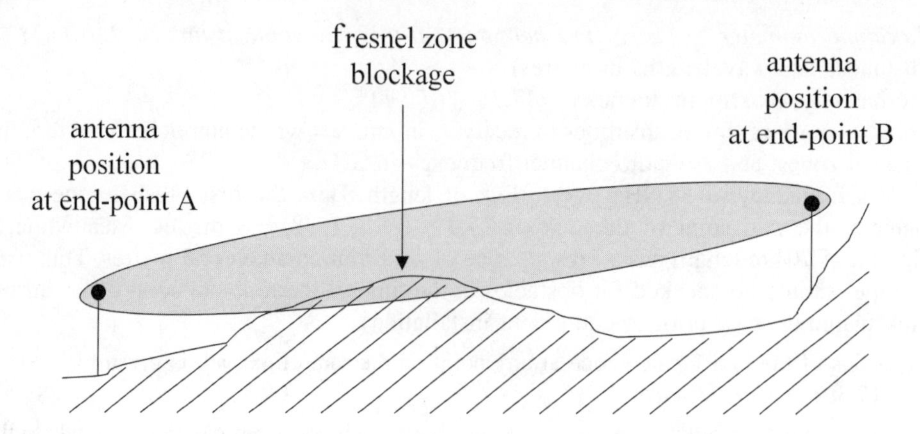

Figure 8.7 Example link cross-section and super-imposed Fresnel zone

8.5 Line-of-Sight Check (LOS-check)

It is customary to conduct a line-of-sight check (LOS-check) prior to the installation of a microwave or millimetre wave radio sytem. A LOS-check involves visiting the proposed installation sites of transmitting and receiving antennas to assess the availability or otherwise of a line-of-sight between the two stations; to determine whether there are any obstacles within the Fresnel zone and to assess any other likely radio interference effects arising from the locality (e.g. nearby high power radio transmitters or large metal objects or areas which might cause harmful reflections).

The following is a list of the tasks usually carried out during the site visit and LOS-check:

1. **End-point coordinates:** determination of the exact coordinates of the two end-points of the radio link — the coordinates are most easily determined by means of a *GPS* (*Global Positioning System*) receiver. The longitude, latitude (or equivalent), as well as the height and height above ground of the intended antenna position should be determined. The exact coordinates are crucial for good radio frequency and coverage planning.

2. **Map location and compass bearing of link:** the bearing of the remote end of the link should be determined, and the position of the link should be indicated on a map. The map should be used to determine possible obstacles, and can also be used for determining clearly visible landmarks, which may subsequently be used to locate the remote site using binoculars.

3. **Line-of-sight:** in the case of short and medium range links, binoculars may be used during good visibility to locate the remote station and to check for a line-of-sight. The remote station may be located by means of the map bearing and/or using the reference landmarks established in point (2) above. However, when using a compass on a building roof, you must be careful to check that the compass is giving a true reading. The electric motors associated with elevator shafts on building roofs can create significant magnetic fields which may cause false readings of compass bearings of up to 20° or even more. The reference landmarks are always valuable — if only as a check.

4. **Correctly identifying the remote location:** in cases when it is difficult to distinguish the particular building intended for the remote station location, one of a number of aids may be used to aid identification. A second person located at the remote site and holding a strobe light (powerful flashing light) will help to locate the right building. Alternatively, on a sunny day a simple mirror can be used for similar effect.

5. **How tall must the mast be?** in a case where a direct line-of-sight is not available between the two end-points, it may be necessary to determine how tall a mast at one end of the link will have to be in order to overcome the blockage caused by an intervening obstacle. The necessary mast height can be determined by using a helium balloon. The balloon is simply flown on a rope at the radio station end where the mast needs to be installed and let out successively further on a rope until it becomes visible by binocular from the remote end.

6. **Checking the Fresnel zone:** having determined that there is a line-of-sight, the Fresnel zone should be checked to be clear of obstacles. In addition, the areas immediately around the two intended antenna positions should be checked to be free of obstacles and potential reflection surfaces. Particularly dangerous are large metal (reflecting surfaces) or other transmitting radio antennas directed at or close to the intended position of the new antenna. While checking for obstacles, you should also consider the likely condition of the link at other times of year and after a few years operation. (Trees grow taller; trees have leaves in summer and not in winter; mounds of snow can accumulate on rooftops in winter; industrial and business parks spawn new high-rise office blocks; passing ships, planes or other traffic, etc.) Obstacles in the near-field (up to about 50 m from the two end-points) can be particularly problematic.

7. **Photographic documentation:** it is valuable to record the findings of the LOS-check on film. A good quality camera, used with a tripod and telephoto lens, ensures that the radio planners back at headquarters can make an assessment for themselves should there have been any traces of doubt about, for example, the likely disturbance caused by a particular obstacle (e.g. intruding overhead power lines).

8. **LOS-check for links longer than about 8 km:** the relatively simple steps described above represent the normal procedure for LOS-checks of *shorthaul* and *medium-haul* radio links (up to about 8 or 10 km). However, 8 or 10 km is about the greatest distance over which you can expect reliably to perform LOS-checks using only a pair of binoculars. Even to see such distances good weather and visibility is necessary (it is no good going out in heavy rain for such an LOS-check). So, you ask, how can I check the availability of a line-of-sight, and how can I check that the Fresnel zone is free of obstacles? The answer is by commissioning a link cross-sectional profile from a geographic agency, or alternatively, by using a computerised radio planning system incorporating a digital map to generate this cross section, as illustrated in Figure 8.7. (For an accurate cross-section, the exact end-points of the link, as determined in step (1) are crucial.)

ITU-R recommendation P.530 provides a complete methodology for determining the effects on the signal of obstructions in the path, as well as possible interference effects caused by multipath fading (e.g. as a result of radio obstacles or reflections off terrain). It provides a detailed methodology for calculating the reliability of links created from multiple individual hops (such *multiple hop* links are used to overcome range difficulties or to circumvent major obstacles). P.530 also provides estimation methods for the reliability of *diversity* configured radio links. These factors also need to be taken into account in the planning of individual radio links. We discuss these and a number of other planning considerations in the remainder of this chapter.

8.6 LOS-Checks for Point-to-Point and Point-to-Multipoint Radio Systems

In principle, both point-to-point and point-to-multipoint systems share the same requirement for a LOS-check for each individual radio link, but the procedure is somewhat different, due to the different nature of the two types of systems.

In the case of point-to-point systems, it is customary to conduct the LOS-check as we previously described for each individual link, during the site survey and planning stage. The LOS-check is carried out by visiting the intended antenna location at both ends of the link and looking towards the remote station to ascertain whether any obstacles are in the direct path or the Fresnel zone. The coordinates determined during the LOS-check are subsequently used accurate radio frequency planning and coordination with other nearby links.

In the case of point-to-multipoint systems, there is no longer an equal significance to the two ends of the radio link. In point-to-multipoint radio systems, the base station site has a greater importance, and radio frequency planning must be done on a base station or cellular basis rather than on an individual link basis. Good planning of the base station is therefore of very great importance. It is important to choose base station sites which have general good visibility to the desired coverage area of the base station 'cell'. In general, prominent locations (e.g. on tall buildings or radio towers) are good locations for base stations.

Since it is usual to choose the sites of the base stations before the exact locations of the remote multipoint (e.g. subscriber) stations are known, it is usually necessary to perform a more general 'LOS-coverage-check'. The following questions might help in the assessment of a site:

- Are all the critical major business areas and important customer headquarters buildings within direct line-of-sight of the base station?
- Are the rooftops of most surrounding buildings visible from the intended base station antenna site?
- Is the surrounding ground geographically concave (a natural amphitheatre where all points are relatively easily seen) or convex (where further away buildings are increasingly hidden by nearer buildings)?
- Are there any major obstacles near the base station or any surfaces which might cause strong signal reflections and/or *shadows* behind the obstacle?
- Are any other base stations visible which might cause interference to or might be interfered with by the new base station?

The same principles as for the formal LOS-check described previously apply. So there must be direct line-of-sight for all likely links and the Fresnel zones should be clear. Of course, it is only first possible to conduct the detailed LOS-check for a particular remote station site, once the exact location has been determined (e.g. once the customer has signed up for WLL service). At this point, a normal LOS-check is performed, as we described in the previous section. Since the link lengths to remote point-to-multipoint sites are usually relatively short (e.g. 5 km), it is usual to try to save time and money by conducting the LOS-check only at the remote site. Indeed, it could be the first part of an equipment installer's duties: 'confirm *line-of-sight* or determine most visible nearest base station'. Such a procedure could avoid the need for a separate site survey and LOS-check visit. In cases where the base station is not visible from the remote site, two main options are available:

- Build a mast on the roof of the remote station to overcome the obstacle.
- Seek a direct line-of-sight to an alternative base station (perhaps using a higher gain antenna to enable greater link range than normally possible).

Where the buildings in the area surrounding a base station site are all approximately the same height (the height restrictions of some city councils leads to all the buildings being the same height), you should consider the length of the longest possible link and the dimensions of the Fresnel zone before deciding the appropriate height of the mast at the base station location. Thus, for example, the maximum radius of the Fresnel zone for a cell of 5 km range operating at 26 GHz is about 3.8 m. So a mast of minimum height 5 m might be advisable!

8.7 Increasing Link Range and Overcoming Obstacles on Point-to-Point Links

The link range of point-to-point systems can easily be extended by the use of active repeater stations as we discussed in Chapter 6 and we illustrate again in Figure 8.8. There is also the possibility of multiple hop links, where a number of activer repeaters and individual links are used.

The possible range of a repeated link is in principle unlimited, provided enough sublinks are used, each operated within the range limitations of the radio system and the local climate (as we discussed in Chapter 7). However, the end-to-end *availability* of multiple-hop links is not the same as the availability of the individual sub-links. We have to consider the *unavailability* of the end-to-end connection. Unfortunately, if all the links are available but one is unavailable, then the end-to-end connection is unavailable. The likelihood is that indeed each link will be unavailable at times when other links remain available, so that there is a cumulative effect of unavailability. In crude terms, if each sublink of a 10-hop-link is planned to be available for 99.99% of the time (unavailable for 0.01% of the time), then the end-to-end link as a whole can be expected to available about 99.9% of the time

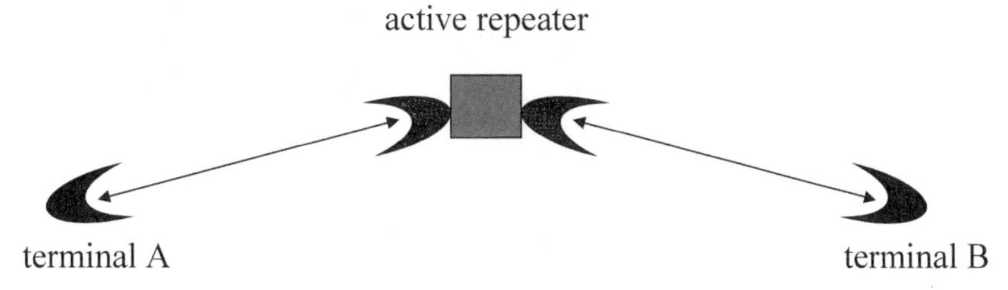

Figure 8.8 The use of *active repeaters* in multiple-hop links to increase link range

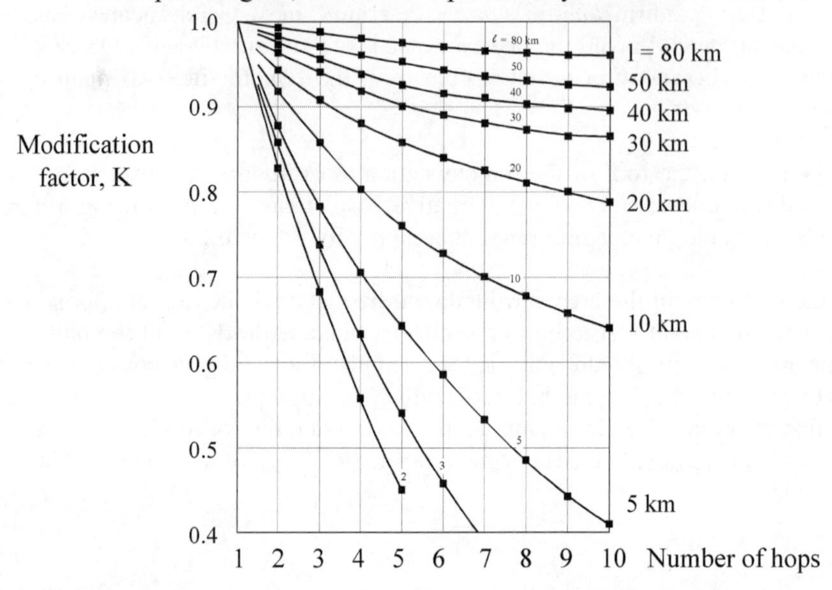

Figure 8.9 Modification factor for 'scaling down' the cumulative unavailability of multiple hop links (ITU-R recommendation F.530) (Reproduced with permission of the ITU)

(unavailable for 10 hops × 0.01% each). (Actually, this crude estimate is not correct, as the weather effects on the sublinks are not 'statistically independent' [as mathematicians would say] of one another. If it is raining on one of the hops, it may not be raining quite as hard in the neighbouring link, but it's likely to be raining. So things aren't as bad as we first made out — luckily some of the unavailability time of the different hops overlaps. So we can 'scale down' the overall unavailability time. The scaling down factor is dependent upon the total number of hops in the complete link, and is presented in a special

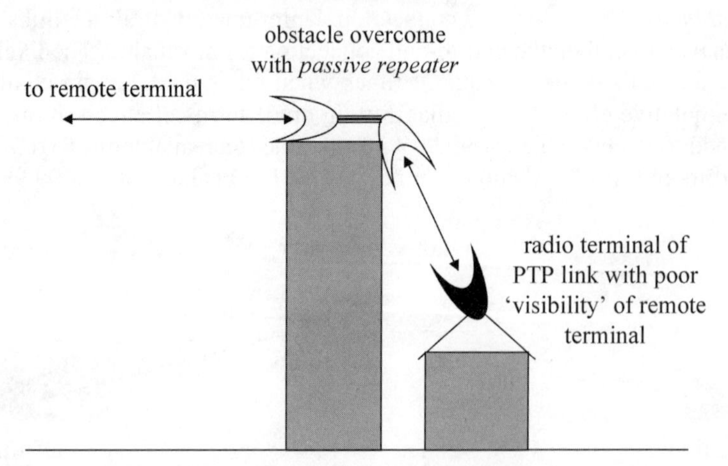

Figure 8.10 Using a passive repeater to overcome an obstacle in the line-of-sight

methodology to predict multiple hop link availability. This is presented in ITU-R recommendation P.530. The relevant diagram is reproduces in Figure 8.9 below.)

Interpreting Figure 8.9 for a link of 360 km, made up of nine multiple hops of 40 km each, we can see that the modification factor (K—for the probability of a given depth of fade or for the probability of a given link unavailability) has value 0.9. Assuming each of the links had been planned for availability of 99.99%, then the cumulative unavailability (prior to applying the modification factor) would be $9 \times 0.01\% = 0.09\%$. The more likely unavailability is determined by multiplying this result by the modification factor K, $0.09 \times 0.9 = 0.08\%$. So the expected availability of the nine hop link will be 99.92% (100%–0.08%).

We can use either active or passive repeaters for overcoming an obstacle in the direct line-of-sight of a point-to-point path (Figure 8.10). While having the advantage of being cheap, passive repeaters (for example back-to-back antennas, as we discussed in Chapter 6) have the disadvantage of reducing the overall range of the link (and quite substantially). Passive repeaters should only be located near to one of the ends of the link (e.g. up to 500 m from one of the terminals) for reliable operation.

Calculating the range of a link containing a passive repeater must be done by considering the link to be a 2-hop link. The signal strength received by the first antenna of the passive repeater should be considered to be the imaginary 'transmitter' output feeding the second (transmitting) antenna of the repeater and thus the second sub-link. The mathematics are quite complex for calculating the maximum range and link availability. The easiest approach is simply to 'check' the viability of the proposed position of a given passive repeater station.

8.8 Overcoming Coverage Problems and Shadows in Point-to-Multipoint Base Station Areas

Range problems and obstacles can also arise when planning point-to-multipoint coverage, but their effects and the means of getting around them are slightly different than those for the point-to-point system solutions we discussed above. In addition, the capacity of the base stations (i.e. the ability of the base station to meet all the bandwidth demands of the remote stations) is an additional problem for the planners of point-to-multipoint systems.

While *omnidirectional* antennas may be used in the base stations of PMP radio systems, as shown in Figure 8.11a, it is nowadays more common to *sectorise* the base station. Sector antennas, as opposed to omnidirectional antennas, broadcast their signals to only a narrower sector area as shown in Figure 8.11b. The benefits of sectorisation are three-fold. First, the same radio frequency may be *re-used* in different sectors, allowing for greater overall capacity in the cell area from the same amount of radio bandwidth. The second benefit is that a sector antenna has a higher gain than an omnidirectional antenna, so that the system range or availability is also increased by sectorisation. The third benefit, is that the more directional nature of sector antennas allows them more easily to be positioned to avoid obstacles and to gain optimum coverage and capacity. An omnidirectional antenna placed at any position on the roof of a tall building is likely to be obstructed by escalator shaft-tops or interfered with by other heavy current electrical rooftop equipment. However, sector antennas can easily be positioned at the individual corners of the building to avoid the obstacles.

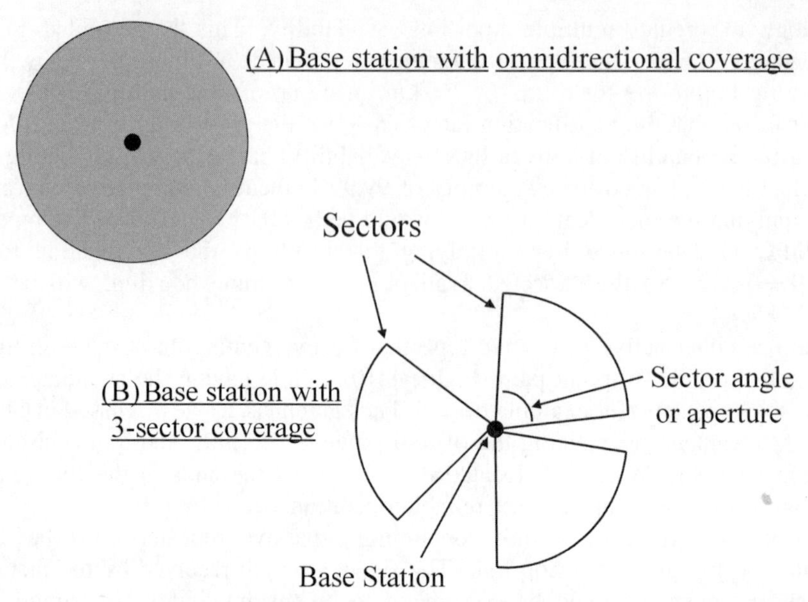

Figure 8.11 Base sector sectorisation

By overlapping neighbouring sectors in urban areas, the radio network planner can gain greater density of coverage (greater availability of bitrate per square kilometre) and simultaneously reduce the likelihood of there being no line-of-sight available to the next base station.

In the example of Figure 8.12, three sectors (originating from three different base stations — A, B and C) have been purposely overlapped. In the triangular-like shaded area which is common to all three sector coverage areas, there are three line-of-sight

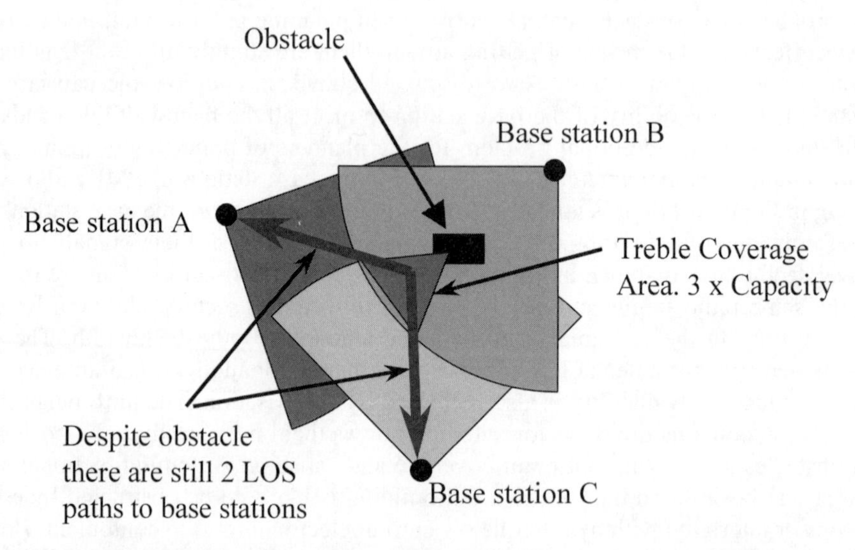

Figure 8.12 Use of overlapping base station sectors to increase network capacity and line-of-sight coverage

possibilities for connecting the remote customers to a base station, so the chances that a given building is obstructed from the nearest base station is greatly reduced.

A second important benefit of the overlapping sectors of Figure 8.12 is the potential for frequency re-use. Given no harmful reflections, it would be possible in Figure 8.12 to use the same radio channel at each of the base stations. Because of the highly directional nature of the remote stations they would only communicate with the base station at which they had been pointed during installation. There will be no interference, and direct interference between the base stations does not occur (even though they appear to be pointing at one another), because although they are all transmitting at the same frequency, this is not the frequency to which their receivers are tuned. The frequency at which they expect to receive (as we discussed in Chapter 2) is the duplex channel freqeuncy. So by *re-using* a single radio channel, we have achieved three times the normal bitrate capacity of the spectrum! We return to the subject of frequency re-use and spectrum planning in Chapter 9.

Where there is a major unavoidable obstacle in one of the sectors of a point-to-multipoint base station, repeaters may be used. Point-to-multipoint repeaters are similar in function to point-to-point repeaters (as we saw in Figure 8.8). Thus, the *shadow* behind an obstacle can be overcome with a point-to-multipoint repeater (either active repeater or passive repeater). A simple point-to-point passive repeater exactly like that of Figure 8.8b could be used to provide coverage to a single remote station in the shadow area. Alternatively, the planner may choose to try to eliminate the whole shadow with a sector repeater (Figure 8.13). Such repeaters are normally active repeaters (i.e. they increase the signal strength).

Sector repeaters such as that illustrated in Figure 8.13 usually come in one of two varieties. The simplest type simple amplifies and re-transmits the received signal at the same frequency. The problem with this type of repeater is the likelihood of *overshoot*. If a given remote terminal has (by chance) a line-of-sight to both the repeater and to the base station, then the signal transmitted directly to the base station may cause interference to the repeated signal. Careful consideration in the placement and alignment of the repeater antennas is therefore necessary.

An alternative type of repeater instead not only amplifies the signal strength, but also moves the signal to a different radio channel in the repeated sector in order to avoid the interference we spoke of above.

Point-to-multipoint repeaters, just like point-to-point repeaters, can also be used to increase the range of a given base station cell. Thus, the total range of the cell in Figure 8.13 has been slightly increased by the repeated sector. We could also imagine a cascade of repeated sectors, one after the other (as long as the equipment design and modulation scheme can cope with the electrical and extra propagation delay consequences of the repeater chain). Such a repeater chain might be an ideal means of providing full coverage along a narrow valley.

8.9 Optimising the Capacity of Point-to-Multipoint Base Stations

The simplest way to increase the capacity of point-to-multipoint base stations is to increase the number of sectors and thus the degree of frequency re-use. *Splitting* the sectors from four 90° sectors to eight 45° sectors will double the available capacity. In practice, though,

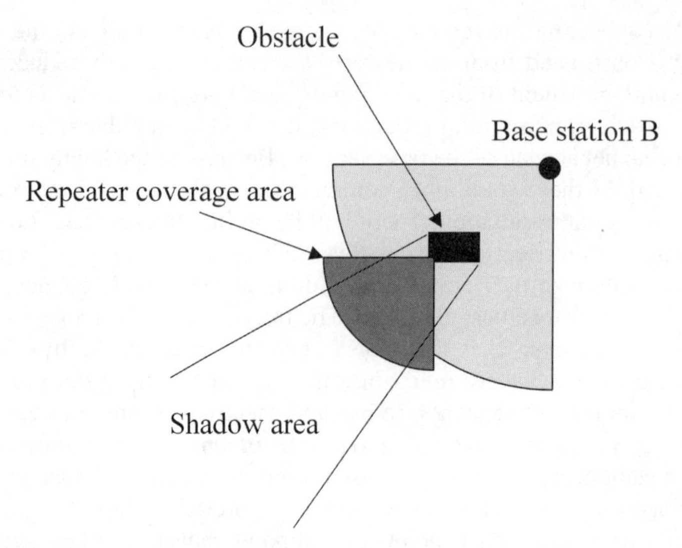

Figure 8.13 Using a sector repeater to 'fill' a shadow area

extra capacity is unlikely to be needed in all of the sectors simultaneously, so it may suffice to split only one of the sectors. In doing so, the radio frequency plan needs to have been carefully considered, so that the newly introduced sector does not interfere with any of the existing neighbouring base stations. In practice, this can only be achieved either manually, by keeping to a strictly repeated frequency re-use pattern (some examples of which we shall discuss in Chapter 9) or by means of a computer planning tool.

8.10 The Significance of the Antenna Aperture Angle in Point-to-Multipoint Sector Antennas

As we discussed in Chapter 3, the angle of opening (or *aperture*) of an antenna is usually quoted in terms of the angle between the 3 dB points of the antenna lobe diagram. In other words, within the aperture of the antenna the gain in all directions is within 3 dB of the maximum antenna gain (Figure 8.14).

Figure 8.14 shows the antenna diagram in the horizontal plane of a nominal 90° sector antenna. Note that the maximum design aperture of the antenna (102°) is greater than the nominal 90°. This ensures that even along the nominal boundaries of the sector (at 90°) the gain of the antenna is at the maximum. On the other hand, it may mean that there is considerable overlap between the sectors of a four 90° sector base station (Figure 8.15). This fact will have to be taken into account when choosing the radio frequencies to be used in the adjacent sectors (the subject of Chapter 9).

Another reality is that the sectors of Figure 8.15 do not have nice 'clean' straight boundaries. The signal of the upper right hand sector of Figure 8.15 can also be reliably received at bearings well beyond even the 'extra regions' of 6° beyond the nominal 90° boundaries, provided the receiver is close enough to the base station (in other words, provided that a much lower sector antenna gain than the maximum is sufficient for the link

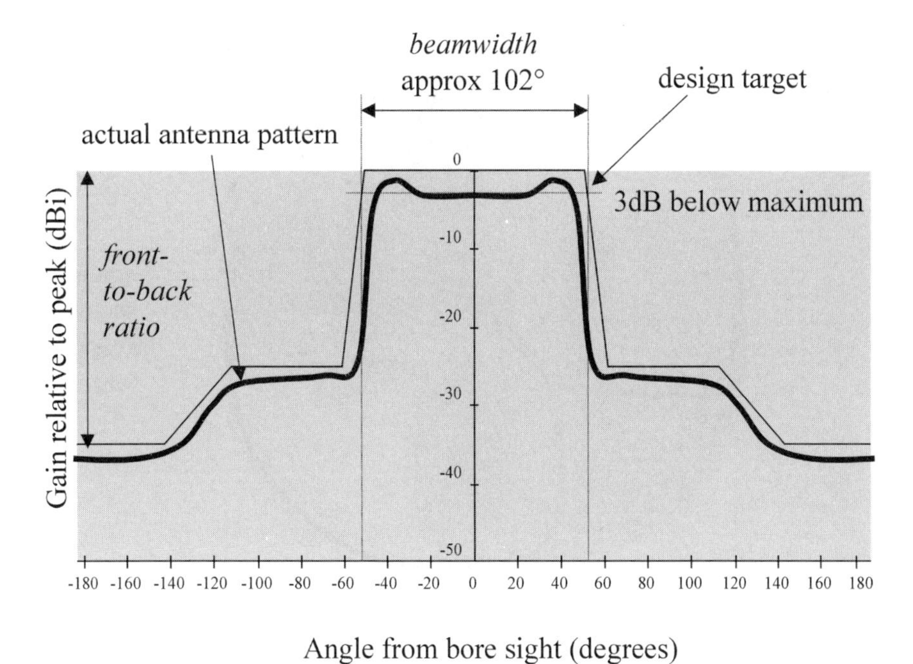

Figure 8.14 Horizontal antenna pattern showing beamwidth and front-to-back ratio

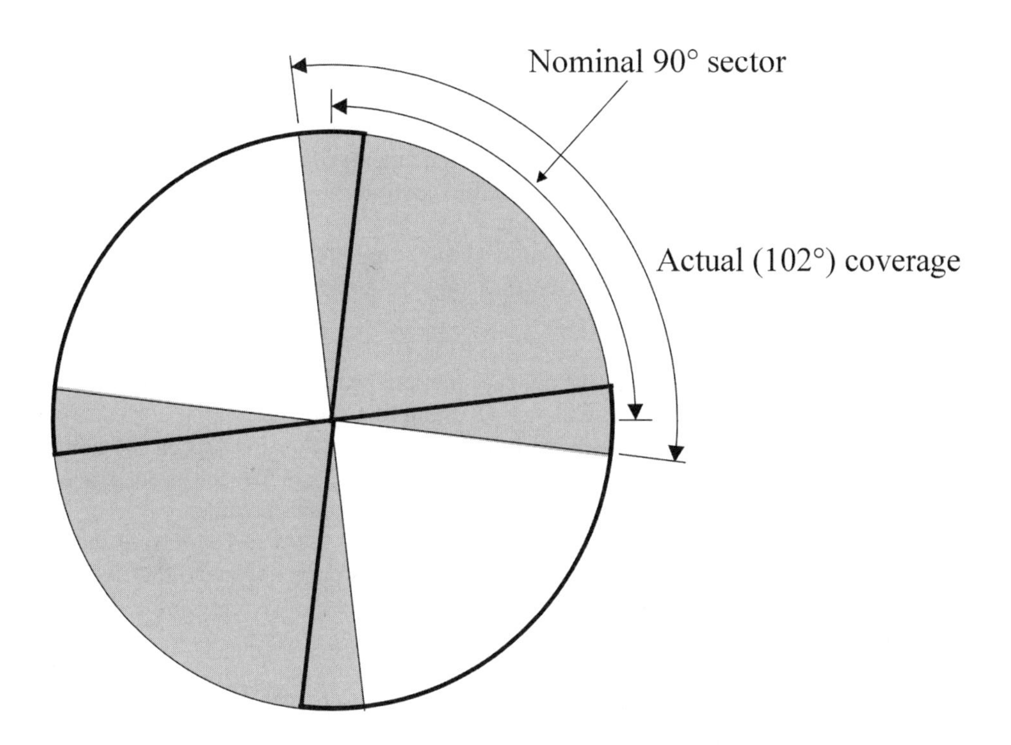

Figure 8.15 In reality, nominal 90° sectors may overlap considerably when considering the antenna aperture angle (antenna gain within 3 dB of maximum)

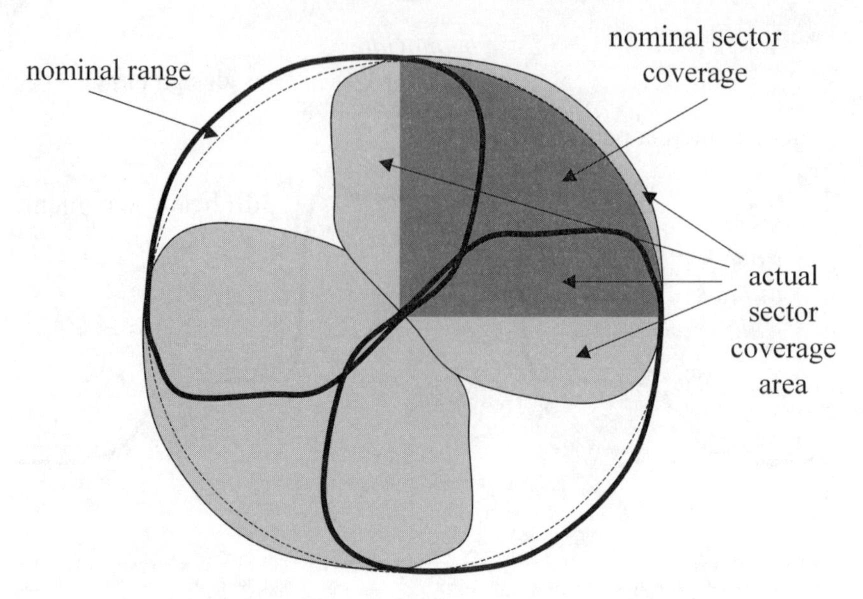

Figure 8.16 Actual sector coverage areas showing *antenna rolloff*

length. Figure 8.16 illustrates (in an exaggerated way) the likely actual overlap areas (area within which 99.99% link availability can be achieved).

So what's the message of Figure 8.16? (Answer) That in reality it is difficult to plan accurately without the assistance of a computer planning tool which records the exact position and orientation of antennas and the detailed antenna radiation patterns. Just because you lie outside the 3 dB *aperture* angle of the antenna does not mean you won't get reliable transmission. At shorter distances than the maximum range, the maximum antenna gain is not necessary for reliable transmission.

So much for the significance of the antenna aperture angle for the horizontal plane, what about the considerations of the vertical aperture angle of a PMP sector antenna? There are two important messages:

- The maximum gain point of the antenna (i.e. the antenna pole [not the steel mounting pole, but the *pole* we discussed in Chapter 3]) should be directed a point corresponding to the top of the building of the most distant remote subscriber located on the horizon. This is important if the maximum range of the system is to be achieved, since the maximum antenna gain must be available to the most distant customer.
- Although the subscribers near to the sector antenna may seem to lie outside the 3 dB vertical aperture angle of the antenna (Figure 8.17), this does not mean that they have unreliable service. (As we discussed above, nearby points do not require the maximum antenna gain of the base station sector antenna.)

It may not appear sensible to orientate the pole of the base station sector antenna to a point at the horizon, particularly when the base station is on a very tall building. It may, indeed, seem far more obvious to try to tilt the base station sector antenna so that most of

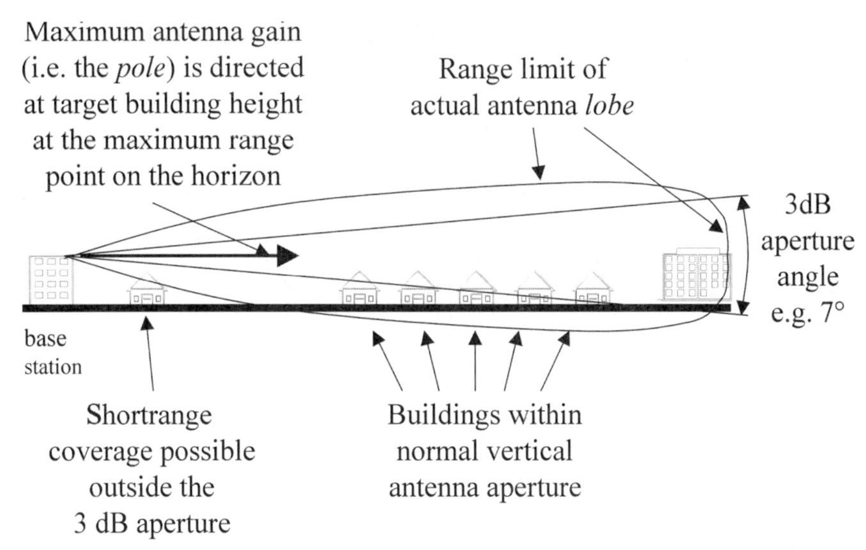

Figure 8.17 The maximum vertical (elevation angle) antenna gain should be directed towards the horizon but nearby points outside the aperture angle still get service because they do not need the maximum sector antenna gain for reliable service

the desired 'target' customer buildings lie within the 3 dB aperture angle of the antenna. However, as Figure 8.18 illustrates, such down tilt of the antenna has a disastrous effect on the range of the system. Now the maximum gain of the sector antenna is not available for the prospective customers at the maximum theoretical system range.

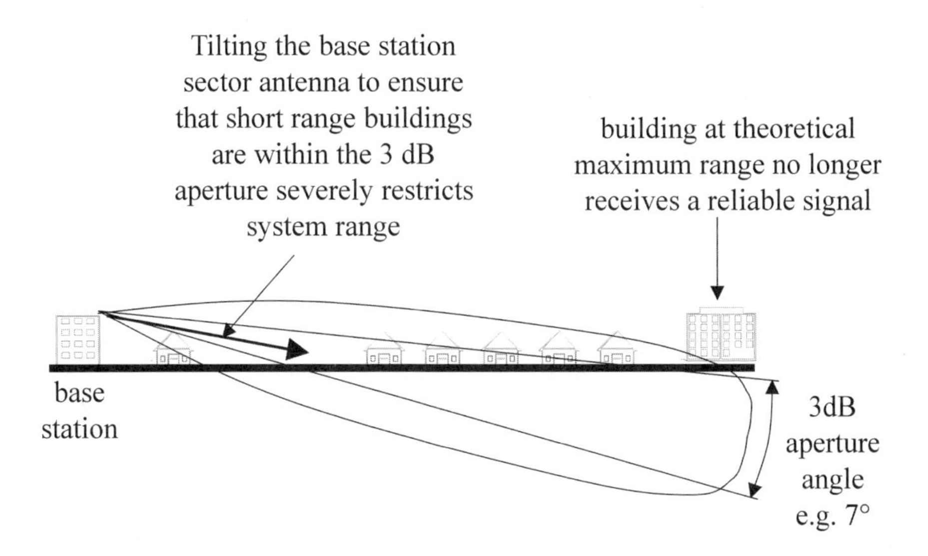

Figure 8.18 Tilting the sector antenna downwards can significantly reduce system range

8.11 The Fading Effects of Multipath Propagation

ITU-R recommendation F.1093 provides a detailed methodology for predicting the effects of multipath propagation. It also describes how the effects of *multipath fading* can to some extent be countered either by counter-measures included in the radio design, or by means of so-called *diversity reception* or *adaptive equalisation.*

Multipath fading is considered by to be the dominant propagation factor affecting radio systems operating at frequencies below 10 GHz. (Above this frequency the fading effects of precipitation (e.g. rainfall) are more limiting on system range, even though multipath fading may still be present.)

Multipath fading is caused primarily by atmospheric effects, but can also be caused by reflections from obstacles, as we discussed earlier in the chapter. Multipath fading leads to problems of signal interference and distortion. The distortion may include one or more of the following variable distortions of the signal amplitude (called *amplitude dispersion*), of the frequency (*frequency dispersion*) and *phase dispersion.* The following counter-measures may either be built-into the system design or planned to be added to the link equipment as part of the link planning phase:

- *Adaptive channel equalisation* (either *frequency domain equalisation* or *time domain equalisation*) is usually added at the radio design phase as a counter-measure for frequency or phase dispersion. As the name suggests an *equaliser* is intended to ensure that the relative frequency or signal phase (i.e. timing) content of a received signal is returned to the original balanced (i.e. equalised level) rather than remaining in the distorted state. Equalisation is carried out prior to *demodulation.*
- *Space diversity (1 + 0 diversity)* is one of the best ways of overcoming multipath fading. The fading is overcome by using two separate receiving antennas (and receivers) which are mounted at the same two end-points, but separated vertically. Multipath fading rarely affects both of the paths simultaneously. The signal improvement achieved by space diversity depends upon the treatment of the two signals by the two receivers. The most common types nowadays employ *hitless switching.* The two signals are demodulated separately and compared with each other to determine which has the lower bit error ratio. This digital signal is then conveyed to the end-user's data equipment. When the second receiver achieves the better signal, then the digital signal feed is changed over in a way which means that the end-user is not aware of the switchover—hitless switching is achieved.
- *Frequency diversity (1 + 1 diversity).* In *frequency diversity* two separate links are operated in parallel and beside one another on different radio channels, but carrying the same input signal. In other words, two receivers and two transmitters are used at each end of the link (four receivers and four transmitters in total). As with space diversity, hitless switching is used to make sure that the end-user receives only the better of the two received signals.

Provided the radio equipment has been designed to be configured in either or both of space diversity or frequency diversity configurations (as well as as a simple *unprotected link*), then the decision to realise the link in a diversity configuration is one taken during link planning.

8.12 System Diversity and Link Hot-Standby Configurations (Protected Links)

There are two main motivations for deciding at the planning stage of a radio link to realise a *diversity* configuration:

- As a counter-measure to multipath fading (frequencies below 10 GHz as we discussed above).
- As a protection against hardware failures (as a so-called *protected link*).

Protected links are those with redundant hardware to ensure disturbance free communication end-to-end despite hardware failures of link equipment. Longhaul and high performance trunk radio links are normally provided as frequency diversity links using hitless switching. Thus most longhaul high capacity trunk radio systems are configured with protection this way.

Shorthaul and medium haul links, on the other hand (i.e. point-to-point links above about 15 GHz), when configured as protected links, are nowadays most commonly configured as *monitor hot-standby* links. Monitor hot standby is a little like space diversity, except that the two receivers might actually share* a single antenna. At both ends of the link there are two transmitters and two receivers. All the receivers are always receiving, two (one at each end of the link) as *master* and two (one at each end) as the *backup* (Figure 8.19).

Figure 8.19 shows half of the equipment operating on the link (only the transmit channel from left-to-right is shown — there is also a duplex channel operating right-to-left, but not illustrated). Transmitter 1 ($T_x(1)$) is active and being received by both receivers, $R_x(1)$ as *master* and $R_x(2)$ as *back-up*. Only the signal of $R_x(1)$ is fed to the end-user. Should the active transmitter fail ($T_x(1)$), then there will be no signal received. Both receivers will notice and the *back-up* receiver ($R_x(2)$) will take control of the situation. It assumes the role of *master*, activates its own transmitter and by so doing is able to instruct the remote transmitter ($T_x(2)$ of Figure 8.19) also to become *active*. (Actually it is usual for both transmitters also to be active, as it takes time for an transmitter to switch on and synchronise. Instead the output signal of the *backup* is greatly attenuated so that the net

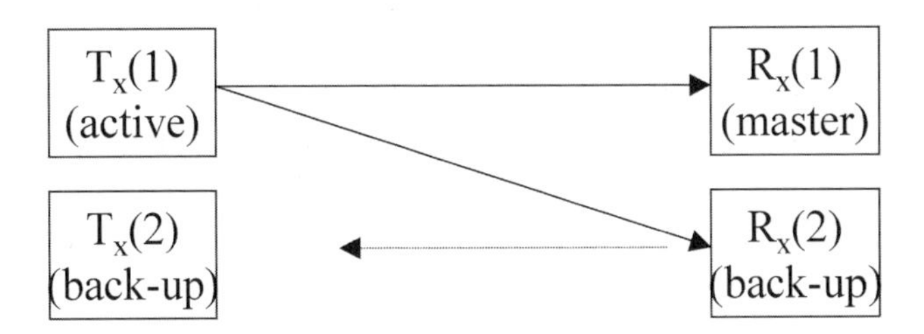

Figure 8.19 Monitor hot-standby (MHSB)

*Using a single antenna for a monitor hot-standby protected link has the advantage that less unsightly steelwork and fewer antennas have to be mounted on the roof. The disadvantage is that the power of the received signal has to be shared between the two receivers, so limiting the possible link range. For protected links of lengths close to the maximum range of the equivalent *unprotected link* it is advisable to use two antennas (like *space diversity*).

output signal is negligible. To 'turn the transmitter on' one simply removes the attenuation.)

8.13 Electromagnetic Interference and Electromagnetic Compatibility

The interference of electrical signals is one of the prime factors limiting the performance of radio systems in general. The subject has become one of major concern, and extensive technical standards have been developed to define the measurement of stray emissions from devices and the sensitivity of devices to interference from ambient electromagnetic radiation (e.g. signal 'spikes' on the power supply, radio signal interference or lightning). Within most countries nowadays, devices are legally required to conform to new *EMI* (*electromagnetic interference*) and *EMC* (*electromagnetic compatibility*) standards. These ensure the device's insensitivity to interference (EMC), and guarantee they themselves do not produce stray emissions (EMI). The commonly known standards are those enforced by the United States FCC (Federal Communications Commission) and ETSI (European Telecommunications Standards Institute).

ITU-R recommendation F.1094 covers the maximum allowable error performance and availability degradations arsing due to interference from emissions and radiation from other sources. These are considerations we must take into account when deciding exactly which radio frequency to use on a particular point-to-point link or within a given sector of a point-to-multipoint base station. We shall return to this subject in Chapter 9.

8.14 Radio Planning Tools

We have discussed in detail a number of factors which need to be considered when planning and surveying point-to-point radio station sites, base station coverage of point-to-multipoint systems and radio link planning prior to installation for both point-to-point and point-to-multipoint systems. No doubt this has made you aware how much simpler the task of planning is when supported with specially designed computer tools recording the exact positions (coordinates and height) of planned radio station sites, and using this information in conjunction with stored digital map information and detailed antenna radiation pattern information to make accurate computer predictions of likely Fresnel zone obstruction, propagation range, reflections and interference. A number of different software companies offer such specialised tools. The best ones also cope with radio frequency planning — the subject of the next chapter.

9

Radio Network Frequency Planning

So that radio systems work reliably and without disturbance, it is critical that radio systems operating in the same geographical or neighbouring areas do not use the same radio frequency. Interference caused by other such radio systems is the largest potential cause of performance problems and therefore deserves careful management. The radio spectrum, as we discussed in Chapter 2, is administered and policed on a worldwide and national level by three levels of regulatory management — the International Telecommunications Union Radiocommunication sector (ITU-R), world regional bodies including CEPT (Conference of European Posts and Telecommunications) and by national regulators and radio communications agencies (including Federal Communications Commission of the USA, Radiocommunications Agency of UK, Regulierungsbehörde für Telekommunikation und Post, RegTP, of Germany). In this chapter we consider in detail how radio interference is caused and discuss the basic methodologies used during the frequency allocation and planning process to try to circumvent it. We discuss radio frequency planning for both point-to-point and point-to-multipoint systems.

9.1 Regulatory Restrictions on the Operation of Radio Systems

Within the radio bands defined by ITU-R (International Telecommunications Union — Radio communications sector), CEPT (Conference of European Posts and Telecommunications), IEEE (Institute of Electrical and Electronics Engineers) and FCC (Federal Communications Commission) in their various recommendations, regulations and specifications, it is up to national regulatory agencies and radio agencies to administer the actual use of the radio spectrum. They do so as they see fit, and there are significant differences in the policies of administration between individual countries. For example, while one country (Germany) may allocate the 26 GHz band for the licensed use by public telecommunications operators use of both PTP and PMP, some countries have allocated the entire band to PTP, while France initially allocated the band for 'private networks'.

The decision by a particular national regulator to allocate a given band for a particular usage determines who may use which bands for what. Usually, the first effect of such a decision is the commencement of development by manufacturers of suitable radio equipment for the particular band and licensed applications. In addition, interested users (public and/or private network operators) may start to apply for allocation of a given quantity of bandwidth in a given geographical region for their exclusive use.

In most countries, the regulatory regime generally allows any individual or corporation to apply for a frequency to operate a PTP (point-to-point) microwave radio system, but PMP (point-to-multipoint) is reserved for the use only of licensed public telecommunications network operators.

Regulation — Obtaining Spectrum for PTP System Operation

Point-to-point frequencies are usually allocated on a point-to-point basis. In most countries, a defined frequency is allocated to each individual link, in response to an individual application. However, in addition, some regulators have chosen to allocate blocks of frequencies to particular large public operators (e.g. mobile telephone operators for the *backhaul* connections of their *GSM* (*Global System for Mobile*) communication base stations to the *Mobile Switching Centres* (*MSCs*). The usual intention of the regulator in making such a block allocation is to promote efficient use of the spectrum and simultaneously offload the radio planning work onto the operator. The operator has an interest to perform careful planning in order that he minimises interference between his links. On the other hand, since his spectrum resource is limited, he also has an incentive to squeeze as many links as he can into the limited radio bandwidth. If he does suffer interference, the responsibility for it and for resolving it lies clearly with him.

Unfortunately, allocating a block of spectrum to an operator for his exclusive use for point-to-point radio network usage does not always lead to efficient use of the spectrum. Experience shows that some operators initially apply for far more spectrum than they really need, thereby ensuring very low interference. Others seem to have applied for spectrum only to prevent others from getting it. A further group of newly-founded operators have applied for radio spectrum, and subsequently resold the spectrum allocation to other more needy operators at considerable profit. For this reason, some regulators have reverted to making only 'unofficial' block allocations; they retain the rights to the spectrum and its planning, but for an initial period have sought to offload radio planning effort onto the operators. Once a particular band has started to run out of capacity, the regulator usually assumes detailed planning control.

Given the complexity of the radio planning task and given the limited spectrum availability and huge, growing demand from telecommunications network operators, it is normal for the regulatory radiocommunications agencies at national level to maintain detailed records of all spectrum usage; the exact coordinates of link and-stations, the capacities of the links and the exact frequencies, bandwidths in use. The archived information often also includes the detailed antenna characteristics (antenna diagram), as well as the transmitted signal power. It is nowadays usually stored in a computer archive and radio planning system. Planning at this level of detail and careful coordination is essential to ensuring minimum interference between operators and simultaneous maximum use of resources.

Regulation — Obtaining Spectrum for PMP System Operation

In the case of spectrum allocated for use in conjunction with point-to-multipoint systems, it is only practicable to allocate spectrum on a geographical zone or nationwide basis. In

Europe, the German and UK regulators were among the first to announce their intentions to allocate spectrum for point-to-multipoint radio system usage by licensed public telecommunications network operators.

In the case of the UK, the initial allocation of spectrum for PMP was in the 3.5 GHz band, made is response to heavy lobbying from Ionica (a public network operator start-up company which had publicised its desire to increase competition for the main dominant network infrastructure owner in the UK (British Telecom) by being prepared to invest in a large scale wireless access network). Following this, in 1995, the UK announced that a further portion of the spectrum (in the 10.5 GHz band) was to be allocated for *broadband* point-to-multipoint. There were to be three nationwide allocations, each of 30 MHz duplex. The spectrum licences were issued during 1996 following a competitive tender to Ionica in conjunction with Scottish Telecom, Mercury Communications and NTL.

In contrast, the process of deciding how to ensure fair distribution of the available spectrum in Germany (to be made available in the 2.6 GHz, 3.5 GHz and 26 GHz bands) was a fairly lengthy one. The initial announcement of the intention to allocate such spectrum to licensed public telecommunications operators was made in December 1995. However, the process of formal allocation of spectrum took until November 1999 to complete. First, the process was held up by the deregulation of the German telecommunications market (infrastructure deregulation — allowing new operators to build their own transmission networks took place in July 1996; the telephone network monopoly was removed in January 1998). Even following deregulation, a considerable amount of time was spent by the regulator in working group debate about the way to divide up the available spectrum, the regulatory constraints which should be placed upon the use of the spectrum and the formal process which would be applied to decide which operators would receive how much spectrum.

The German regulator opened the spectrum for tender according to a two-stage process. In the first phase, interested parties were required to declare their interest in PMP spectrum, stating the preferred radio band, the geographical regions of interest and the amount of bandwidth required. Based on these applications, the German regulator started to allocate spectrum from the PMP bands (2.6 GHz, 3.5 GHz and 26 GHz) in the geographic areas which were not over-subscribed. In the over-subscribed areas, a second competitive tender phase was organised by the regulator. The decision as to which operator would receive the spectrum in each defined metropolitan region was then made according to a set of pre-defined decision criteria.

In several other European countries, the German model for allocating the available spectrum is being adopted. Thus, for example, a similar two-stage allocation process seems likely to be adopted in Austria and Switzerland. In contrast with Germany, though, the second stage competitive tender stage will be based upon a simple auction (over the Internet) of the spectrum to the highest bidder.

Typically, the regulatory agency or radio communications agency levies charges for the use of licensed band spectrum. The fees usually comprise a one-time charge and an annual fee based upon the number of links and/or base stations, the amount of spectrum used and the geographical coverage area. These fees vary widely from country to country. Appendix 11 provides a summary of PTP and PMP spectrum charges in a selection of countries.

Regulatory Constraints on Radio Spectrum Usage

When spectrum is allocated for a given PTP radio link, the operator is assigned a particular radio channel and the maximum allowed transmitter power. In addition, it is normal to specify at which end of the link which of the duplex channels (i.e. upper band or lower band) is to be used as the transmission channel. Since the allocation of spectrum for such PTP usage is made in response to a detailed application which includes detailed link end coordinates (including height) and the specific antenna characteristics, there is not much scope for change of the configuration (e.g. upgrading the bitrate) or the type of equipment without re-application.

In the case of PMP spectrum allocation, the European regulators are tending to follow the model initially set by the german regulator. The spectrum is allocated in Germany according to a zonal scheme. The operator is authorised to use a given number of pre-defined radio channels (e.g. 2×28 MHz) within the defined geographic zone, and has the freedom to conduct the radio planning of individual base stations and sectors within this area: allocating the individual radio channels to individual sectors as he sees fit. However, to ensure that there is not undue interference with other operators in neighbouring areas, the identical radio channels are not allocated in the immediate neighbouring zone. In addition, the operator is obliged by his spectrum licence to maintain a maximum signal strength at a distance 15 km from the edge of the authorised geographical coverage zone. The signal strength at this point, or more correctly the *power spectral density* (or *power flux density*, *PFD*) is limited to a given maximum value at this point.

Maximum *Power Flux Density* (PFD_{max}) at given range beyond the licensed PMP coverage area:

$$PFD_{max} = X \text{ dBW/(MHz.m}^2\text{) at distance } d \text{ km}$$

Radio band (GHz)	Distance beyond authorised coverage zone (d in km)	Maximum Power Flux Density X at this point (dBW/MHz.m^2)
3.5	15	−122
26	15	−110

The Power Flux Density (PFD) in dB_m (during good weather — the 'worst' case, since the maximum power is achieved during good weather conditions) can be calculated from the following formula:

$$PFD \text{ (in dB}_m) = P_t + G_t - 20 \log_{10}(D) - 10 \log_{10}(B) - 71$$

$$PFD \text{ (in dBW)} = PFD \text{ (in dB}_m) - 30$$

where P_t = transmitter output power in dB_m, G_t = transmitting antenna gain in dB_i, D = distance from base station in km at which PFD is measured, and B = radio channel bandwidth in MHz.

To illustrate the calculation of the power flux density as above (for the purposes of checking the radio network plan), let us assume a sector transmitter power of $P_t = 18\,dB_m$, a sector antenna gain of $G_t = 18\,dB_i$ (circa 45° sector) and a radio channel

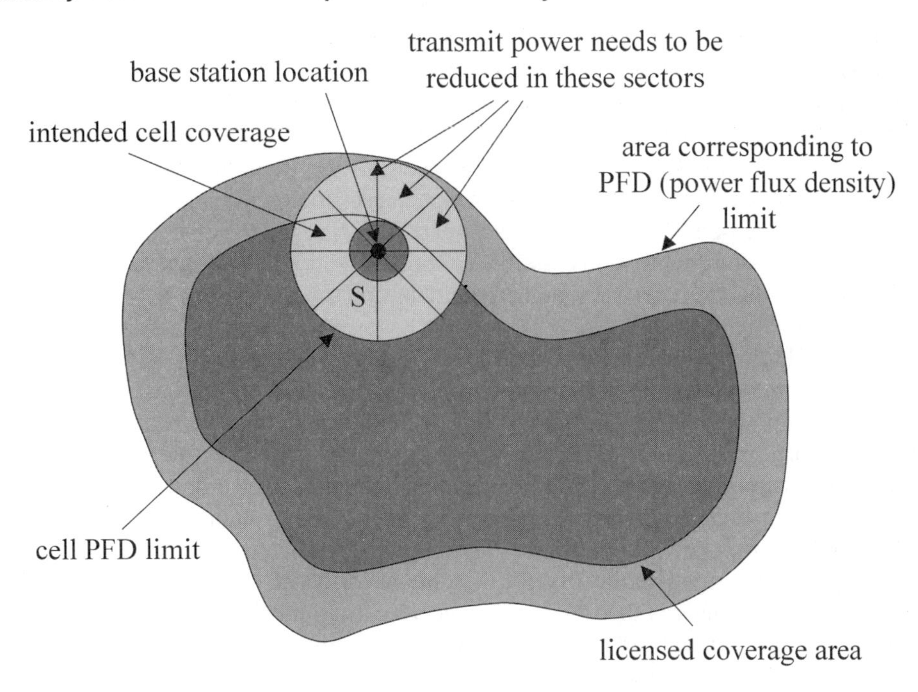

Figure 9.1 Power flux density limitation at 15 km beyond licensed PMP coverage area. (Note: subscriber terminals in sector S [pointing towards the licence area boundary] may also need to limit their transmitter power output)

bandwidth of 28 MHz. Let us also assume that the normal range of the base station is intended to be 5 km, and that we are interested in the maximum PFD at a point 15 km beyond the intended coverage range (see Figure 9.1). At this point $D = 5 + 15 = 20$ km. From the above, the maximum PFD in dBW is $18 + 18 - 20 \times \log(20) - 10 \times \log(28) - 71 - 30 = -105$ dBW. This is a value which exceeds the maximum permitted power flux density (-110 dBW at 26 GHz and -122 dBW at 3.5 GHz). So what is the consequence, you ask? The consequence is that you may not use the maximum available transmitter power of 18 dB_m during good weather conditions. The maximum transmitter power (given the above system parameters during good weather) at 26 GHz will be 13 dB_m and at 3.5 GHz only 1 dB_m.

There are two methods in practice by which radio network operators can conform with such power flux density limitations imposed by regulators. The first alternative is to manually configure the output power of the transmitter to the maximum value permitted. Unfortunately, however, such permanent limitation of the output power will also greatly reduce the signal power available during bad weather, and thus greatly restrict the range of the base station and the availability of individual customer links (connections), as we discussed in detail in Chapter 7). For this reason, it is preferable to use an automatic means to adjust the transmitter output power. Such a means is provided in radio systems which offer *Automatic Transmit Power Control* (*ATPC*).

As the name suggests, a radio system which incorporates ATPC will automatically adjust the output power of the transmitting radio. The control is achieved by conducting a permanent measurement of the received signal level at a given point remote from the transmitter (for example, at the most remote subscriber station within a given base station sector). During good weather, the received signal will be relatively strong, so that not so much transmitter signal power need be output for reliable transmission. However, as soon as the system becomes affected by bad weather (for example, caused by the fading effects of heavy rain), then the received signal will weaken. At this point in time, a message is sent from the remote station to the base station which causes the transmitter output power to be progressively increased, in order that an adequate *fade margin* (reserve of signal power) is maintained, so that reliable transmission continues without disturbance. Once the bad weather has passed, the received signal strengthens and the transmitter power may be once again automatically reduced.

In fact, ATPC may be used in either or both of the downstream (adjusting the base station transmitter power as above) and upstream directions (i.e. adjusting the remote stations output power). ATPC in the upstream direction may also be required in order to fulfill regulatory spectrum usage constraints. A remote subscriber terminal near to the base station and transmitting towards the licence area boundary (see Figure 9.1) may indeed generate an even higher power spectral density at the 15 km boundary (due to a higher antenna gain). Fortunately, ATPC in the *upstream* direction is usually designed and built-into the system to improve the overall system performance. ATPC, when used in the upstream direction, ensures that the signals received by the base station from the various remote subscriber terminals are all roughly of the same signal strength. This allows each of the signals a fair and equal chance of good reception and low interference. We return to ATPC later in the chapter.

Having considered the minimum requirements placed on radio network planners by radio regulation agencies, we now go on to discuss in detail the considerations which go into good radio network planning. These considerations affect regulators and radio network operators alike, when the prime interest is to ensure maximum efficiency of spectrum usage and simultaneous minimum interference betweeen neighbouring radio systems.

9.2 Interference between Radio Systems

Perhaps the most important duty undertaken by a radio network planner is the coordination of radio frequency usage in order that adjacent radio systems do not interfere with one another. Interference results at the receiver of a radio link, when that receiver is unable to distinguish between the wanted (i.e. the carrier) and the unwanted (interfering) signals. Such destructive and irreversible interference occurs when two incoming signals are nearly similar in frequency, polarisation and signal strength. There are three main types of interference which the radio network engineer normally considers during network planning:

- co-channel interference;
- cross-polar interference; and
- adjacent channel interference.

Co-channel interference is caused by an unwanted (disturbing) signal using the same radio channel (frequency) and polarisation as the wanted signal. Provided the wanted signal is much stronger than the unwanted signal, then the receiver will still be able to distinguish between the two and receive the wanted signal correctly. The relative ability of the receiver to distinguish between co-channel interferers is usually defined in terms of the minimum *carrier-to-interferer* signal power ratio, which must be maintained so that good reception is possible. This ratio is known as usually denoted with the terminology C/I (carrier/interferer ratio) and defined either as an absolute value (e.g. 20) or in dB (the dB value equals 10 $\log_{10}$ of the absolute value; so for absolute value $C/I = 20$, $C/I = 13$ dB). The capability of the receiver to distinguish between co-channel interferers is sometimes referred to as the co-channel discrimination of the receiver. This is the minimum value of C/I at which good reception is possible with the receiver.

Cross-polar interference results from an unwanted signal of the same radio frequency but opposite polarisation (e.g. horizontally polarised rather than vertically polarised, etc.) disturbing the wanted signal. Again, the interference is characterised in terms of the relative strength of the two signals (carrier-to-interferer, C/I and quoted, as above, as an absolute ratio or in dB). The capability of a receiver to receive the wanted signal correctly despite the interference is defined to be the cross-polar discrimination of the receiver. This is the minimum value of adjacent channel C/I at which good reception is possible with the receiver.

Adjacent channel interference results from an unwanted signal using the next higher or lower radio channel frequency disturbing the wanted signal. As in the two cases above, the interference is characterised in terms of the relative strength of the two signals (carrier-to-interferer, C/I and quoted, as above, as an absolute ration or in dB). The capability of a receiver to receive the wanted signal correctly despite the interference is defined to be the adjacent channel discrimination of the receiver. This is the minimum value of cross-polar C/I at which good reception is possible with the receiver.

As you can predict from our earlier discussion in Chapter 6, it is generally easier to discriminate between radio signals of different polarisations rather than signals of the same polarisation and the same (*co-channel*) or *adjacent channel* frequencies. So it is not a surprise to learn that the cross-polar discrimination is greater than the adjacent channel discrimination, which in turn is greater than the co-channel discrimination (for the smaller the C/I value required, the smaller is the required difference in power between wanted and unwanted signals for good reception). Typical values are −25 dB for the 'receiver cross-polar discrimination' (interferer signal strength before the antenna may be stronger*), −10 dB for *adjacent channel discrimination* and 13–15 dB for *co-channel discrimination* (with 4QAM modulation; a higher separation of 20–22 dB may be necessary for more sensitive 16QAM modulation).

9.3 Coordination Contour and CIR Contour Diagrams

For reliable operation of a radio system it is necessary to ensure that we retain sufficient carrier-to-interference signal strength ratios, as we have just discussed. For the practical purposes of planning networks it is valuable to plot *CIR* (*Carrier-to-Interferor Ratio*) *contour diagrams*. Figure 9.2 is an example of such a CIR contour diagram. The diagram

*As we discussed in Chapter 6, the cross-polar discrimination of the antenna is typically much higher ($C/I = -35\,dB_c$), but the manufacturing and installation tolerances limit the values which should be used for radio frequency re-use planning to around −20 or $-25\,dB_c$.

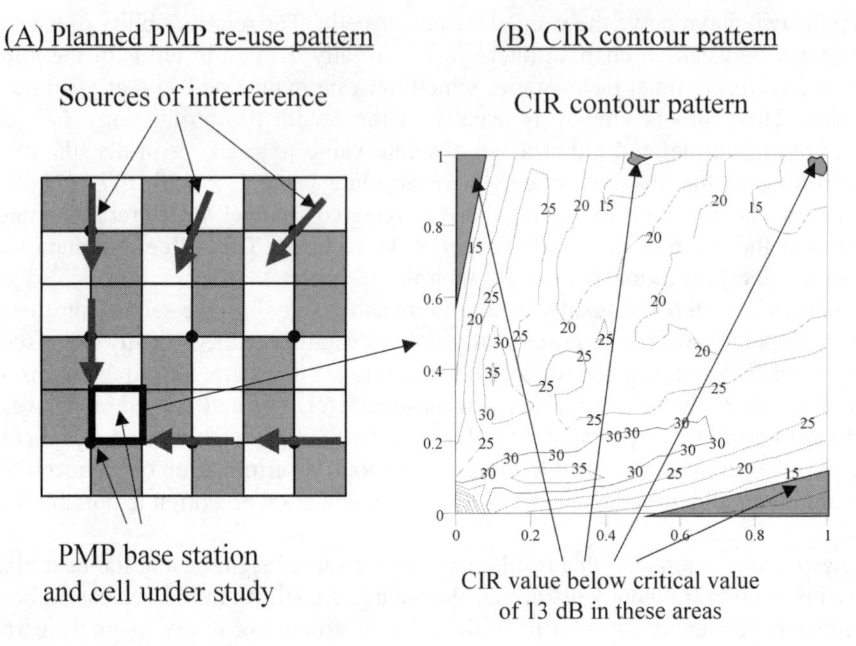

Figure 9.2 Carrier/Interference Ratio (CIR) contour diagram of a PMP system. (Note: the diagram assumes subscriber terminal antenna of approximate 7° beamwidth. (A larger beamwidth would result in the receipt of more interferer signals and so reduce the CIR)

shows the calculated value of the CIR (either in dB or in absolute ratio value) for each theoretical point at which a given radio channel frequency might theoretically be used.

In the example of Figure 9.2b we have chosen to show the CIR contour pattern of a given sector of a PMP (point-to-multipoint) *frequency re-use pattern*. Figure 9.2a shows nine adjacent base stations (represented by dots) of such a PMP frequency re-use pattern, and the pattern of four 90° sectors transmitted from each base station. The shading of the sectors (represented by squares) corresponds to the actual radio frequency used in the sector (there are two separate channels used in total — a two-frequency re-use pattern as we shall discuss later in the chapter). The CIR contour diagram is plotted for the upper right hand sector (i.e. north east) of the lower left hand base station 'cell'. The frequency planner is considering using the radio channel frequency in this sector corresponding to the lighter-shaded sector squares and has plotted the CIR contour diagram assuming the use of this frequency. The arrows show possible sources of strong interfering signals, caused by the overshoot (as we shall also discuss later) of signals from remote subscriber terminals directly into the base station under study.

Assuming that the contour values of Figure 9.2b are quoted in dB, and assuming a required minimum CIR of 13 dB is needed for reliable operation, we can see from Figure 9.2b how most of the sector area will be 'free of interference' (by this we mean that the interference level is lower than the 'threshold' value at which the radio receiver is no longer able to 'ignore' it). There are, however, four areas along the boundary areas of the cell (labelled with arrows in Figure 9.2b) where the minimum CIR value is not achieved. In these areas, there will be unacceptable interference and radio *coverage* will not be possible. We shall discuss these problems for both PTP and PMP systems later in the chapter and consider possible work-around solutions.

You may notice in Figure 9.2a how the 'light' and 'dark' shading of the individual sectors of the base stations is not constant. In the top row of base stations the upper right hand sector (i.e. north east) is always 'light'. The same is true for the lowest row of base stations. However, for the middle row, the upper right hand sector is always 'dark'. This 'flipping' of the cell pattern between adjacent rows is done on purpose. It has the benefit of reducing the interference. This should be clear from the fact that two of the base stations (in the middle row of Figure 9.2a) do not cause interference to our sector under study (as they would if their 'top right' sectors were light-shaded). There are no 'interference arrows' arising from these two cells in Figure 9.2a. As a result, the contour pattern of Figure 9.2b is not symmetric about the diagonal lines of the sector under study.

It is is an important observation to note that the geometry of the cells and the radio channel *frequency re-use* pattern is critical to the overall performance of a radio network. Perfect symmetry sometimes works to our disadvantage, as we shall see several times in this chapter.

The calculation of all the possible sources of interference is necessary before an accurate CIR contour diagram (sometimes also called *interference diagram*) can be plotted. We shall consider in the remainder of this chapter a number of the main geometrical effects leading to interference or to the resolution of interference.

Of course, the main sources of interference which need to worry the frequency planner are unwanted transmissions received directly over line-of-sight paths from adjacent base stations (as in Figure 9.2a). But in addition, depending upon the radio frequency band of system operation, it may also be necessary to consider interference or carrier signal attenuation resulting from reflected, diffracted and/or refracted signals. Such signals may emanate from other terrestrial radio systems, or even signals originating from satellite radio systems. Urban environments are prone to *reflections*. Long links, particularly operating at frequencies under 10 GHz may be subject to *refractive* signal path interference. Links in mountainous areas are prone to *reflection* and *ducting* (a combination of reflection and refraction) and large areas of water are prone to reflection and refraction due to water vapour. Detailed prediction methodologies are presented in ITU-R recommendation P.452.

9.4 Frequency Planning and Re-use in PTP Networks

The main consideration in the planning of the *re-use* of frequencies in a network of PTP (point-to-point) radio links is the problem of overshoot (Figure 9.3). The basic problem is that it is impossible to operate two links directly one-behind the other, in the same orientation, using the same frequency without causing interference. Figure 9.3 illustrates the problem.

By changing the orientation of the links in Figure 9.3 and operating them instead in a 'back-to-back' configuration (with both 'A-ends' at the middle point, as shown in Figure 9.4) the disturbance of the second link can be eliminated.

Figure 9.4 illustrates the possibility of 'back-to-back' operation of two radio links using the same duplex radio channel. By making sure that the two lower band channels are used for transmission from the mid-point outwards (i.e. at each of the respective 'A-ends') we avoid the possibility of interference.

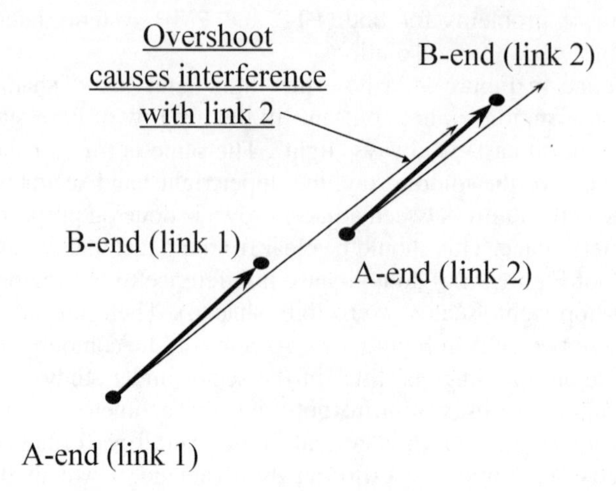

Figure 9.3 The possibility of overshoot must be considered in the frequency re-use planning of point-to-point networks

The radios at the 'B-ends' expect to receive at the lower band frequency (full arrows of Figure 9.4) and transmit at the upper band frequency (dotted arrows). Meanwhile, the 'A-ends' transmit in the lower band (full arrows) and receive in the upper band (dotted arrows).

The overshoot of the lower band signals beyond the B-endpoints does not cause interference as there are no further receivers behind the B-ends. Meanwhile, the overshoot of the upper band signals (from the 'B-ends' past the 'A-end' midpoint and towards the other 'B-end') is also not a problem. There is no possibility of interference as the receivers into which the signals overshoot are also 'B-ends'. They are thus only capable of reception of a lower band signal anyway. We have not eliminated the overshoot. We have simply 'disarmed' its capability to interfere.

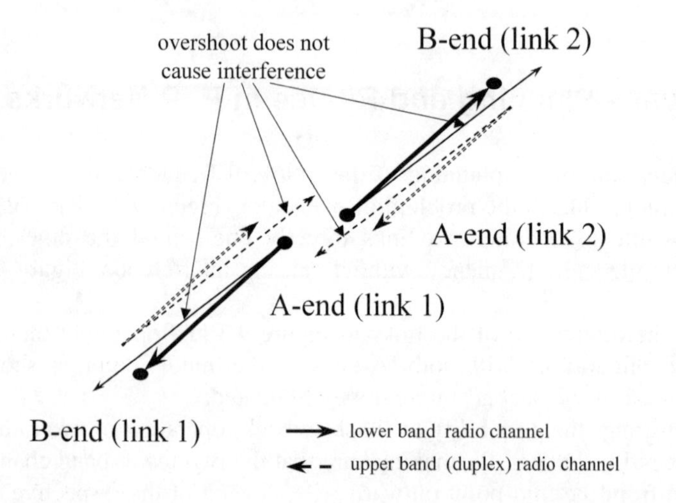

Figure 9.4 Operation of 'back-to-back' links is possible using the same duplex radio channel

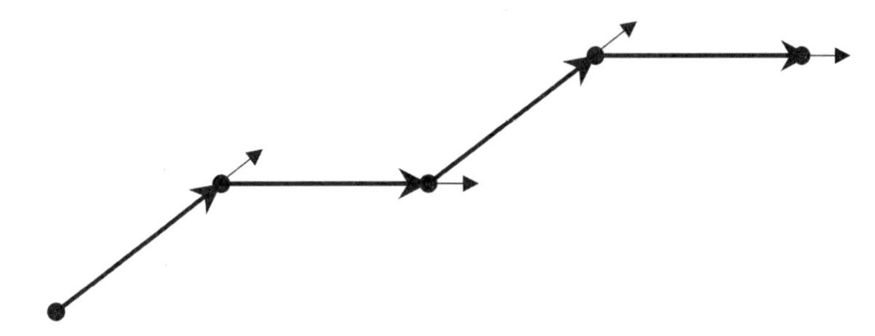

Figure 9.5 Zig-zag layout of multiple hop links to enable re-use of a single radio channel

As an alternative (or complement) to the technique of using 'back-to-back' operation of the frequencies to reduce the power of interfering signals, we can also choose carefully the orientation of individual links to one another. The zig-zag cascaded links of Figure 9.5 cause less interference to one another than a simple 'straight-line-cascade' would cause.

The zig-zag configuration of Figure 9.5 is particularly effective in reducing interference when high gain (i.e. highly directional) transmitting and receiving antennas are in use. Even so, the situation is not quite as straight-forward as the simple diagram of Figure 9.5 may suggest. We may also have to consider the interference caused by *side-lobe* transmissions or *side-lobe* receptions. Particularly during good weather, a strong sidelobe signal might easily have the strength to bridge the gap straight from point A to point C (as in Figure 9.6). In real networks, it is therefore normal to calculate all the overshoot, side-lobe and other 'stray' transmissions in the calculation of overall interference. Given the complexity of this task, it is normal to use a specialised computer software tool to perform it, and to suggest the radio channel which should be used for a proposed new link addition to an existing radio network.

9.5 Coordination Distance and Area

ITU-R defines the *coordination distance* to be that distance at which other 'adjacent' radio systems operating on the same channel can be considered to be negligibly affected by interference. Similarly, the *coordination area* is an area around a point-to-point base station or satellite earth station outside of which interference can be considered negligible.

It is also relatively common practice to plot *coordination contours* or *interference contours* which identify on a map the strength of likely interference, as we saw in Figure 9.2.

In particular, ITU-R recommendations F.1095, 619 and P.620, set out methodologies which are recommended to be adopted by national radio planning and coordination administrations for determining coordination areas and distances when planning terrestrial radio applications and/or satellite applications to share the same radio band. Such coordination is important in order that different network operators do not disturb one another either within a given country, or across the borders of neighbouring countries.

It is important to calculate the coordination distances between existing and planned radio systems and stations as accurately as possible. The most accurate anaylsis is by calculation of the interference caused by the two radio systems to one another, considering the actual transmitter power and the exact details of the antenna pattern.

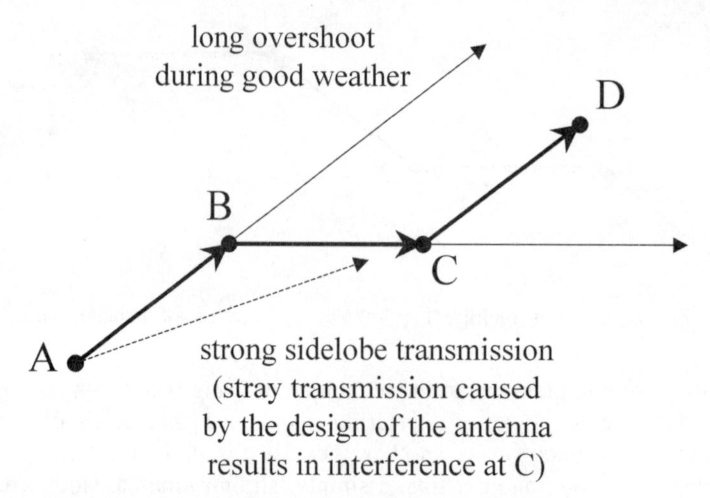

Figure 9.6 Interference can be caused by sidelobe transmission, particularly during good weather (angle of main sidelobe [angle ABC] is exaggerated)

The coordination distance is the distance at which the interfering signal power (I) is equal to (or less than) the critical value required to meet the required *Carrier-to-Interference* (*CIR*) ratio value. The actual value of I at a given distance from an *interfering station* is given by the following equation:

Interfering signal power at coordination distance d (ITU-R recommendation F.1095)

$$I = P_T + [G_R - D_R(\theta)] - L(d) + [G_T - D_T(\theta')]$$

where I = interfering signal power in dBm at distance d from interfering station (to resolve for the coordination distance d above, we must substitute here the value of maximum tolerable interfering power considering the desired carrier signal power level, C, and the minimum value of CIR);

P_T = maximum transmitting power (dBm) of the interfering station within the *reference bandwidth* (i.e. the bandwidth used by both stations);

G_T = gain in dBi of the transmitting (interfering) antenna;

G_R = gain in dBi of the receiving antenna (of the *interfered-with station*);

$D_T(\theta')$ = antenna discrimination in dB of the transmitting antenna (i.e. the antenna gain of the transmitting antenna where the interfered-with station is an angle θ' from the main pole);

$D_R(\theta)$ = antenna discrimination in dB of the receiving antenna (i.e. the antenna gain of the receiving antenna where the interfering station is an angle θ from the main pole);

$L(d)$ = total path loss due to the distance d between the transmitting and receiving station (during good weather).

The calculation of interference and coordination distance is most easily and accurately carried out using computer software tools. The coordination distances and area for a given existing radio station has a form similar to Figure 9.7a.

Since it is (unfortunately) not always possible to calculate the coordination distance accurately, and since sometimes not all the information required for the calculation is

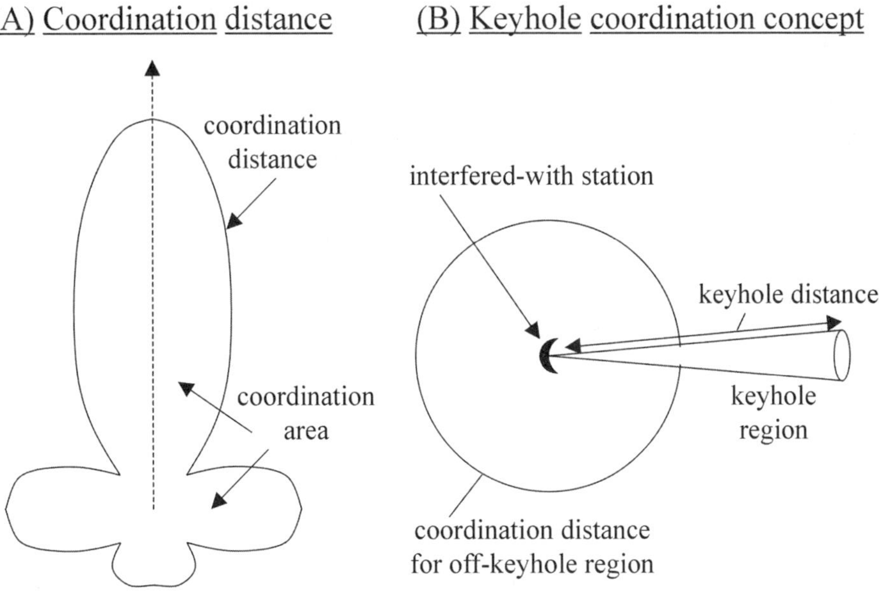

Figure 9.7 Coordination distance and coordination area

available (e.g. one national administrator may not have detailed antenna information from an operator in a neighbouring country), ITU-R recommendation F.1095 also defines the concept of a *keyhole coordination region* (Figure 9.7b). This concept can be applied to microwave and other radio systems with highly directional antennas. The keyhole region corresponds to the main pole of the antenna, and the keyhole distance corresponds to the main antenna gain value. Meanwhile, in other directions (i.e. in the off-keyhole region) a shorter coordination distance (than the keyhole distance) may be used. The off-keyhole distance, however, needs to include a 'margin of safety', and is thus greater than the accurately calculated coordination distance would be.

9.6 Radio Planning of PTP Links using Repeaters

Where point-to-point radio paths are composed of multiple-hop links employing radio repeaters, each of the individual hops needs to be considered to be an independent link using the relevant radio channel when conducting the radio network planning. Even though each of the hops of the multiple hop link are carrying the same signal, there is still the possibility of interference, as we discussed earlier in the chapter (in Figures 9.3, 9.4 and 9.5).

Planning Tools

To ease the complex task of assigning radio channels to individual radio links in point-to-point radio networks there are a number of commercial computerised planning tools available on the market. Typically, such planning tools comprise computer software, which must usually be supplemented with the following database information:

- *digital map* information (giving geographical information, heights, coordinates, etc. in a digital format — such databases are usually available on a nationwide or regional basis and are usually very expensive: the most sophisiticated digital maps also include land usage, building height information and even aerial photographs in digital format).
- *antenna pattern* information (usually supplied by the radio equipment manufacturer).

Once the tool is installed and configured, the radio planner needs to input the end-point coordinates (latitude, longitude and height) of established links, together with the radio channel and bandwidth used. The end at which the upper band or lower band radio duplex channel is used for transmission must also be accurately recorded (as we discussed earlier in Figure 9.4).

When planning a new point-to-point link, the radio planner inputs the proposed end-point coordinates. The coordinates are usually obtained by means of a global positioning system receiver used during the LOS-check and site survey, as we discussed in Chapter 8. Normal GPS provides accuracy of coordinates to within 40 metres (where more accurate coordinates are required *differential GPS* has to be used). The planner also inputs the required radio bandwidth of the link. The computer planning tool can then be used to calculate and output the following information:

- the appropriate radio channel for use on the link, and which duplex channel (i.e. upper band or lower band) should be used at which end of the link;
- the geographical cross-section of the link (i.e. whether line-of-sight is available between the two proposed end-points);
- the profile of the Fresnel zone and whether there are any geographical obstructions within it. The system might also be capable of indicating any likely building obstructions (if a building height database is available);
- the likely interference patterns and contours (often in the form of a map showing relative signal strengths).

9.7 Frequency Re-use in Point-to-Multipoint Radio Networks

In PMP networks, the *re-use* of radio frequencies and channels has to take account in particular of the likelihood of interference caused by adjacent sectors at the same base station or by neighbouring base stations. The ability to re-use frequencies in PMP networks is more restricted when planning PTP (point-to-point) networks due to the large *beamwidth* of the sector and omnidirectional antennas used at PMP base stations.

Typically, the radio frequency planning for PMP networks takes place at two administrative levels. The regulator first allocates blocks of frequencies (spectrum) within a given region to a particular operator (so-called *coordination*). As we discussed earlier in the chapter, the regulator usually makes constraints on the power spectral density of the residual signal at a given *coordination distance* beyond the authorised coverage area (i.e. outside the *coordination area*). Once the operator has received his block of spectrum, it is usually up to him to plan his own frequency re-use within his licensed geographical

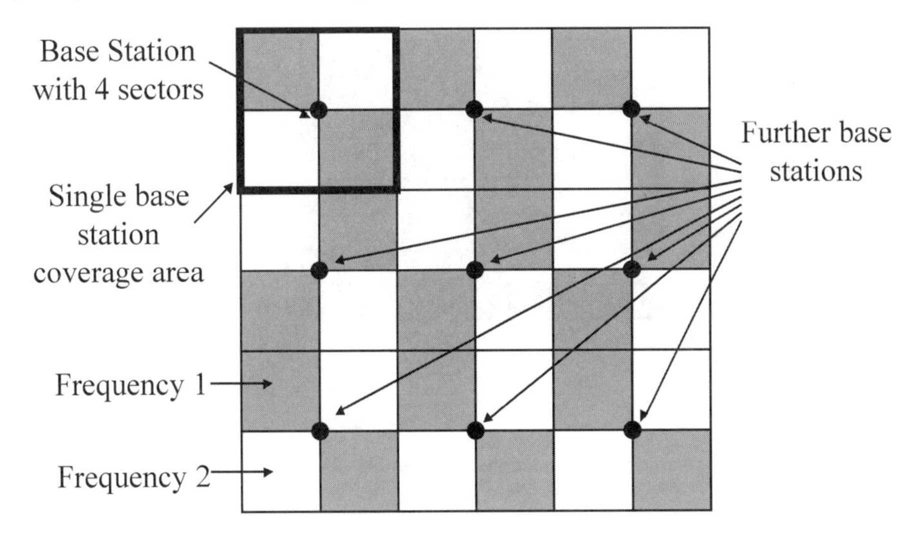

Figure 9.8 Two-frequency re-use pattern

coverage area. In the remainder of the chapter we discuss the main considerations of PMP frequency re-use planning.

Basic Frequency Re-use

To minimise the possibility of interference it is normal to apply a strict *re-use pattern* of either 2, 3 or 4 frequencies or sector frequency combinations. Figure 9.8 illustrates a possible 2-frequency or 2-sector-frequency-combination re-use pattern.

Each base station in Figure 9.8 is transmitting to four separate 90° sector areas: north east, south east, south west and north west. The individual sector areas are illustrated by a pattern of four squares per base station. Actually the coverage of the sector antenna is not a convenient square, but a more complicated shape more like Figure 8.16 of Chapter 8. For the purpose of initial consideration and illustration of the principles of frequency re-use planning of PMP networks, we shall assume somewhat simplified coverage patterns during the whole of this chapter. As with point-to-point radio planning, the actual reality is best predicted by means of a computer planning tool.

(As an aside, you may wonder why the single cell (4-sector) pattern of Figure 9.8 is not repeated exactly. In the first and third rows the upper right hand (north east) sectors are 'light', while in the second row they are 'dark'. The reason for this, as we discussed briefly in conjunction with Figure 9.2, and will return to again later in this section, is to reduce interference.)

Where each sector uses a single radio channel, then the pattern of Figure 9.8 will indeed be a 2-frequency pattern. However, it is also possible that more than one radio channel could be used in each of the sectors. In this case it is useful instead to think of a 2-frequency-combination pattern. Thus, the darker sectors (labelled 'frequency 1' in Figure 9.8) could correspond to the use of the channels 1, 3, 5 and 7 of Figure 9.9 (we call this 'sector pattern A' in Figure 9.9) while the lighter sectors of Figure 9.8 (labelled 'frequency 2') could correspond to the channels 2, 4, 6 and 8 of Figure 9.15 (sector pattern B of Figure 9.9).

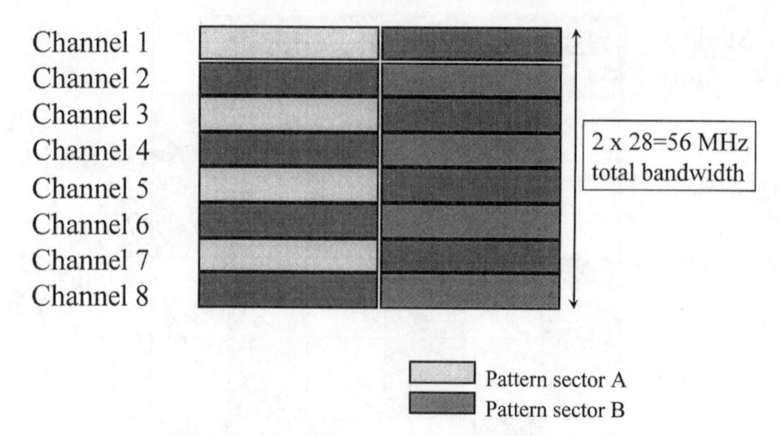

Figure 9.9 Fixed frequency combination patterns — a possible allocation of 4×7 MHz per sector in a 2-sector-combination frequency re-use pattern

The advantage of allocating the frequencies in the fixed manner illustrated in Figure 9.9 is that it ensures the appropriate separation of channels from one another to minimise interference. Thus, the scheme of Figure 9.9 avoids the operation of adjacent channels in the same sector.

The problem with 2-sector frequency re-use is that interference between adjacent channels is inevitable along the fringe areas between adjacent sectors. This is unavoidable, due to the overlap both of the spectrum masks of the adjacent sectors areas and due to the 'imperfect' nature of the antenna pattern (Figure 9.10). (We discussed the spectrum mask in Chapter 2 and the overlap of the sector coverage in Chapter 8.)

A partial solution to the problem of interference is achieved by the 'flipping' of the frequency re-use pattern as we saw in Figure 9.2. This reduces the number of neighbouring interfering base stations by careful selection of the geometry.

Another partial solution to the problem of interference along sector boundaries is to redirect remote subscriber antennas to other adjacent base station sites, as illustrated in Figure 9.11. This solution can always be considered when a particular remote station is suffering interference when directed at the normal 'home' base station.

Another way to reduce interference is by using more radio spectrum. Figure 9.12 illustrates possible frequency re-use combination patterns for three-sector re-use and four-sector re-use patterns.

Figure 9.12 shows three-sector and four-sector patterns comprising a nominal 28 MHz frequency allocation per sector made up of four 7 MHz channels per sector. As in Figure 9.9, the aim is to avoid adjacent channels in the same sector. But in addition, we can now also aspire to avoid adjacent radio channels not only in the same sector but also in the adjacent sectors (at least to some extent). Thus, for example, in the four-sector pattern of Figure 9.12, we can try to plan for patterns A and C to adjoin one another, and patterns B and D. In this way we eliminate the possibility of interference along the corresponding sector boundaries, as we show in the vertical boundaries between the sectors A:C and B:D of Figure 9.13a.

Unfortunately, the four-sector repeat pattern of Figure 9.13a does not remove the adjacent channel interference along the horizontal sector boundaries A:D and B:C, but at

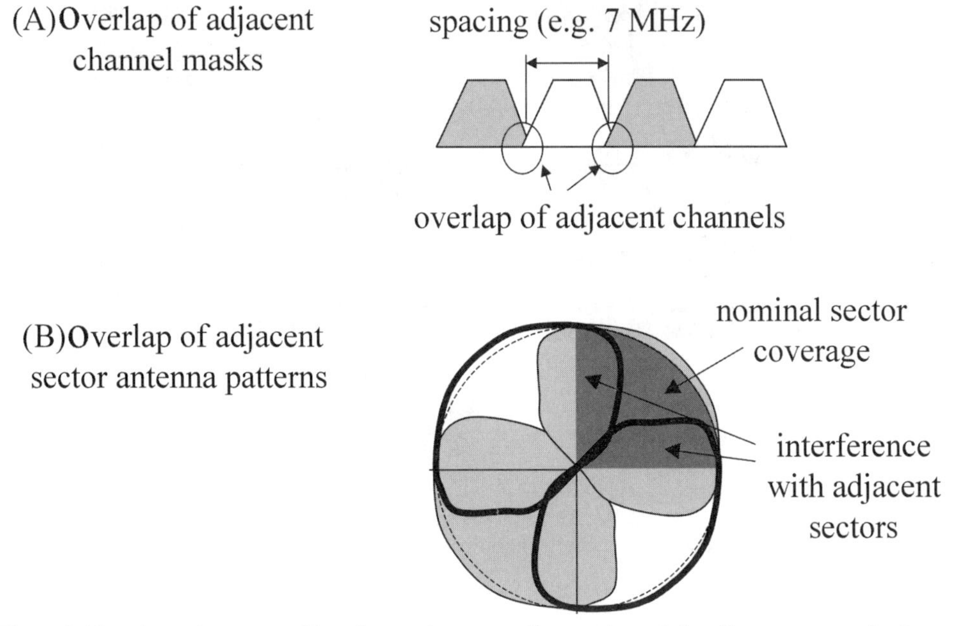

Figure 9.10 The main causes of interference between adjacent channels in adjacent sectors of a 2-sector frequency re-use pattern

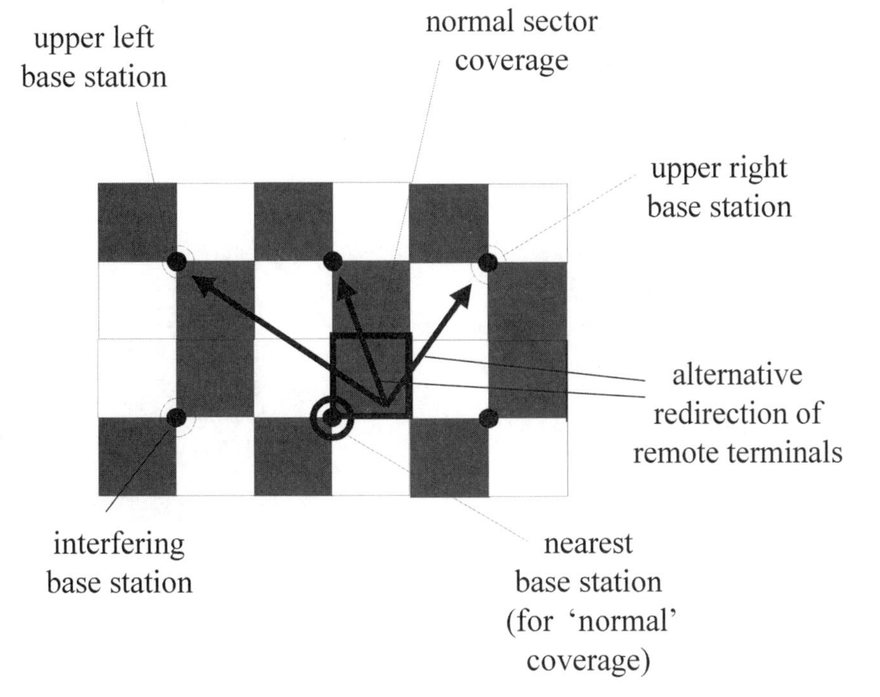

Figure 9.11 Re-directing subscriber terminal antennas to avoid the interference along the boundaries of the adjacent sectors of the nearest base station

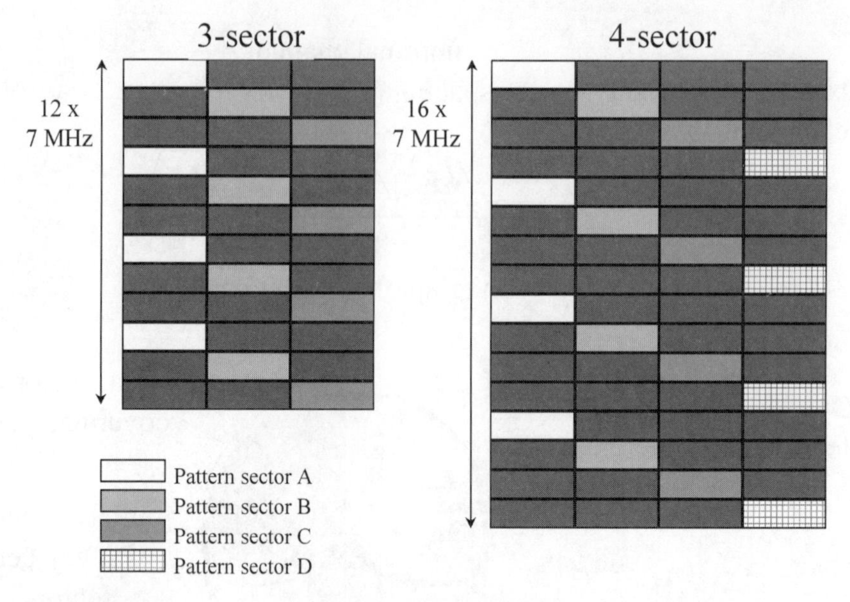

Figure 9.12 Fixed frequency re-use combination patterns for three-sector and four-sector patterns

least the pattern is an improvement upon the pattern of Figure 9.13b, which might perhaps have arisen from over-simplistic frequency planning.

Actually, the pattern of Figure 9.13a is also not the optimum. Figure 9.14 shows an even better pattern of the sectors and the allocation of the individual radio channels to the various 'sector patterns'. In Figure 9.14, we have eliminated the adjacent channel interference along adjacent sector boundaries entirely.

The main lesson to be learned from our discussion leading up to Figure 9.14 is that the best frequency re-use pattern is not always the first simple plan you think up. In addition,

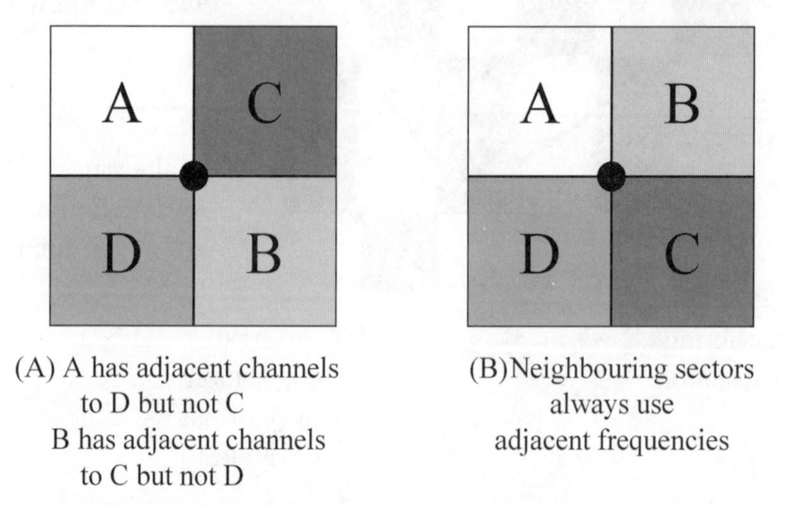

Figure 9.13 Planning of adjacent sectors

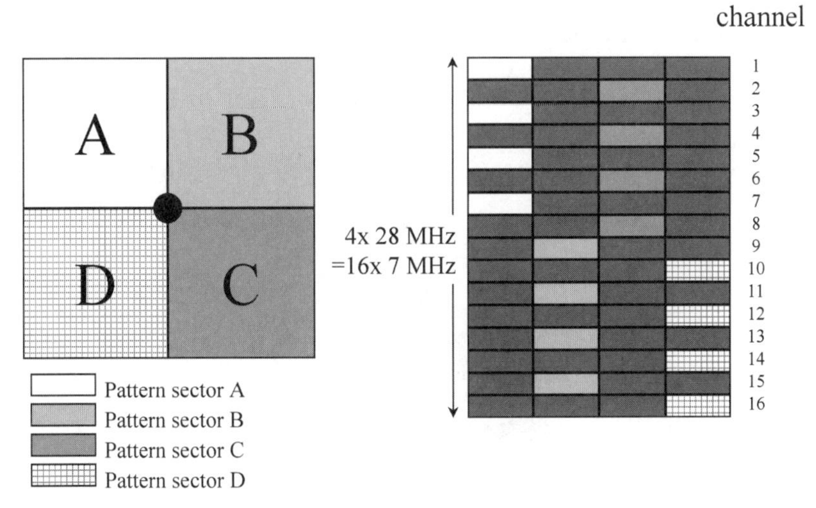

Figure 9.14 Four-sector pattern avoiding adjacent channel interference between adjacent sectors

the likelihood in practice is that only very few base stations in a real network actually would require the full four 7 MHz channels of capacity we show in Figure 9.14. It may be that five 7 MHz channels are required in sector A and only three channels in sector C. Why not use the spare channel from sector C in sector A, you ask? Why not indeed? Because maybe such *frequency re-use* does not fit in with the wider network frequency plan. The complexity of the problem of determining the best frequency re-use pattern quickly exceeds the capability of the human brain. As with point-to-point radio network planning, point-to-multipoint *frequency re-use planning* is best performed using a computer tool designed specifically for the purpose. Only with computer assistance can you expect to approach a near-optimum usage of the radio spectrum.

9.8 Further Considerations for PMP Frequency Re-use Planners

In this section we consider in general terms a number of other difficulties which may face radio network planners of point-to-multipoint networks, and also some of the 'tricks' they might use to overcome them.

Obstacles Cause Interference by Reflection and Create Shadows

Large obstacles causing strong reflections near the base station can be a major source of interence and poor coverage or *shadow*. In the extreme case 2-frequency re-use pattern illustrated in Figure 9.15, the top right-hand sector has been plunged into shadow while the bottom left-hand sector is suffering in interference. The best solution to such a problem is to choose another base station site, at least for the sector antennas serving the top right sector. Other solutions, as we shall discuss next, are to use signal polarization to help alleviate the interference and base station repeaters to help eliminate the poor coverage areas in the shadows.

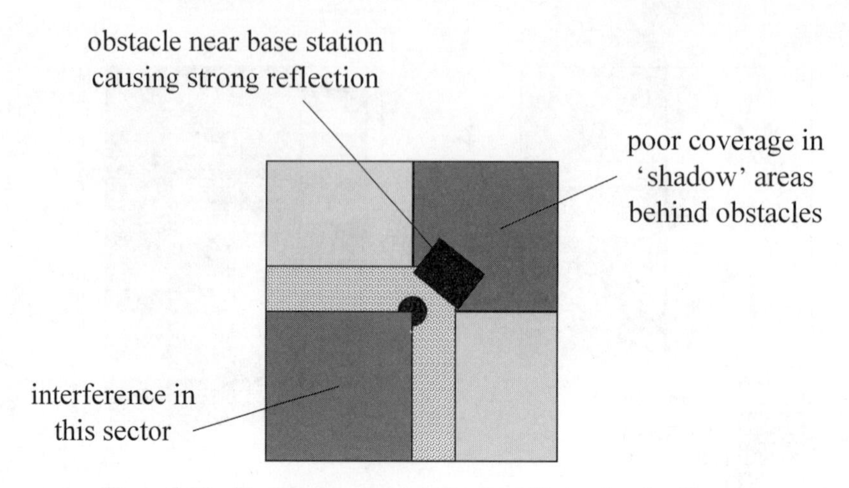

Figure 9.15 The *shadow* and *interference* effects of major obstacles

Using Vertical and Horizontal Polarisation to Improve Frequency Re-use

In cases where a PMP network operator only has limited spectrum, say only sufficient for a 2-sector frequency re-use pattern and/or where the operator encounters interference along sector boundaries, due to reflections or whatever, one possible recourse is to employ both the horizontal and vertical channel polarisations.

In the case of the interference caused by reflection (as in Figure 9.15), one could resolve to use the horizontal polarisation of the channel in the upper right sector and the vertical polarisation in the lower left sector. Since the cross-polarisation discrimination (as we discussed earlier in the chapter) is better than the co-channel discrimination, we will save ourselves at least some of the interference in the bottom left-hand sector.

Alternatively, the operator may choose to make a four-sector re-use pattern from his two allocated frequencies by choosing to use both vertical and horizontal polarisations of each channel separately (Figure 9.16).

Overcoming Shadows by Using Repeaters

Shadows like those of Figure 9.15 can also (to some extent) be dealt with by means of active and passive repeaters, though this can introduce significant complication to the frequency planning process. The use of passive repeaters boosts the original signal and re-transmits it to fill the shadow on the other side of the obstacle. But passive repeaters can introduce overshoot difficulties (as Figure 9.3). Active repeaters help eliminate this problem, but require extra frequency allocations in the sector.

Increasing Cell Capacity by Adding Radio Channels or Splitting Sectors

The two easiest ways to increase the capacity of a base station cell are:

- first to increase the number of radio channels used in each sector;

- subsequently to increase the number of sectors by splitting sectors into ones with smaller sector angles.

Thus, in practice the sectors of Figure 9.14 may start each using one 7 MHz channel. In other words, channel 1 in sector A, channel 2 in sector C, channel 9 in sector B and channel 10 in sector D. As the demand in sector A grows, so channels 3, 5 and 7 are added progressively. Once all of the channels are used up, it becomes time to split one or more of the sectors.

Frequency planning considerations, as well as the demand for capacity, need to be taken into account when deciding how many of the sectors to split and into what reduced angles to split the sectors. Figures 9.15a and 9.15b illustrate how, in the case of a two-sector frequency re-use pattern, the splitting of a single 90° sector into two 45° sectors is not possible, since two adjacent sectors then require to use the same radio channel combination (sector frequency combination). Instead it is necessary to convert two 90° sectors at the same time (Figure 9.17c). Additionally, it might be necessary to change the frequencies of other sectors even though these sectors are not split in the process (Figure 9.17d). Alternatively, by splitting a single 90° sector into three 30° sectors, it is possible to retain all the existing arrangements in other sectors (Figure 9.17e).

Where both the horizontal and vertical polarizations of each of the radio channels are in use (as we discussed in Figure 9.16), it is prudent to consider when the radio channel frequency and polarizations are allocated to individual sectors how the sector splitting will be carried out. This can help save a lot of field service effort during the process of splitting -- caused by the need to visit a large number of remote (typically customer) terminals in order to re-install antennas for the opposite polarisation.

If the upper two sectors (upper left and upper right) of Figure 9.18a are both initially set in vertical polarisation (and the bottom two sectors in horizontal polarisation), then a sector of all four sectors into a $6 \times 60^{\circ}$ pattern is relatively straight forward provided the new 60° sectors share the same central horizontal sector boundary as the 90° sectors of Figure 9.18a (this is illustrated in Figure 9.18b). In the case of Figure 9.18b, all the remote subscriber terminal antennas will retain the same polarisation. This saves a considerable amount of

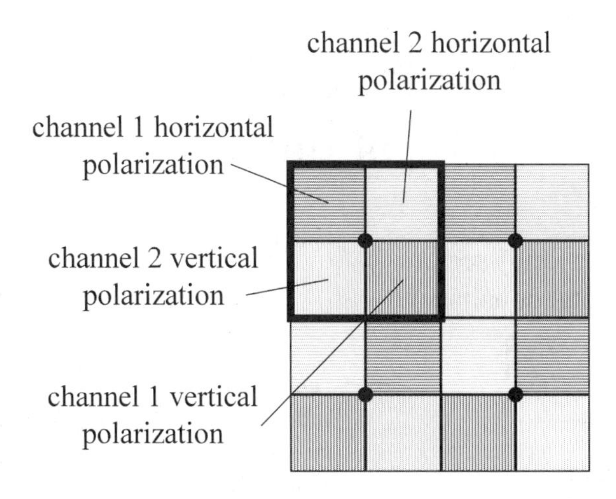

Figure 9.16 Using vertical and horizontal polarisation to improve frequency re-use from two-sector to four-sector

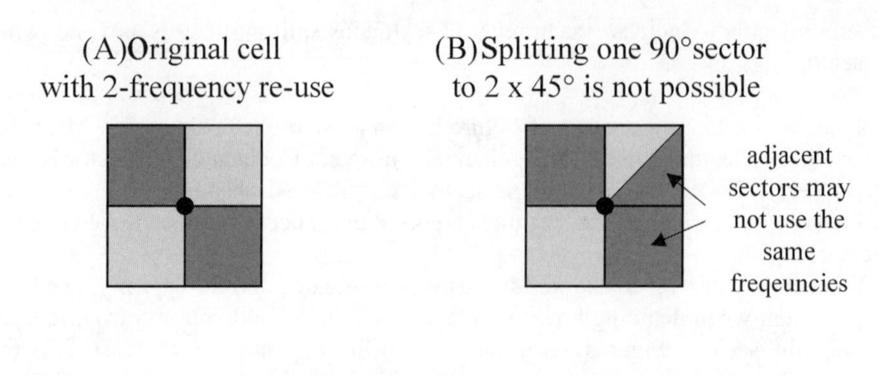

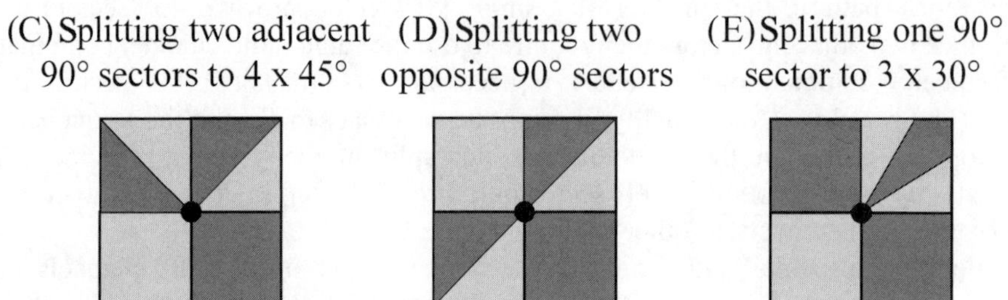

(C)Splitting two adjacent
90° sectors to 4 x 45°
(D)Splitting two
opposite 90° sectors
(E)Splitting one 90°
sector to 3 x 30°

Figure 9.17 Radio channel allocation changes caused by splitting sectors

field service effort and thus money, because the change of antenna polarisation from horizontal to vertical typically involves a manual re-mounting of the antenna. The antenna has physically to be rotated through 90° and then re-mounted or replaced by the antenna of the opposite polarisation.

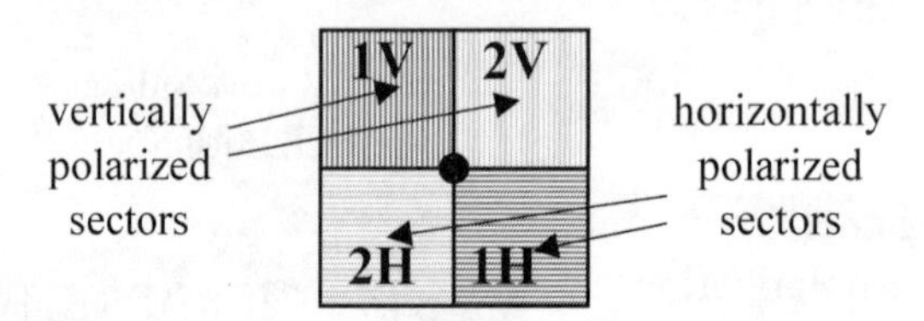

Figure 9.18 Splitting base station sectors of different polarisations

It is usually cheaper and easier to change the polarisation of the sectors at the base station than to visit many (perhaps hundreds of) customer sites to re-arrange the antennas. Figure 9.18c by contrast shows a less fortunate situation, where the 60° sectors share the central vertical boundary, so that about one-sixth (17%) of all remote terminal antennas will have to be visited and manually re-mounted or replaced.

Further frequency planning considerations are:

- how much field service effort is required to change the radio channel frequencies in the base station sectors and at the remote terminals (does the equipment, for example, need to be changed at either the base station site or the remote terminal sites). (It is becoming increasingly common for equipment to be reconfigured across a range of radio channels by means of a software command, possibly issued from a remote network management system. Even so, if the new planned frequency of a given sector lies outside the channel range of the currently-installed equipment, then it might be necessary to swap the radio units both at the base station sector site and at all the remote terminal sites);
- how the newly split sector fits into the established pattern of base station cells, and what interference problems arise directly as a result;
- transmitter output power adjustments necessary in the newly formed sectors due to the use of sector antennas of higher gain (halving the angle of the sector will add 3 dB gain to the antenna, provided the vertical beamwidth remains the same).

Not all sector splits are equally good. Just because you can work out the geometry (so that new adjacent sectors need not share the same frequency, and so that the field service effort of changing radio channel frequencies and polarisations is minimised) does not mean you have a good frequency plan. The interference patterns of specific split sector angles interfere more with the existing network than other angles. Analysis shows, for example, that the better split of a single cell or a relatively small number of isolated 90° cells within an established repeat pattern of 90° sectors conforming to the previous pattern of Figure 9.8 is into 30° sectors and not into 45° sectors. We attempt to explain why with the assistance of Figure 9.19.

Comparing the two different geometries of Figure 9.19a and Figure 9.19b, we see that the new 45° sectors of Figure 9.19a need a greater portion of the sector area to be changed to the alternative frequency. In addition, the new sector marked in the figure is likely to be interfered with by two other established base station sectors (as compared with the one of Figure 9.19b).

The Effect of Imperfect Cell and Sector Alignment

Surprising though it may at first seem, the misalignment of the sectors from the perfectly geometrical re-use patterns which we have discussed so far is actually to our advantage, as we see in Figure 9.20.

Consider first the bottom row of three cells in Figure 9.20, and in particular the top right hand cell (light shaded) of each cell. Along the lower horizontal edge of the cell, subscriber terminal antennas will all be aligned directly to the west, in order to point directly at their respective base stations. Now, because of the perfect geometry of the three cells, all three base stations lie in a direct line behind one another, so that subscriber terminals in the

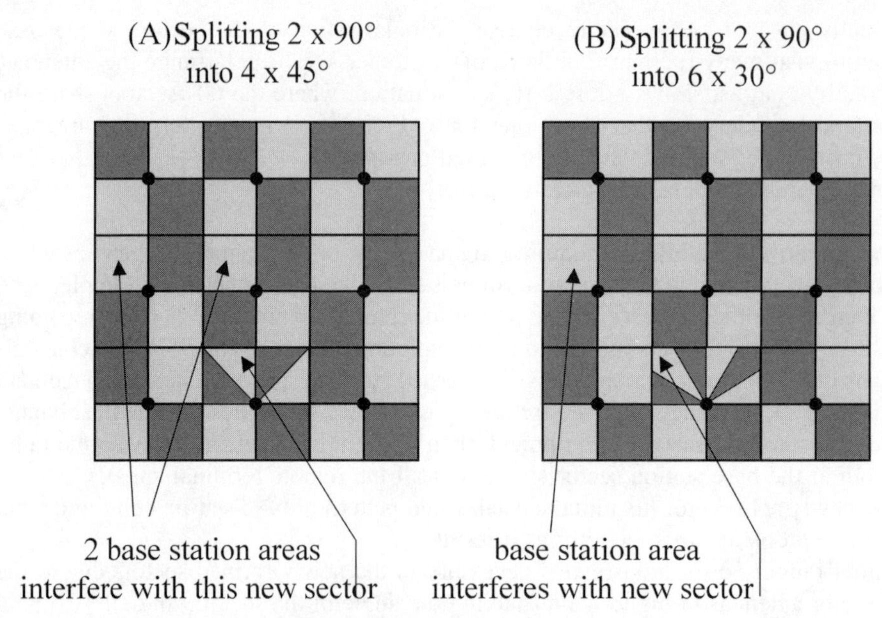

Figure 9.19 Splitting cells within an established frequency re-use pattern

furthermost right of the three cells are likely to disturb both of the other (i.e. middle and leftmost base stations), as the arrows illustrate.

In contrast, the upper row of three cells of Figure 9.20 are arranged in a higgledy-piggledy fashion (as is likely always to be the case in practice). Fortunately, this configuration is more advantageous. In our example, the two right-hand cells are rotated

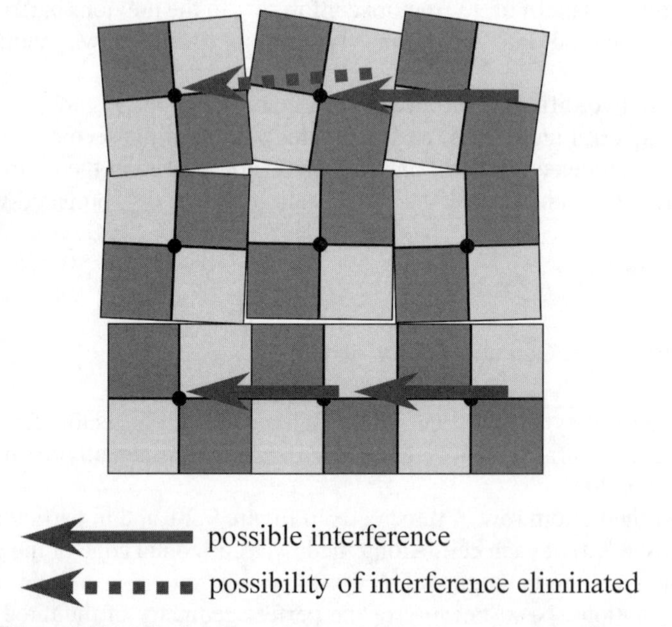

Figure 9.20 Misalignment of sector boundaries reduces interference!

slightly to the right and the left-hand cell is rotated slightly to the left. Now, along the line of the unbroken arrow (from the right-most cell to the centre cell) there is still the possibility of interference, but the number of subscriber terminals affected and the severity of the interference is only comparable to that between the right pair of cells in the lower row. Meanwhile, however, because the left-most cell of the upper row is now slightly tilted to the left, it is no longer subject to inteference along the horizontal boundary under study. In effect we have reduced the possibility of this type of interference by about 50%. Imperfection has worked to our advantage!

The Effect of Higher Modulation

The first effect of higher-order modulation, as we discussed in Chapter 4, is to increase the number of bits/second of user bitrate carried per baud (in other words, per Hertz) of radio bandwidth. However, this 'gain' in capacity is not without a 'cost'. The 'cost' is the reduced sensitivity of the receiver and the greater susceptibility to interference. This is illustrated in Figures 9.21 and 9.22, which we shall explain next.

Both Figures 9.21 and 9.22 show the relative signal strength which must be attained at the receiver location, in order to achieve a given *Bit Error Ratio* (*BER*), given a particular minimum tolerable *CIR* (*Carrier-To-Interference*) ratio. Figure 9.21 shows a number of plots for a radio system operating using 4-QAM (*Quadrature Amplitude Modulation*). Figure 9.22 shows the equivalent plots for a similar radio using 16-QAM.

Thus, when using 4-QAM modulation and assuming a minimum tolerable CIR of 13 dB and a target BER of 10^{-10}, the signal strength relative to background noise (E_b/N_o — the energy per bit compared with the *noise floor*) must be at least 11 dB. The minimum CIR of 13 dB corresponds to an absolute CIR of 20, or in other words a *coordination distance*

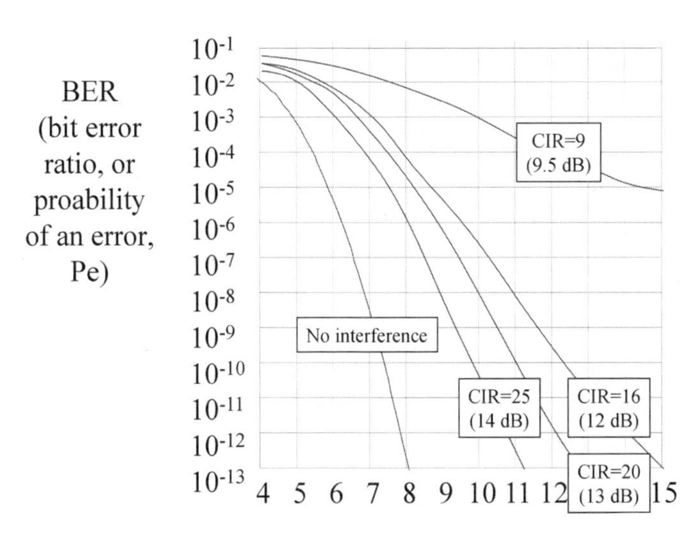

Figure 9.21 Required signal strength at receiver given BER target and minimum tolerable CIR value (for 4-QAM modulation)

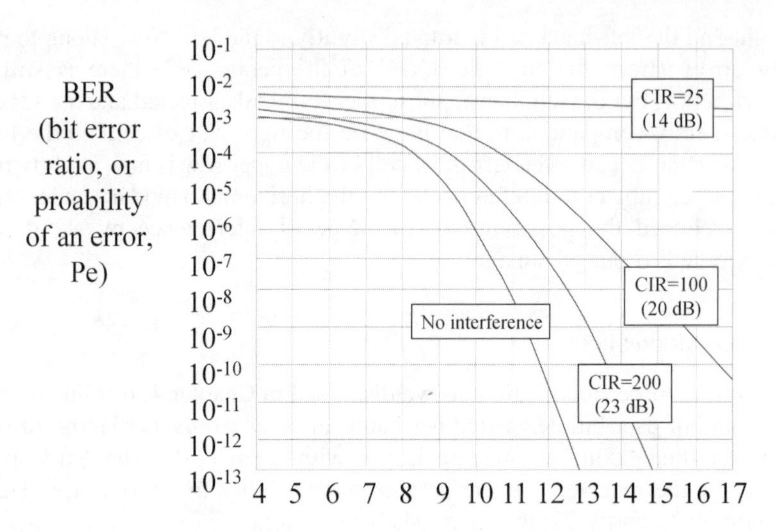

Figure 9.22 Required signal strength at receiver given BER target and minimum tolerable CIR value (for 16-QAM modulation — compare with Figure 9.19 for 4-QAM)

factor of 4.5. Thus, the next (interfering) base station using the same frequency must be at least four-and-a-half cell radii away.

At 16-QAM the frequency re-use range is much greater. As we can see from Figure 9.22, a BER of 10^{-10} is unachievable using 16-QAM unless a minimum CIR of around 23 dB is achieved. This is equivalent to a *coordination distance* factor of 14.

In practice, it may not be possible to design radio networks with coordination distance factors as high as 14 (23 dB). The two-frequency re-use pattern introduced in Figures 9.2 and 9.8 in this chapter, and used as the basis for much of our discussion is a pattern which cannot tolerate such a high distance factor. In this case, it is educational instead to return to the CIR contour diagram (Figure 9.2). Figure 9.23a is a repeat of Figure 9.2, showing the CIR contours which correspond to the now-renowned upper left-hand base station sector. Figure 9.23a has been shaded in the areas where the CIR value does not meet the minimum tolerable value of 13 dB (distance factor 4.5). This is the area in which 'coverage' is not possible. In other words, subscribers located in these areas and pointed to the 'home' base station will suffer intolerable interference.

The coverage area achieved when using 16-QAM is considerably smaller than that possible with 4-QAM, as Figure 9.23b shows. In Figure 9.23b, the shaded area corresponds to the much higher minimum CIR requirement of 16-QAM of 23 dB (distance factor 14).

So what are the morals of our story about higher modulation?

1. Radio systems using higher modulation are more susceptible to interference.
2. Therefore greater coordination distance factors should be used designing frequency re-use schemes.
3. In areas where only isolated base stations are to be deployed (where there are no nearby sources of interference), or in 'hot spots' in the immediate vicinity of base station sites

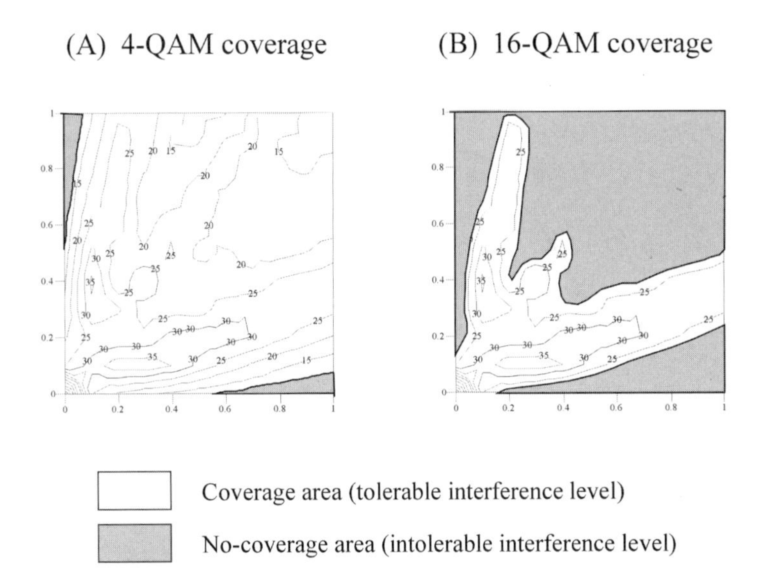

Figure 9.23 Higher modulation may result in a reduced geographical coverage area

(where coordination distance factors can be achieved on a 'microcell' basis), higher modulation may usefully increase the possible system capacity.

As a final thought, it is educational to consider the effect of increasing the modulation (to ever higher modulation schemes) on the overall capacity of a radio network. As we have learned, there are two counter-productive influences at work:

- the higher modulation increases the bit/s per Hertz of radio bandwidth, but;
- reduces the possible geographic coverage area, because more area is lost due to interference (Figure 9.23b).

Figure 9.24 shows the relative overall capacity of a radio system when designed to operate at a number of different possible higher modulation levels. It can be seen that the optimum modulation (i.e. that achieving the best overall capacity over a given geographic area using a limited amount of spectrum) is in this example around 8-QAM.

The Percentage Geographic Coverage of a PMP Frequency Plan

From shaded CIR-contour diagrams such as Figure 9.23, we are easily able to visualise and calculate the overall percentage geographic coverage of a PMP frequency plan.

In reality, the coverage area of Figure 9.23a (repeated in Figure 9.25a) can still be improved upon. In other words, even customers in the grey-shaded areas can be provided with service. This is achieved by directing the remote subscriber antennas in these areas to neighbouring base stations rather than to the normal 'home' base station. Figures 9.25b and 9.25c show the coverage possible from the 'upper left' and 'upper right' base stations. By combining all three diagrams (Figures 9.25a, 9.25b and 9.25c), near full coverage can be achieved.

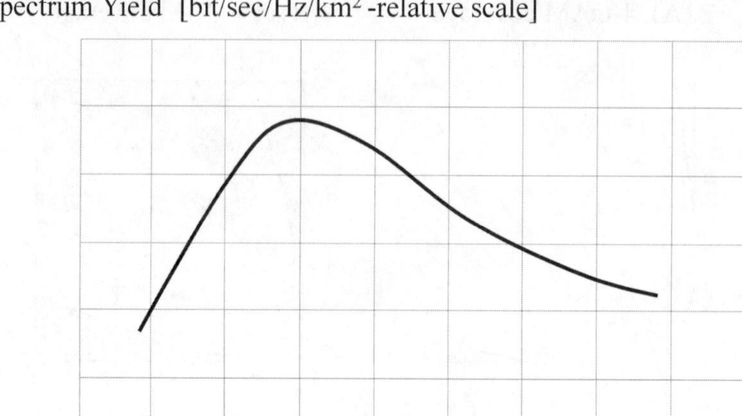

Figure 9.24 The effect of the modulation scheme on overall network capacity

Our various tricks so far have already achieved nearly 100% coverage; but the tricks are not quite exhausted! As Figure 9.26 illustrates, the use of high gain antennas at the remote sites brings even more coverage. (The more directional and thus more *discriminating* antenna is able to 'ignore' some of the interference, but in addition is necessary in order to achieve acceptable link *availability* over the greater distance to the neighbouring base station.)

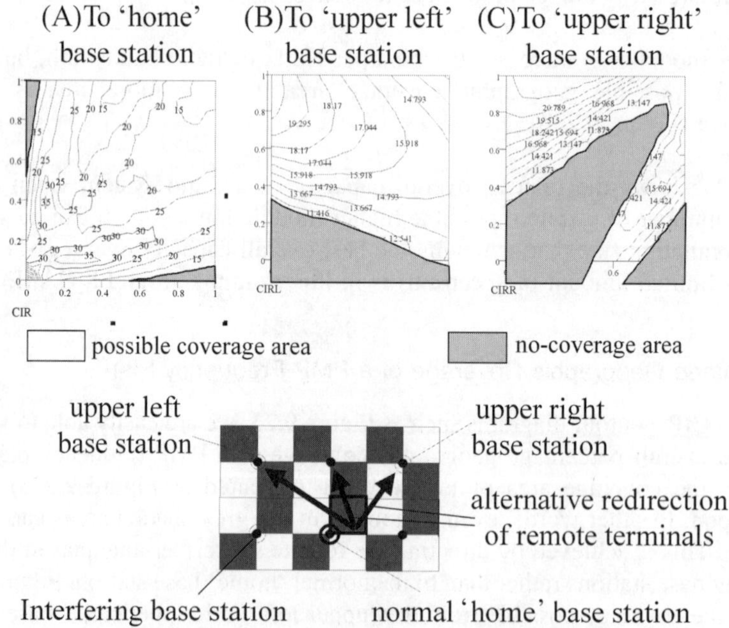

Figure 9.25 Improving geographical coverage in sector fringe areas by directing remote subscriber antennas to neighbouring base stations

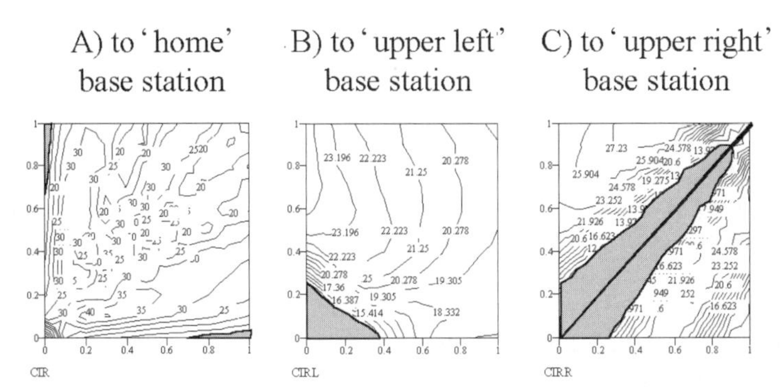

Figure 9.26 Reducing the effects of interference by using a more *discriminating* (i.e. higher gain or more directional) antenna (to 'mask out' the source of the interference)

9.9 Frequency Planning Considerations of ATPC

As a general principle we can say that the use of *ATPC* (*Automatic Transmit Power Control*— as we learned about in Chapter 6) is a good thing for both PTP (point-to-point) and PMP (point-to-multipoint) radio systems. ATPC aids the frequency re-use planning of a network, by increasing the amount of spectrum re-use which is possible, but can make the task of planning much harder, since the signal powers from the different transmitter sources are constantly changing.

From a frequency re-use point-of-view, ATPC can be thought of as *reducing* the transmitter power from the maximum during periods of good weather, thereby reducing the amount of signal overshoot and so th range of possible interference. During bad weather, the transmitter power is increased again, as necessary, to overcome the effects of signal fading.

In a PTP (point-to-point) radio system, ATPC is applied by measuring the received signal level at both ends of the connection and using this to adjust the transmitter output power accordingly, in order to maintain a small fading margin between actual received signal strength and minimum detectable signal strength.

The use of ATPC is sometimes required by the technical standards for a particular PTP radio band, but this is not always the case. Unfortunately, where ATPC is not prescribed for a particular band, and where this band of spectrum is to be shared between a number of different operators in the same region, then it does not make sense unilaterally to apply ATPC. This is because by using ATPC an operator will be a 'good citizen' and tend to reduce his disturbance to the nearby links of other operators, but himself will suffer worse interference. The 'case' for not requiring ATPC in some bands (usually the *shorthaul* bands) is two-fold: first, it allows the equipment to be less complex, and therefore cheaper; and second, the interference calculations between adjacent links are easier for the radio frequency planner. The calculations are made simpler by the constant, pre-defined output power of each transmitter. In consequence, the interference caused by one link to another can be exactly determined simply by knowing the geographic distances and link orientations.

In PMP (point-to-multipoint) systems, ATPC applied in the *upstream* direction (i.e. from remote subscribers to the base station) is almost unavoidable. ATPC in the upstream direction helps to ensure that all the remote signals arriving at the base station are of nearly the same signal strength. In this way all the signals have an equal chance of good reception and none are subject to undue interference or masking by the other 'competing' signals.

In the *downstream* direction of a PMP system, it is not so critical to overall radio system operation that ATPC be applied. One reason which might necessitate the use of downstream ATPC is the regulatory restriction of power signal density (as we discussed at the beginning of the chapter) at a given distance beyond the base station location. By using maximum signal power at all times, the power spectral density may exceed the maximum permitted value particularly during good weather. In this case, downstream ATPC could be used to reduce the base station transmitter power. However, downstream ATPC is relatively complex for the radio system designer to build into the system and for the frequency planner to cope with.

The PMP radio system designer has to conceive a methodology by which he can determine the appropriate transmitter power for downstream ATPC. His problem is that the signal is received by many different remote stations, all of which will monitor different received signal strengths, depending upon their distance from the base station and according to their local prevailing weather. Maybe a remote subscriber station relatively distant from the base station on one side of the sector is not subject to the same degree of fading due to a local thunderstorm that is causing a nearer remote station in another part of the sector to be 'off-air'.

For the PMP frequency re-use planner, downstream ATPC can complicate things, because now the interference calculations between adjacent base stations are much more complex. No longer can we rely on the geometrical distances between base stations as the guarantor of given CIR values. Now we also have to consider the possibility that the output power of one base station could be greater than the output power of ist neighbour. In this case, we increase the possibility of interference.

9.10 How much Spectrum does a Radio Network Operator Need?

For a nationwide backbone or backhaul network *interconnecting Base Transmitter Stations* (*BTSs*) to *Base Station Controllers* (*BSCs*) and *Mobile Switching Centres* (*MSCs*), mobile telephone network operators typically use three or four channel frequencies (e.g. three or four times 7 MHz (4 × 2 Mbit/s) channels) plus perhaps one or two 28 MHz channels for high capacity links (16 × 2 Mbit/s). (This equates to around 56 MHz or 84 MHz total allocation to the operator.) A quite dense network of point-to-point links (of many thousand links) could be built in a a given country using such an allocation within one or more of the shorthaul bands (e.g. 23 GHz, 26 GHz or 38 GHz CEPT [European] bands).

In the case of point-to-multipoint (PMP) networks the spectrum requirement depends upon a large number of factors, including the band of operation, the system range (affected by the local climate), the required density and overall capacity of connections. A rough rule-of-thumb is to plan on three or four-times the spectrum required in the most densely populated *sector*. Thus, if the radio network operator requires 28 MHz capacity in a given sector (of given angle and range) then its 'ideal' overall spectrum allocation request should be for 84 MHz or 112 MHz of spectrum (corresponding to three-frequency and four-frequency re-use patterns, respectively). The four-frequency re-use pattern, as we discussed earlier in the chapter is necessary to overcome the difficult realities of real frequency planning. Where such a 'generous' spectrum allocation cannot be made available to the operator (for example, where only $2 \times$ frequency allocation is available), it will be necessary to deploy different signal polarizations. This we also discussed earlier.

9.11 Guard Bands

Guard bands are segments of spectrum laid aside and left unused by radio frequency planners. The most common use of guard bands is by radio regulatory agencies, to provide 'separation' between the spectrum allocations of different radio network operators in the same geographical region. The guard band dramatically reduces the possibility of adjacent channel interference of one operator by the other, immaterial of how well or badly each of the two operators' radio frequency planning has been carried out.

Part III

Applications, Network Integration and Management

10

Radio Applications and Network Integration

The quality of a telecommunications network as perceived by the end-user depends upon the end-to-end performance of the network and the customer's 'application'. Only when the network and application have both been tuned to work in harmony with one another can the end-user expect good end-to-end performance. Optimum performance relies upon the right choice of components to make up the network and to support the chosen application as well as upon the correct tuning of the application to assist best possible operation of the network components. This is a task for a specialised type of network and application and for a specialised service-design engineer. The task is given a special name: 'network integration'. In this chapter we consider the different types of radio systems and the applications to which they are suited. We then go on to consider a number of the general considerations which a network integration engineer must take into account when designing the radio part of the network and incorporating it into the network as a whole. In subsequent chapters, we will go on to consider different types of radio and fixed wireless access systems and the specific applications for which they are intended.

10.1 Choosing the Right Radio System Type for a Particular Application, Range and Bandwidth

Figure 10.1 illustrates the basic main types (classes) of telecommunications radio systems, showing the relationship between the operating frequency and the range and bandwith of the system. The range of the system, as we discovered in Chapter 8 is very dependent upon the mode of signal propagation (whether by line-of-sight, or by *skywave*, *troposheric scatter* or some other means of *over-the-horizon* propagation). However, the different types of propagation are not all equally suited to specific applications, as we shall discuss briefly in the next few sections. As is clear from Figure 10.1, most fixed wireless access systems are based upon microwave radio systems. They are thus generally only suited to line-of-sight propagation and have restricted range. On the other hand, networks with significant capacity (and thus large numbers of customer connections) can be developed in this band. The restricted range actually helps us increase the overall connection density within a given spectrum bandwidth.

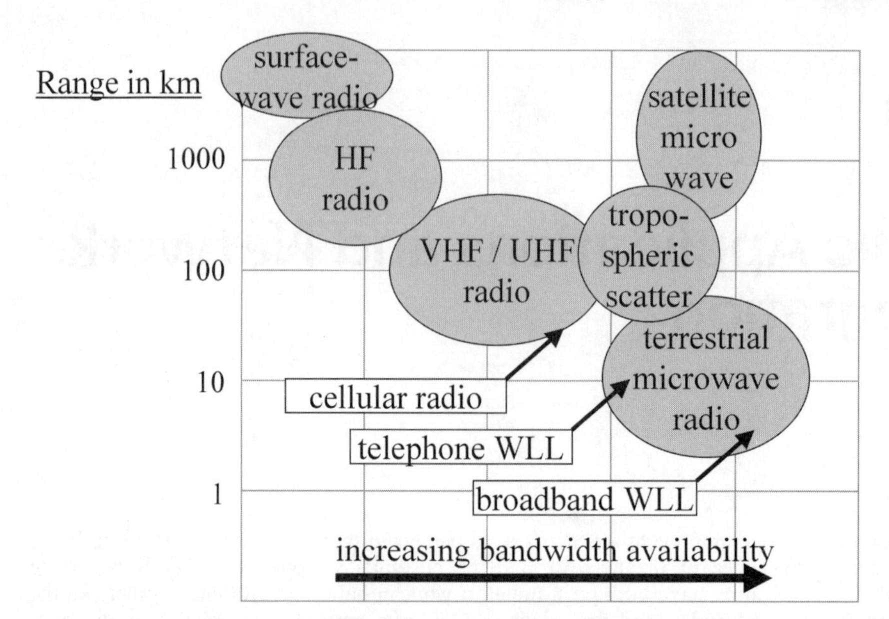

Figure 10.1 Different classes of radio system, their operating frequency, range and bandwidth

Surface-Wave Radio Systems

Surface-wave radio systems have a good range when using relatively low frequency radio waves. (In this context low frequency is the range 50 kHz–2 MHz). Surface wave radio is typically used for broadcast radio transmissions, particularly for public radio stations, maritime radio, and navigation systems. Broadcast, surface wave radio transmitters are usually very tall radio masts, several hundred feet high, which combine high power transmitters with an omnidirectional lobe pattern. The high power enables domestic radio receivers to be relatively unsophisticated, and therefore cheap. The omnidirectional pattern allows them to transmit over a wide area.

High Frequency (HF) Radio

Radio waves in the frequency band from 3–30 MHz propagate over greater distances in a *skywave* form than they do as a surface wave. For *skywave* propagation, a directional antenna is set up to transmit a radio wave at a specific angle to the horizon. The signal is transmitted via the earth's ionosphere (a region about 200–350 km above the earth's surface). Layers of the ionosphere which have different physical properties cause the signal to be refracted and eventually reflected back to the earth's surface, an effect we discussed in Figure 7.5 of Chapter 7.

Unfortunately, high frequency skywave radio is very sensitive to weather conditions in the ionosphere. Time-of-day and sun spot cycles can break up reliable service, giving rise to fading and interference. Nonetheless, high frequency radio using skywaves was one of the earliest transmission methods employed for international telephone service. As early as

1927, the first commercial radio telephone service was in operation between Britain and the United States, many years before first transatlantic telephone cable was laid in 1956.

Very High Frequency (VHF) and Ultra High Frequency (UHF) Radio

At frequencies above about 30 MHz, neither skywave propagation nor surface-wave over-the-horizon propagation is possible. For this reason, the VHF and UHF radio spectrum is used mainly for line of sight radio transmission systems. *Just-over-the-horizon* transmission is also possible, but only if the antennas are elevated clear of the electrical *ground effects* of the earth's surface.

VHF and UHF radio systems are becoming increasingly common as the basis of a large number of applications:

- local radio stations;
- citizens band (CB) radio;
- cellular radio (mobile telephones);
- Radiopagers;
- Broadcast TV.

The advantage of VHF and UHF radio is that much smaller antennas can be used. It has made possible the wide range mobile communications handsets, some of which have antennas only a few inches long. A drawback is the restricted range of VHF and UHF systems, although this has been turned to advantage in cellular radio telephone systems, where a number of transmitters are used, each covering only a small *cell* area. Each cell has a given bandwidth of the radio spectrum to establish telephone calls to mobile stations within the cell. Adjacent cells use different radio bandwidth, thereby preventing radio interference between the signals at the cell boundaries. The fact that the radio range is short also means that a non-adjacent cell can *re-use* the same radio spectrum without chance of

Figure 10.2 Frequency re-use in non-adjacent sectors of cellular radio networks — the restricted range helps to increase the capacity available from a restricted spectrum bandwidth

interference, so that very high radio spectrum utilisation can be achieved. Figure 10.2 illustrates the re-use of radio spectrum, as achieved in *cellular radio* networks for portable telephone sets. Cells marked with the same shading are using identical radio bandwidth (a four-frequency re-use pattern is shown).

Typical bands used for cellular radio network applications are the 450 MHz, 900 MHz, 1800 MHz (Europe) and 1900 MHz (USA). Within these bands, large scale pan-European and pan-continental mobile telephone networks have appeared since the early 1990s and now are able to support many millions of customers within each domestic national network. Allocating less than 100 MHz of bandwidth in total, many countries have licensed two, three or even four separate mobile telephone operators to share the bandwidth. Yet, each of the operators may support more than a million individual mobile telephone users.

The reason so many customers can be supported in so little bandwidth is due to three main factors. The first is that the restricted range enables frequency re-use, as we discussed above. The second reason is that the connection speed made available to each user is restricted (typically 8 bit/s to 10 kbit/s). This is adequate for a telephone call, but rather slow for modern data or Internet communications. The third reason is that each customer only uses a connection (i.e. makes a call) very occasionally. For higher bandwidth applications, there is simply not enough spectrum available in the VHF or UHF bands to support simultaneous use by many different customers.

Broadcast TV applications use large channel bandwidths in the band (typically 70 MHz per channel), but the same signal is received simultaneously by many different TV receivers. However, as television viewers increasingly expect more channels and 'choose-what-you-view', the limitations of the VHF/UHF band have made it impractical to consider it for future TV broadcasting. Broadcast TV is slowly migrating to satellite and cable TV. Meanwhile, the VHF/UHF bands are being gradually freed up to make more spectrum available to meet the ever-growing demand for mobile telephone services.

Microwave Radio

Microwave is the name given to radio waves in the frequency above 1000 MHz (1 GHz). The prefix *micro* is in recognition of the very short wavelength (of the order of one centimetre). Microwave systems have historically been widely used as high capacity, point-to-point transmission systems in telecommunications networks (such as high-capacity trunk telephone network connections between major cities), and for backhaul connections in mobile telephone networks (between mobile radio Base Transmitter Stations (BTSs) and the backbone Base Station Controllers (BSCs) Mobile Switching Centres (MSCs). Increasingly, microwave spectrum is being allocated to fixed telecommunications network operators for applications such as telephony *Wireless Local Loop* (*WLL*), *fixed wireless access*, *broadband wireless access*, and similar applications.

The high frequency and short wavelength of microwave radio allows high capacity radio systems to be built using relatively small but highly directional antennas. The small scale yields benefits in terms of cost, installation and maintenance.

Microwave antennas are operated in a line-of-sight mode, over distances between 3 and 70 km apart, depending upon the radio frequency in use, the radio propagation conditions and by the type of system (PTP, point-to-point or PMP, point-to-multipoint).

Tropospheric Scatter

It takes *tropospheric scatter* to make over-the-horizon communication possible. Radio waves appear to be reflected by the earth's atmosphere. It is not a well understood phenomenon and various explanations have been offered. Perhaps the easiest to grasp is the notion of radiowave reflection (commonly called *scatter*), caused by irregularities in the troposheric region of the earth's atmosphere. The scatter occurs in a *common volume* of the earth's atmosphere, corresponding to the region 'visible' to both the receiving and transmitting tropospheric scatter antennas (Figure 10.3). The scatter angle is the angle of path deviation (or the *reflection angle*) caused by the scatter effect.

Large dish-shaped antennas are used for tropospheric scatter systems, together with microwave radio frequencies in the range of 800 MHz–5 GHz. Communication distances achievable with tropospheric scatter radio systems are 100–300 km. Their main drawback is the bad and continually varying signal fading that is experienced. In the United Kingdom, the tropospheric scatter method of *trans-horizon radio* transmission used to be a standard method of communicating with offshore oil drilling platforms in the North sea, but it is tending to be supplanted by small-dish satellite communication, and the spectrum is being freed for other uses.

Satellite Systems

Satellite transmission provides an excellent means of long distance communication, either around the globe or across difficult terrain. It also provides an effective means of *broadcasting* the same signal to a large number of receiving stations.

The type of satellites most commonly used in telecommunications networks, called *geostationary* satellites, orbit the earth directly above the equator at such a height that they travel once around the earth's axis every twenty four hours. Because both the satellite and the earth move at the same speed, the satellite appears to be geographically (or geo) stationary above a particular location on the equator.

When used for telecommunications purposes, a geostationary satellite is equipped with microwave radios and antennas, which allow line-of-sight radio contacts between the satellite and other microwave antennas at *earth stations* on the ground. Specific spectrum

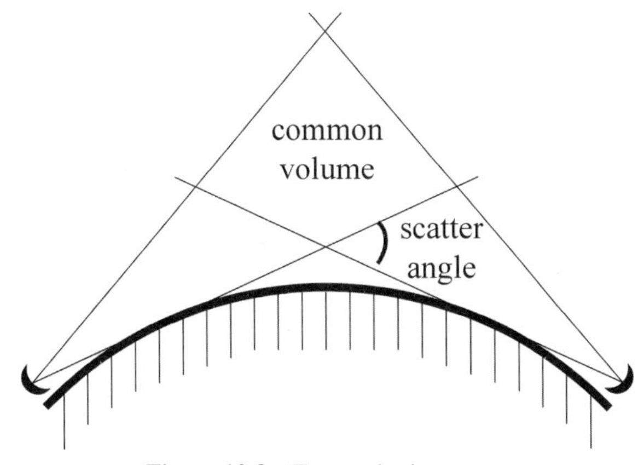

Figure 10.3 Tropospheric scatter

bands are allocated by ITU-R, regional and national regulatory bodies for *earth-to-space* and *space-to-earth* communication.

Communication between two earth stations can then be established by a tandem connection consisting of an *uplink* from the transmitting station to the satellite, and a *downlink* from the satellite to the receiving station. On board the satellite the uplink is connected to the downlink by a *responder* (receiver) for each uplink and a *transmitter* for each downlink, and because the two normally work in pairs, they often designed as a single piece of equipment usually referred to as a *transponder* (Figure 10.4).

The most important class of satellite radio system used in access networks (and in broadcast networks) are *VSAT* (*Very Small Aperture Terminal*) systems. VSAT systems appeared during the 1980s. As the name suggests, the antennas of VSAT systems are small dish antennas (typically 60–300 cm diameter). The small dish (i.e. relatively low gain antenna compared with historical 12 m and 18 m diameter *earth station* antennas) is made possible by using high power transmitters and high power satellite transponders.

VSAT is commonly used for high bandwidth corporate communications, for example for high speed international data communications networks. Satellite communication is particularly effective in providing highspeed links to remote destinations within very short planning and installation lead times.

Since the basic radio systems used in satellite communications are similar to terrestrial microwave, similar principles of operation and planning apply, although there are special considerations to be made regarding propagation through the earth's atmosphere.

The main drawbacks of satellite communication are the relatively high costs (associated with manufacturing the satellite and launching it) and the significant signal delays resulting from the time it takes the signal to travel up to and back from the satellite, 40,000 km above the earth's surface. In the case of telephone conversation, this propagation leads to a rather annoying one second delay between talking and hearing the listener's response. Even

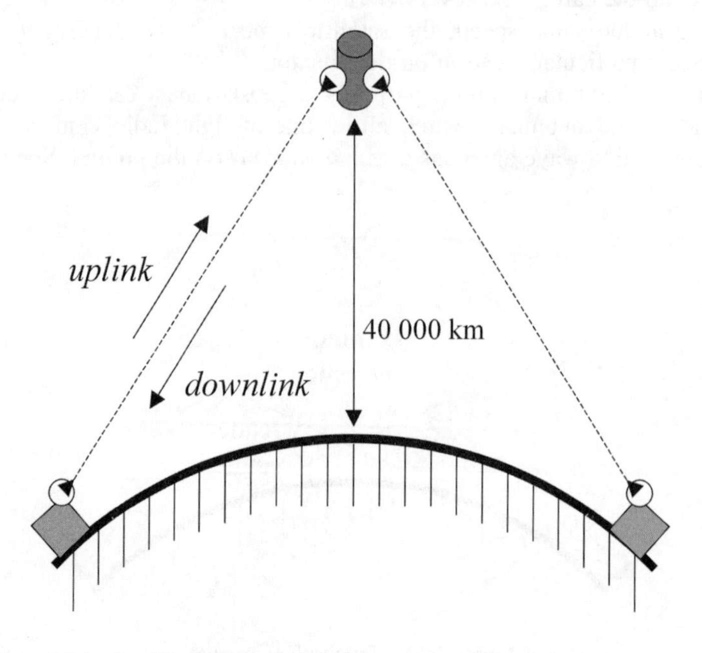

Figure 10.4 Satellite *uplinks* (*earth-to-space*) and *downlinks* (*space-to-earth*)

worse, sometimes you even hear an echo of your own voice, half to one second after talking. The talker's voice has to be carried 80,000 km to the listener (taking around $\frac{1}{4}$ to $\frac{1}{2}$ second). If the listener responds immediately, then his response comes back another $\frac{1}{4}$ to $\frac{1}{2}$ second later (i.e. one second in all). However, in normal conversation, the listener responds almost as soon as you stop talking. When you don't hear a response, you are sometimes tempted to start talking again: "did you hear" . . . meanwhile the response comes back, and both are speaking at once.

Echo is normal on a telephone conversation. The speaker's voice is 'echoed' by the listeners telephone and returns to the speaker's earphone. Normally, the echo is immediate, so the speaker thinks he is hearing his own voice as he speaks . . . and hearing it in the earphone confirms to him psychologically that the telephone is not 'dead'. The problem in the case of satellite communication is that the echo might be returned between a half and one second later. This is very offputting for the speaker, and requires special techniques of echo *cancellation* to be applied, as we shall discuss later in the chapter.

Propagation delay can also be a major problem for data communications in cases where the data application has not been specifically designed to cope with the delay. We shall also discuss this later.

10.2 Network Integration

Choosing the right type of radio equipment for a given application is critical to good network performance. But just as critical is the *integration* of the radio sub-network into the network as a whole.

In the remainder of this chapter, we study the broad considerations which must be taken into account by network integration engineers. In particular, we shall consider in turn, with particular reference to the implications for the planning of fixed wireless access networks:

- the end-to-end network *transmission plan*;
- the *interfaces* used to connect the fixed wireless access system to the main *backbone* network and to *Customer Premises Equipment* (*CPE*);
- the *synchronisation* plan;
- the *network signalling* and/or *data protocols* used to carry specific *network services*;
- the effects of *propagation delays*;
- *network management* and *service monitoring*;
- the need for *accounting*;
- the effects of *network failures* (and how to build robust networks).

10.3 Network Transmission Plan

A formal network *transmission plan* sets out the necessary network design rules and equipment planning and commissioning rules a telecommunications network. The transmission plan is intended to maximise the quality of connections and communications made across the network, minimising the effects of signal losses, noise, interference, distortion and the like. ITU-T (the standardisation sector of the International Telecommunications Union) has set out a series of recommendations governing the design of international public telecommunications networks. In particular, ITU-T recommendations G.101 and G.100 set out the transmission plan and the related terminology, respectively. The plan includes design guidance on:

- the overall signal *loudness* (i.e. signal strength or volume) and *loss plan* at all points through the network (this ensures that the listener can hear the talker, and vice versa);
- ensuring the electrical *stability* of individual circuit connections;
- the minimisation of signal *distortion*;
- the minimisation of *crosstalk* and *interference* (corruption of the signal caused by the disturbance of other signals being carried in parallel across the network);
- the limits on acceptable *noise* disturbance;
- (for digital circuits) the maximum allowed *Bit Error Ratio* (*BER*), *jitter* and *quantisation distortion*;
- the maximum line lengths (actually, maximum line lengths are more commonly quoted in the specifications applying to particular types of cables and interfaces used for connecting *Customer Premises Equipment* (*CPE*) to the network);
- the control of *sidetone* (sidetone is the ability of a talker to hear his own voice in the earpiece while he is talking. With no sidetone, the talker gets the impression that the phone is 'dead'. Too much sidetone causes the speaker to drop his voice, so that he can no longer be heard);
- the limits on acceptable signal *propagation times* (excessive propagation times manifest themselves as a delay in transmission, a potential cause of slow data throughput and response times, or of unacceptable gaps and pauses in conversation);
- the control of signal echo (echo is like sidetone, but with a delay of more than an eighth of a second before the speaker's voice is heard by him again. It must be countered by *echo suppression* or *echo cancellation*, as we will discuss later).

In addition to ITU-T recommendation G.101, which sets out the transmission plan for international networks, recommendations G.121 and G.122 set out the corresponding details for national networks forming part of an international network. These recommendations ensure that national public telecommunications networks are designed in such a way as to be able to be interconnected to make international networks meeting international quality standards. In effect, the end-to-end allowances of acceptable quality degradations are 'shared out' and allocated to the various parts of the connection. Provided each part meets its own quality goals, then the end-to-end connection will also be acceptable.

ITU-T recommendation G.121 covers signal loudness and signal losses along the course of the connection. This ensures that the listener hears the talker, not too loud and not too soft, and that the losses in both directions are roughly the same. Recommendation G.122 covers the guidelines for ensuring circuit stability. Other recommendations in the G-series detail other specific aspects of circuit design and control, as well as the procedures for circuit commissioning.

Most national publicly licensed telecommunications operators are obliged by law and telecommunications regulators to conform to the ITU-T recommendations. In addition, individual *interconnection standards* (agreed within and part of the official telecommunications regulation of a particular country) may further sub-divide the national network 'allowances' into separate parts, corresponding to *long distance* and *local network* parts.

We discuss each of the main transmission plan quality parameters in turn.

Signal Loudness and Network Loss Plan

Figure 10.5 illustrates the network model used to define the signal loudness and network loss plan. The overall loss from talker's mouth to listener's ear is called the *Overall Loudness Rating* (*OLR*). The value is recommended by ITU-T to have an optimum value of 10 dB, and should not exceed 29 dB. As part of the *Access Network* (*AN*) of Figure 10.5, a wireless access sub-network must not exceed its allocation of the overall allowed end-to-end loss (or other parameter).

The overall loudness rating is subdivided into the losses incurred in the sound-to-electrical signal conversions, which take place in the microphone and the earphone (termed, respectively, the *Send Loudness Rating* (*SLR*) and the *Receive Loudness Rating* (*RLR*)), and the network loss.

It is standard to design networks to have a network loss of between about 0 and 3 dB. Typically analogue networks were designed to have a network loss of about 3 dB plus 0.5 dB for each switching stage in the connection. This nominal small loss was designed to ensure the stability of the circuit (a subject we shall cover presently) but without causing too much loss in the strength of the signal. Loss of signal strength is generally a 'bad thing' because the signal becomes more prone to disturbance from 'background sources' of electrical noise. Modern digital networks are no-loss networks (i.e. 0 dB).

As also shown in Figure 10.5, the transmission plan may also define precise network loss and other transmission parameter values at a number of specific interconnection points along the course of a connection. These standard points are:

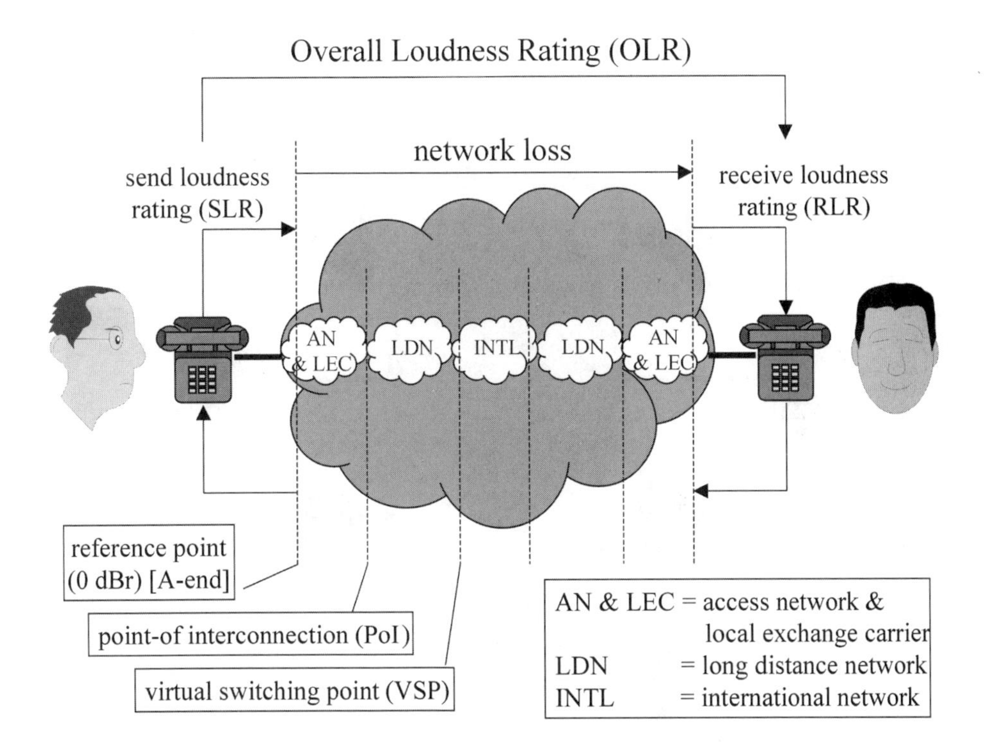

Figure 10.5 Basic transmission plan network model for signal loudness and network loss

- the 0 dBr reference point at which the customer's connection line (the *access network*) is assumed to join the main local switching network;
- the point-of-interconnection (typically defined by national regulations) where two public operators networks are interconnected within a given country (typically the point of interconnection between a *local exchange carrier's* network and that of the *long distance network* operator;
- the *Virtual Switching Points* (*VSP*) defined by ITU-T recommendations for the interconnection of national gateway exchanges to the international network.

The send loudness rating of a given telephone microphone can be measured using standardised test equipment (an *Intermediate Reference System* (*IRS*) as defined by ITU-T recommendation P.48), as can the receive loudness rating of a given telephone earpiece. If the resulting electrical signal emanating from the microphone, or the sound wave emitted by the earpiece is not loud enough, then can be corrected by a variable amplifier (as shown in Figure 10.6). The use of such amplifiers allows an optimum signal loudness to be achieved at all points along the course of the connection (Figure 10.5).

If the signal reaching the 0 dBr reference point from the speaker on the left-hand side of Figure 10.6 is too weak, then it will become subject to distubance from noise during its transmission through the network. In this case the signal should be amplified by the amplifier A_T. If, on the other hand, the signal is too strong, then it might be distorted during transmission, because the dynamic range of the transmission equipment is not great enough to carry such a strong signal. Worse still, it might cause circuit instability (see the next section). In this case, the signal should be attenuated at position A_T. (Actually, it is most

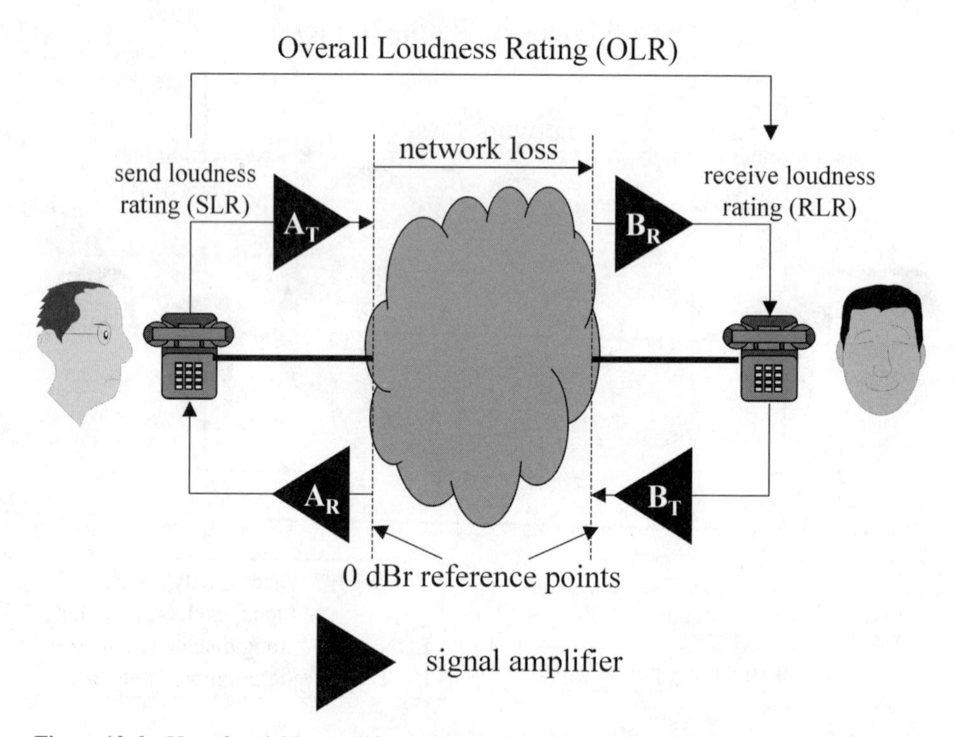

Figure 10.6 Use of variable amplifiers at transmit and receive ends to optimise signal loudness

common to use fixed gain amplifiers and variable attenuators (or *pads*) at the positions of the variable amplifiers (A_T, B_R, B_T, A_R) shown in Figure 10.6. This allows for signal attenuation as well as amplification.)

In a similar manner, it is helpful to have a variable amplifier/attenuator available at the receiving end (B_R on the right or A_R on the left) to boost or attenuate a signal incoming from the network to the optimum level required by the telephone earpiece.

Where the telephone handset does not include a variable amplifiers/attenuator for signal volume, the job of signal loudness correction will normally have to be carried out in the access network equipment (in our case, probably by the radio terminal or base station equipment). Not all access network equipment offers this capability. (Indeed most modern digital equipment is designed to assumed a standard 'no loss' criterion.) It is not critical to have such variable amplification/attenuation, but it is handy to have it on a per-customer basis when a specific customer complains of signal loudness problems.

Circuit Stability

Particularly in analogue telephone networks, and in mixed analogue/digital networks (including modern *integrated services digital networks*, in which old two-wire analogue telephones are used), special precautions need to be made to ensure the electrical stability of the network as a whole. A particular problem arises in at the point where the two-wire telephone access line is converted to a four-wire signal format (two wires each for *transmit* and *receive* 'pairs') for transmission across the main part of the network. This may be at a point in the wireless access network. The device used at this point might be either a *hybrid convertor* (two-wire analogue to four-wire analogue convertor) or an *A/D convertor* (two-wire analogue to four-wire digital convertor). Figure 10.7 illustrates such a conversion device, and the signal reflections (signal *return path* or *echo*) it can cause.

Each line of Figure 10.7 representing a *transmit* or *receive* connection is actually a pair of wires (i.e. is a two-wire connection), even though it is shown as a single line. Thus, the telephone illustrated is connected by a two-wire connection to the *hybrid* or *analogue-to-digital convertor*, which in turn is connected to two two-wire connections (four wires in total), one *pair* each for transmit and receive.

Unfortunately, it is not possible to build perfect devices for two-wire to four-wire conversion. The problem lies therein, that the two-wire connection carries both transmit and receive signals, which are to be separated onto separate pairs for the four-wire (long haul) part of the connection. The problem arises when some of the receive signal (from the long haul receive pair) finds its way back onto the transmit pair (as shown by the semicircular arrow leading from the receive pair directly across the two-wire/four-wire convertor to the transmit pair).

The signal *return* happens whenever the *balancing impedance* is not matched *exactly* to the impedance (i.e. *resistance*) of the telephone. In practice, this means that some of the receive signal always finds its way back onto the transmit pair. So what, you say? The problem is, that this stray signal can lead to problems of circuit *instability* or signal echo. Neither can be tolerated.

A further potential source of circuit instability or echo is the possible return path between the telephone earpiece and microphone (shown in Figure 10.7 with a semicircular 'return' arrow beside the telephone handset).

Circuit instability can arise in the four-wire part of the circuit (between the two two-wire-to-four-wire convertors at either end of the main part of the connection), when there is a net amplification in the loop [formed of receive and transmit pairs]. Imagine in Figure 10.7 that the strength of the 'receive' signal *returned* on the transmit pair is as strong or stronger than the intended *transmit* signal. Imagine also, that it is amplified only slightly by the network part of the connection, and that a significant signal *return* happens once again at the A-end of the connection (shown by the dashed semicircular arrow in Figure 10.7). In this case, the twice-returned 'receive' signal is now adding to the original 'receive' signal, in effect amplifying it. However, if an amplified form of the 'receive' signal now results on the *receive* pair at the B-end, then an even stronger reflected signal will be returned. This causes even more amplification. The cycle continues, causing ever more amplification with each 'lap' of the four-wire part of the circuit until the circuit becomes unstable and the signal volume saturates the electronics. At this time, the circuit is singing with a loud tone, in slang commonly known as 'feedback'.

To avoid circuit instability, the network designer must ensure that there is signal attenuation in the potential return path, so that the return signal is weakened with each lap rather than strengthened. In the analogue world this was achieved with the nominal small network loss of about 3 dB in each direction on the four-wire part of the network. In the digital world, this is achieved by more efficient two-wire to four-wire analogue to digital convertors which achieve near optimum separation of *receive* and *transmit* signals. Even so, network operators still have to beware of the possible echo return path through the telephone handset. I have myself encountered such feedback on noisy 'open line' foreign exchange dealing desks, where a number of separate telephone lines are presented to a broker

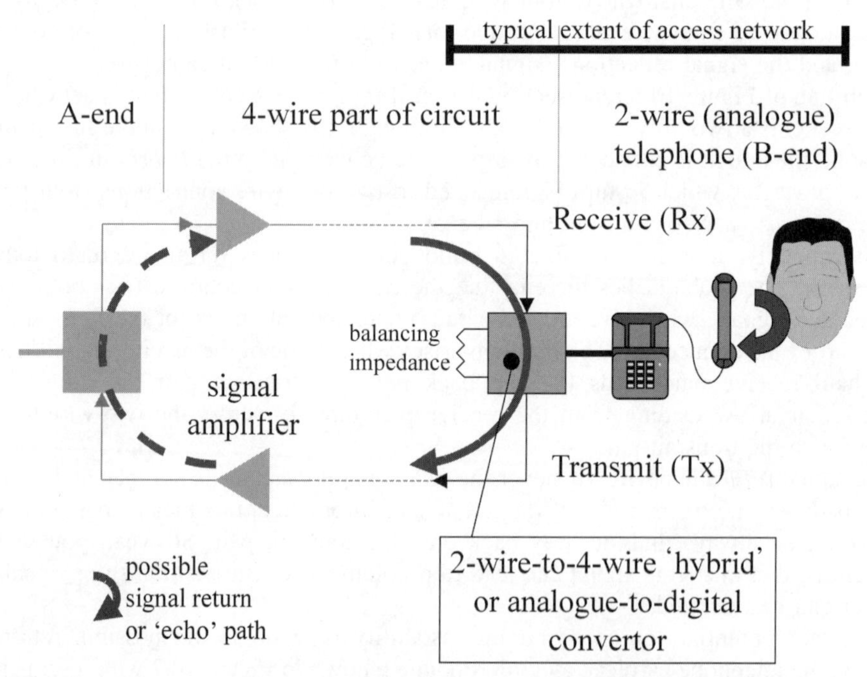

Figure 10.7 Two-wire to four-wire conversion and the problems of signal reflections

on loudspeakers, and the broker shouts his 'bids' and 'offers' down a microphone rather than using a normal telephone handset.

In short, circuit instability is achieved by sticking rigidly to the *signal loudness* levels defined by the network designer's *transmission plan.*

In fully digital networks, with digital telephone handsets (i.e. ISDN telephones rather than two-wire analogue telephones), there is no requirement for the two-wire-to-four-wire convertor which is the major source of circuit instability and signal echo. Even so, there still needs to be a certain amount of signal loss of the audio signal (i.e. volume reduction between microphone and earphone) to prevent the audio echo return path between earphone and microphone (Figure 10.8) from being established.

Signal Distortion

Signal distortion arises most commonly when the signal strength is too great for the dynamic range of the equipment being called on to carry or transmit it. For example, a loudspeaker 'overloaded' by a powerful amplifier causes distortion. In the same way, over-loud signals applied to a telecommunications circuit can also lead to distortion. For voice signals, distortion leads to problems of understanding. For data signals, distortion can lead to the incorrect interpretation of incoming data.

Signal distortion is best avoided by sticking rigidly to the signal loudness transmission plan as discussed above. Even so, some distortion may be unavoidable, due to the 'imperfect' nature of real equipment. In this case, it is sometimes possible to reverse the distortion by means of an *equaliser*. Some radio equipment offers the possibility for such *equalisation*, though in other systems it may not be necessary. The need for it depends upon the quality of the system, the dynamic range (i.e. the bandwidth or bitrate of the carried signal), and the type of radio modulation used. Higher bandwidths or bitrates will tend to require equalisation more than lower bitrates.

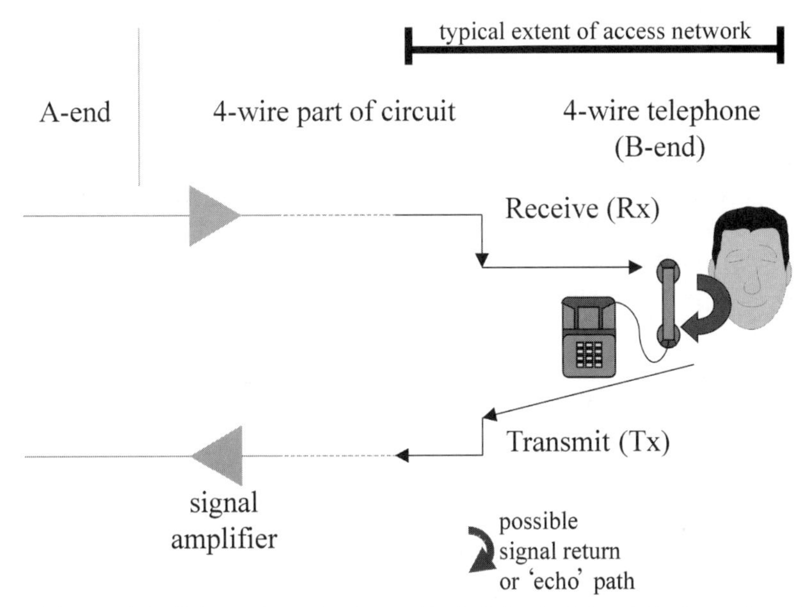

Figure 10.8 Echo and signal return paths in fully digital or fully four-wire networks

Crosstalk and Interference

Crosstalk and *interference* are unwanted signals induced in a given circuit or connection by other adjacent telecommunications lines. As with signal distortion, the prime cause of crosstalk and interference is the presence of an unduly strong signal — this time on an adjacent cable, probably running parallel with the first. There are obviously strong parallels between this type of cable-induced crosstalk and interference and the interference caused by other radio signal sources, and the negative effects are similar — causing signal corruption. The resolution is, however, slightly different. Radio interference, as we have seen, is avoided by careful radio frequency planning, to maintain a tolerable *Carrier-to-Interference Ratio* (*CIR*). Meanwhile, crosstalk and cable interference is minimised by adjusting the signal loudness levels and network losses defined by the *transmission plan*.

Circuit Noise

Circuit noise, like crosstalk, is caused by the induction of signals onto the transmission line or communications circuit by another nearby electric field. In the case of noise, the induction is caused either by nearby power cables or by *background noise* which originates from heavy electrical plant, electric motors or other atmospheric and cosmic sources of strong electric fields.

Noise is best minimised by ensuring that the signal strength at all points along the connection is never allowed to fade to a volume level comparable with the surrounding noise. Thus, a minimum *Signal-to-Noise* (*S/N*) ratio of signal strengths is maintained throughout the connection. This ensures that the wanted signal is still perceptible amongst all the background. If the signal becomes too weak in comparison with the noise, amplification is then unfortunately of little value, because it boosts the wanted signal and noise equally. It is thus difficult to remove noise without affecting the signal itself, though there is some scope for removing noise which lies outside the dynamic range (frequency spectrum) of the signal itslef by simple filtering. This can slightly improve the signal-to-noise ratio (S/N).

Particularly in radio receivers, where the wanted signal is at its weakest, it is important for radio designers to ensure that it is not subjected to unnecessary sources of noise (for example, originating from high power components which themselves are part of the receiver circuitry). Some radio designs include *Low Noise Amplifiers* (*LNAs*), which are specifically designed to amplify the very weak signals received by the antenna while adding the mimimum amount of noise.

As with crosstalk, interference and circuit instability the best 'cure' for noise disturbance is avoidance by good planning and sticking rigidly to the signal loudness levels set out in the transmission plan and the operating limits defined by the radio designer.

Bit Error Ratio (BER) (Digital Circuits)

Most modern telecommunications networks are digital networks. There are three main reasons for this:

- the advent of large scale computer chip development and production has made digital technology very cheap;
- digital networks are better suited than analogue networks for carrying large volumes of

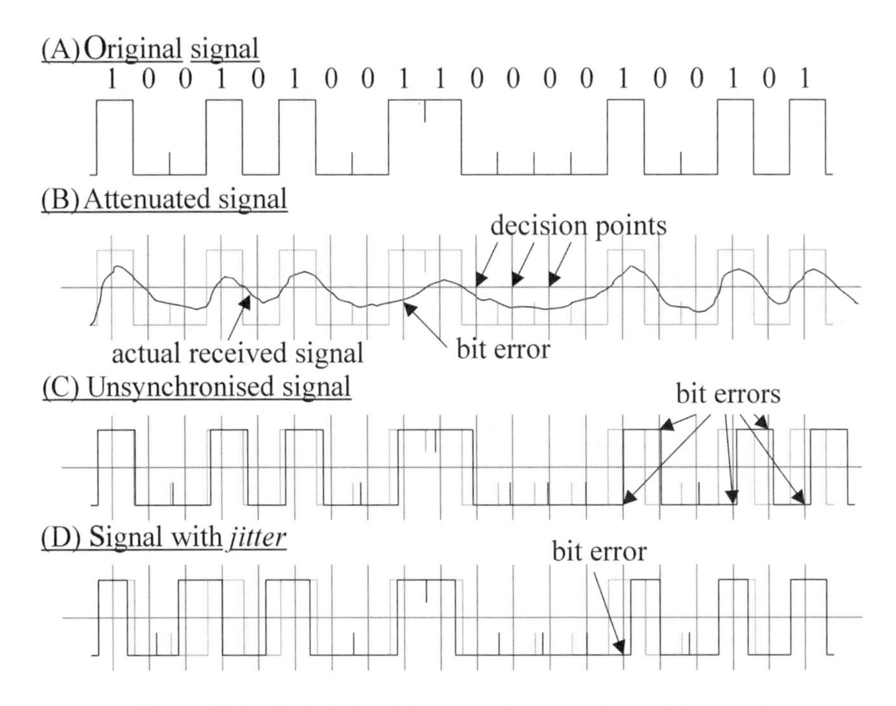

Figure 10.9 Digital signal format and causes of signal degradation

data and computer information, and;

- digital networks paradoxically provide better quality of transmission for voice, audio and other analogue signals, since the distortion caused during transmission can be virtually eliminated.

The manifestation of a poor quality digital transmission circuit is the appearance of *bit errors* in the received signal. Figure 10.9 illustrates a typical digital signal and the various possible main causes of bit errors.

Figure 10.9a illustrates a typical digital signal, comprising a long string of electrical pulses (on/off; voltage/no voltage or different voltage/current values) corresponding to information *bits* of value '0' or '1'. This is the representation of a numerical value expressed in *binary code*. The number of *bits* we require to express a given numerical value depends naturally upon the magnitude of the number, and the accuracy with which we wish to express it. The number of bits we send per second (the *bit rate*) governs how many numbers we can send per second. The numbers we send can represent alphanumeric computer information (according to the ASCII-code, which defines a binary number value to each letter of the alphabet, to each of the numbers and the mathematical signs and punctuation marks. Alternatively, by means of *Pulse Code Modulation* (*PCM*), which we shall discuss later, we can also represent analogue signals in a digital binary form.

The exact information represented in Figure 10.9a is the binary value 10010100110000100101B (the 'B' indicates the number is *binary*). Alternatively, maybe this is five separate 4-bit values 1001B, 0100B, 1100B, 0010B, 0101B (in so-called *hexadecimal* code this would represent the *decimal* values 9, 4, C, 2, 5). The stream could also represent two and a half *bytes* of information (8 bits each). The first bytes would have the *hexadecimal values* 94 and C2. (The *ASCII* computer code (*American Standard Code for Information Interchange*) assigns an 8-bit value to the most commonly used alphanumeric and symbol values. For less commonly used symbols a second *byte* of code (16 bits) is used.)

Now, if any of the bits of the digital signal of Figure 10.9a are corrupted during transmission, then the incorrect value ('1' instead of '0' or '0' instead of '1') may be received. Each incorrect value is called a *bit error*. Since there are only two allowed values of signal which the receiver may detect, it is possible to eliminate some of the signal distortions caused during transmission. Let us assume that the electrical pulses corresponding to bits of information are: 0 volts for value '0' and 5 volts for value '1'. If the receiver detects a value of 3.5 volts, then we could fairly safely assume this was intended to be 5 volts rather than 0 volts, and that the signal pulse had suffered attenuation or distortion during transmission. Due to our digital coding and detection, we are able to *regenerate* the original digital signal exactly.

The capability to regenerate a signal without errors is not unlimited. Once the signal pulses of a digital signal are sufficiently corrupted or distorted, our binary detection means will make the 'wrong' detection. Thus, in our example above, if the signal further deteriorated to 2.4 volts (from the original 5), the detector might decide (based upon a 2.5 volt 'decision criterion') that the original value was '0'. This would be a bit error (see the 9th bit of Figure 10.9b, where such a bit error has led to the conversion of the original values 94, C2 into the received values 94, 42.)

The quality of a digital transmission line is usually quoted in terms of the proportion of received bit errors to the total number of bits transmitted during the same period. This value is called the *Bit Error Ratio* (*BER*). Typical values of BER measured in modern digital networks range from around 10^{-6} (1 error in every 1 million bits sent) to 10^{-12} (1 error in every 1 billion [US 1000 billion] bits sent). For high quality voice telephone communication, values of BER even as low as 10^{-3} are adequate, and 10^{-5} gives quite good quality. However, for most types of modern data communications (including frame relay, internet protocol and ATM (asynchronous transfer mode)) much higher quality of transmission is a pre-condition for reliable and secure transmission. A maximum BER of 10^{-9} is usually set as the target for modern data networks.

So much for the target maximum number of errors which we can tolerate. What are the main causes of bit errors? These are illustrated in Figures 10.9b, 10.9c and 10.9d.

Figure 10.9b illustrates bit errors emanating from excessive attenuation of the signal pulses. In particular, the ninth bit has suffered a bit error because at the point in time where the receiver made its 'decision', the received signal strength was the 'wrong side' of the decision threshold. This cause of bit errors is prevalent when the transmission line distance between the digital signal source and the receiver is too great. Such errors can be reduced by reducing the length of the connection (cable or radio path), or by *regenerating* the signal at an intermediate point and by retransmitting the regenerated signal.

Figure 10.9c illustrates bit errors emanating from a lack of *synchronisation* between the signal transmitter and the signal receiver. (This might be a lack of synchronisation between a radio transmitter and a radio receiver in the same sub-network. Bit errors of this type occur when different sub-networks are not synchronised to one another, e.g. the customer premises equipment sending data into the radio access network is not synchronised with the radio network.) In the case of Figure 10.9c, the transmitter is sending bits at a rate which is slightly slower than the rate at which the receiver is *clocked* to make its decisions. As a result, by the time the 15th bit arrives at the receiver, the detector has already made its (incorrect) decision based upon the value of the tail-end of the 14th bit. This first bit error is subsequently followed by many more. Keeping all the transmitters and receivers of the radio part of the network in *synchronisation*, as well as synchronising the radio sub-network with all the other components and sub-networks making up the network as a whole, is therefore crucial. We return to this subject in more detail later.

Jitter (Digital circuits)

Figure 10.9d illustrates bit errors emanating from *jitter*. Jitter is the result of variations in pulse lengths; these occur either when a particular pulse (or part of a pulse) takes longer to process in the digital circuitry than surrounding pulses or when the pulse propagates through the network faster or slower than surrounding pulses. The effects of jitter may also cause *wander* over a longer period of time. Signal jitter has lead to an error in the detection of the 15th bit of the pattern in Figure 10.9d.

Quantisation Distortion (Digital Circuits)

The effect of bit errors is to cause direct corruption of data signals or *quantisation distortion* of voice or other digitally-coded analogue signals. Figure 10.10 will help us to understand quantisation distortion by explaining how we digitally-code analogue signals. The coding takes place by means of *Pulse Code Modulation* (*PCM*).

Figure 10.10 illustrates the conversion of an analogue signal into a digital one by means of *Pulse Code Modulation* (*PCM*). Figure 10.10a shows the original analogue signal and

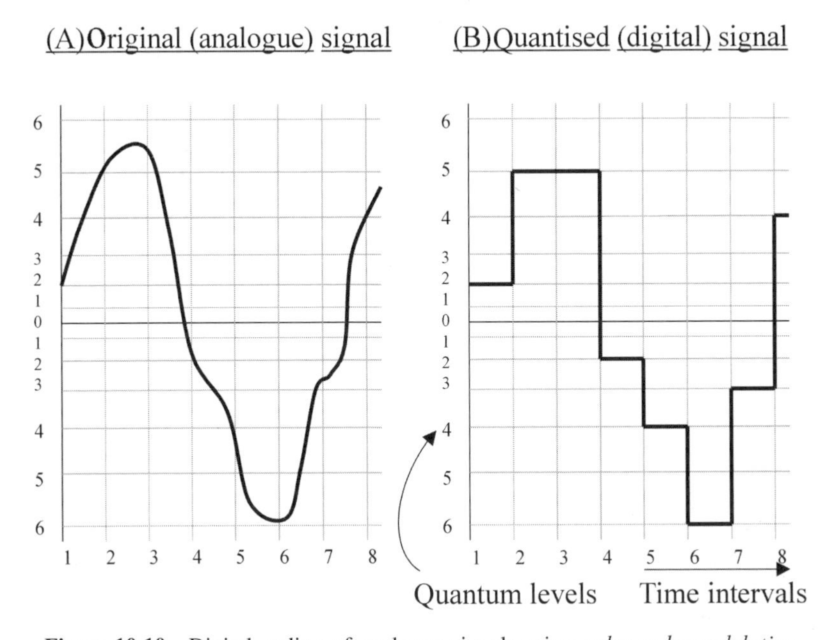

Figure 10.10 Digital coding of analogue signals using *pulse code modulation*

the 'framework' of signal amplitude levels (quantum levels) and intervals in time which determine the digital signal code values. At each regular interval in time (typically each 125 μs, equivalent to a *sampling rate* of 8000 Hz), the signal amplitude value is compared against the pre-determined quantum levels. The nearest *quantum level* value is taken to have the same amplitude as the original analogue signal. This quantum level has a numerical (or *digital*) value. By transmitting only the digital values for each interval in time, we are able to reconstruct a close approximation of the original signal at the receiving end (Figure 10.10b). Of course, the reproduced signal is rather 'square' in nature, and does not exactly match the original. The difference between the original and the 'square' reproduction is termed *quantisation noise*.

The more quantum levels that are used to approximate the original analogue signal value and the more frequently we sample the signal, so the less inaccurate is the reproduced signal. Thus, for example, it is normal to use 256 quantum level (corresponding to an 8-bit binary value per sample and quantum values between -127 and $+128$) and a sampling rate of 8000 Hz for a standard telephone channel. Thus, the standard telephone channel bit rate of 8000×8 bits per second, or 64 kbit/s. High fidelity music stored on Compact Disc (CD), on the other hand, uses 32-bit sampling. (As an aside, you may notice that the quantum levels at low signal amplitudes are more closely spaced. This is done on purpose, since it has been found that this improves the human perception of the reproduced signal.)

Network transmission plans specify limits on the additional quantisation noise which may be added by a given portion of the network, or sub-network. This is quoted as a value defined in *quantisation distortion units* (*qdu*). One qdu is equivalent to a distortion of the digital signal of one unit value caused during transmission.

Quantisation noise is clearly generated at the point in the network where the original signal is converted from analogue to digital (e.g. in a digital telephone handset or at an analogue-to-digital two-wire-to-four-wire convertor). However, further quantisation noise can also be introduced during the path through the network. There are three main causes:

- Transmission bit errors (as discussed previously: while it is normal for data transmission to talk of the BER, for voice or audio transmission engineers refer instead to the qdu value of the connection).
- Signal *compression* (for example, the compression of the signal to a lower bit rate by means of *ADPCM* (*Adaptive Differential Pulse Code Modulation*) or *CELP* (*Code Excited Linear Prediction*)). Compression reduces the network backbone capacity (i.e. the bit rate required to carry a signal) at the cost of increased quantisation noise.
- Other forms of signal processing (for example, *echo cancellation*, as we shall discuss later).

Maximum Line Lengths

Figure 10.11 illustrates the different network interfaces relevant to wireless access networks and equipment, including those for connecting a radio access network to a backbone switching or transmission backhaul network, as well as those for connecting customer premises equipment (sometimes these different interfaces are referred to as network interfaces (including *Network-Network Interfaces* (*NNI*) and *Network-Node-Interfaces* (also *NNI*) and *User-Network-Interfaces* (*UNI*)).

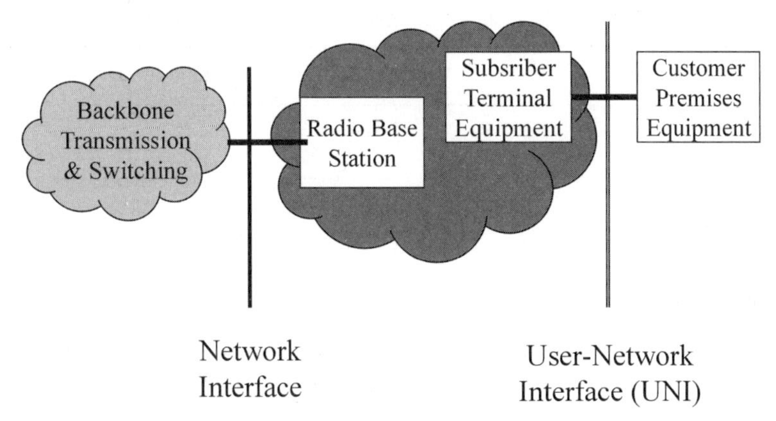

Figure 10.11 User and network interfaces to radio access (sub-)networks

All types of interfaces are subject to cable length limitations. These limitations arise, amongst other things, from the defined electrical power to be used when transmitting across the interface, upon the type of cabling defined and upon the bit rate or bandwidth of the connection. Generally, however, there are a small number of different categories of equipment, for which similar limitations apply. We summarise these as a rough planning guide in Table 10.1.

Non-conformance with line interface specification will almost certainly lead to problems.

Another thing to note, especially in customer premises installations, is that it is becoming standard to install *structured cabling* systems in office buildings. These are intended to provide for multi-purpose wiring for all possible forseeable telecommunications services which could be required in the building. By so doing, the building occupier is spared the annoyance and expense of permanently having to install new cables each time a new telephone extension or local area network port for a personal computer is needed. The common types of cabling used are *shielded* (*STP*) and *unshielded twisted pair* (*UTP*) cable, either *Category 5* (*Cat5*), *Category 6* (*Cat6*) or, most recently, even *Category 7* (*Cat7*). The common patch panels and patch cables use RJ-45 plugs and sockets. RJ-45 is an 8-pin rectangular connector. Beware: just because the cable and the plug from the wireless *terminal station* fits into the building structured cabling of the customer does not mean that the interfaces on plug and socket are compatible with one another! We return to this subject later.

Sidetone

Sidetone is the name given to the signal which you hear in the earpiece as you speak into the microphone of a telephone. The signal is usually a weak ‘return’ signal. In effect, you are listening in the earpiece to what you are saying in the microphone. Since it is normal, even when not telephoning, to be able to hear yourself as you speak, so the human user of a telephone *expects* to get sidetone when speaking on the phone. Without any sidetone, the talker gets the impression that the phone is ‘dead’. With too little sidetone, the talker raises his voice, so that the volume in his ear is brought up to the level he is used to hearing as he

Table 10.1 Line length limitations of line interfaces

Physical interface type	Line length limitation
Network Interfaces	
G.703 (64 kbit/s, 1.5 Mbit/s, 2 Mbit/s, 34 Mbit/s)	Several kilometres before regeneration, coaxial cable version (75 ohm) has slightly greater range than twisted pair version (120 ohm)
SDH (synchronous digital hiearchy) STM-1 (155 Mbit/s) electrical	Generally, the electrical interface is only used as the 'local' interface to connect to optical line termination equipment located in the same location as the radio base station
SDH (synchronous digital hiearchy) STM-1 (155 Mbit/s) optical	The optical interface will be either *multimode*, *metropolitan monomode* or *monomode*. The longest range (many kilometres) is achieved with monomode cable. Multimode and metropolitan monomode interfaces are for metropolitan networks, up to about 15 km
User-Network Interfaces	
G.703 (64 kbit/s, 1.5 Mbit/s, 2 Mbit/s)	Several kilometres
X.21 (commonly used for n*64 kbit/s, leaselines)	Around 100 m
V.24 (serial interface to data equipment with 25-pin D-plug)	Around 15 m
Ethernet 10 base T	Around 100 m twisted pair cable
Analogue telephone pair (a/b)	600 ohm cable impedance – several kilometres with fairly narrow guage twisted pair cable

talks. With too much sidetone, the talker lowers his voice, so as not to be distracted (at which point the listener complains that he cannot hear, of course).

Achieving the right amount of sidetone is really a matter for telephone handset and radio *terminal station* designers, though this level could be influenced by signal return paths introduced in network terminating equipment (e.g. a subscriber radio terminal) which is connected to the telephone.

Propagation Delay and Echo

Propagation delay and echo, as we discussed in conjunction with Figure 10.7, can lead to severe line quality and communication problems for both voice and data communications. In any network where the one way propagation delay is more than about 30 ms, it is advisable to make provision for of *echo cancellation* measures. (Such measures may be built-in to a wireless access network by the equipment designer, but when not, may have to be added by means of additional 'external devices' afterwards.) Echo cancellation removes the adverse quality effects of echo. The problem is that the echo is practically unavoidable: there is bound to be some sort of reflection back from the listener's end to the talker. Provided the echo is not delayed, i.e. the network *loop delay* (propagation time for two passages across the network, there and back) is not greater than 60 ms, then the echo is perceived as sidetone (see above). Experience shows that longer loop delays are perceived by humans as a line degradation. The talker gets the impression he is 'talking in a pipe' or 'talking in a hall'. At even greater propagation delays (e.g. 40 ms one-way), the talker becomes aware of what common language calls an 'echo'.

10.4 Network Interfaces — To the Backbone Network and Customer Premises Equipment

For two pieces of telecommunications equipment to interwork properly with one another (e.g. the wireless access sub-network with the backbone network, or the wireless sub-network with the end-users equipment), they must be connected to one another by means of a compatible interface. And just because the plug fits the socket does not mean the interface is the same one! To define the interface requires more than just the definition of the physical connector. Telecommunications standards have historically defined two different classes of interfaces: those used in public telephone networks, either for voice connections or for *leaselines*; and those used for data communications.

In general, leaseline and voice telephony interfaces are designed for the 'transparent' carriage of the signal. Thus, the same signal appears at the output end of a connection as went in at the input. Between input and output can be thought of as being an empty 'pipe'. The signal goes in, takes a little while (at its own speed) to propagate through the pipe, and then emerges at the other end, unchanged other than by signal attenuation, distortion, etc.

In contrast, data signals are more rigidly monitored and controlled during transmission by data *protocols*. The protocol has a function a bit like a human language: it checks that the 'listening' computer is ready to receive the signal before the 'talking' computer is allowed to 'speak'. Afterwards it checks the 'listening' computer 'heard' the message and understood it, otherwise the 'talker' is asked to repeat it. Data protocols are defined in detail by international standards, which have to set out the entire 'etiquette' of computer conversation.

In addition to the human language, audio signal format or data protocol used for communication *during* the 'conversation phase' of a connection, a further protocol or *signalling* mechanism is used during the *call set-up* phase to allow the caller to indicate to the network the telephone number or other network address information (e.g. Internet address) to which a connection should be established by the network (Figure 10.12).

We shall not describe in detail all the various signalling and data protocols and the reasons for them here. However, to appreciate the most common interfaces in radio access networks, it is worthwhile briefly to explain that most modern *signalling stacks* or *protocol stacks* (the definitions of the various procedures going to make up the interface as a whole) are arranged in *layers*. Each layer has a specific function within the interface, as Figure 10.13 illustrates.

Figure 10.13 illustrates an office data communications *Local Area Network* (*LAN*) connected to a *wide area* public telecommunications network, and the various interfaces between the various pieces of equipment, including the *layer* functions undertaken by each.

The computer itself has a communications *protocol stack* which comprises all three of the protocol layers (*physical layer*, *link layer* and *network layer*). Meanwhile, the LAN wiring and hub only participate at the physical layer. The *LAN bridge* comprises *physical* and *link* layers, while the *router*, like the computer, contains all three layers. So what are these layers and what do they do?

The physical layer specification defines the type of medium used for the transport of the communications message (e.g. twisted pair cable, coaxial cable, fibre optic cable, etc.). The specification includes the electrical characteristics of the cable (e.g. allowed impedance, etc.), as well as describing the voltages, currents and signal shape (the so-called *line coding*) that shall be used to represent the 1's and 0's of the digital signal. The physical layer specification may also define the actual connectors (plugs and sockets) that should be used to connect cables to the equipment (but perhaps surprisingly, the connector is not always standardised). The physical layer thus provides for carriage of the bit stream.

The *link layer* defines a set of rules for passing information along one link of a multi-hop data connection. The rules define when the 'talker' can talk and how the 'listener' can confirm that he has heard, or ask for a repeat. In the case of a LAN, the link layer also contains 'address information' (like the address on an envelope), which ensures that the information carried over the 'shared' medium gets to the intended destination computer or other device in the same LAN. This *link address*, however, is not sufficient to define a remote location, only reached by a *wide area network*. For this purpose the *network layer protocol* and a corresponding *network address* is used. The address (the desired destination of the information sent by the computer) is attached to the information by the computer and interpreted by the router. Thus, these two devices include *network layer* protocol functions,

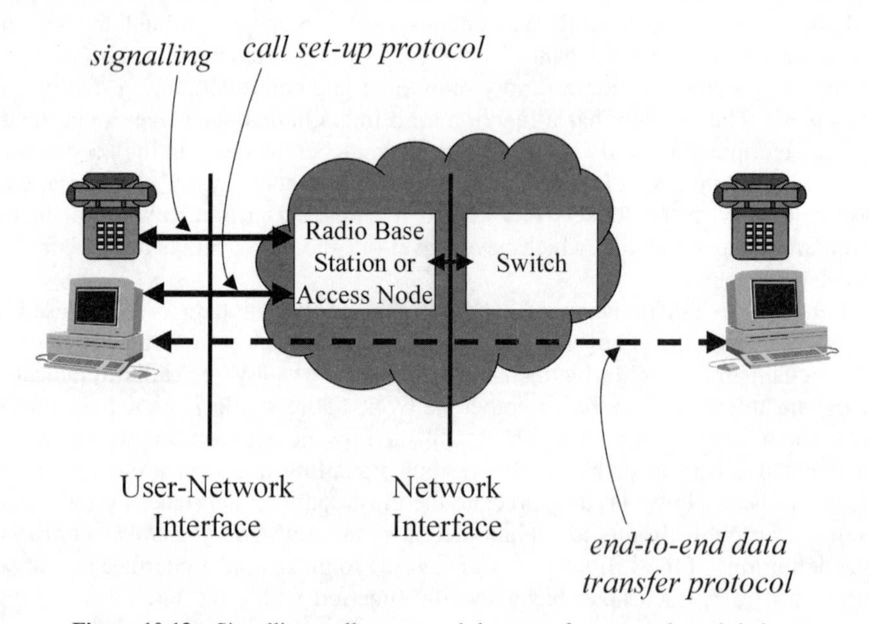

Figure 10.12 Signalling, call set-up and data transfer protocols and their purpose

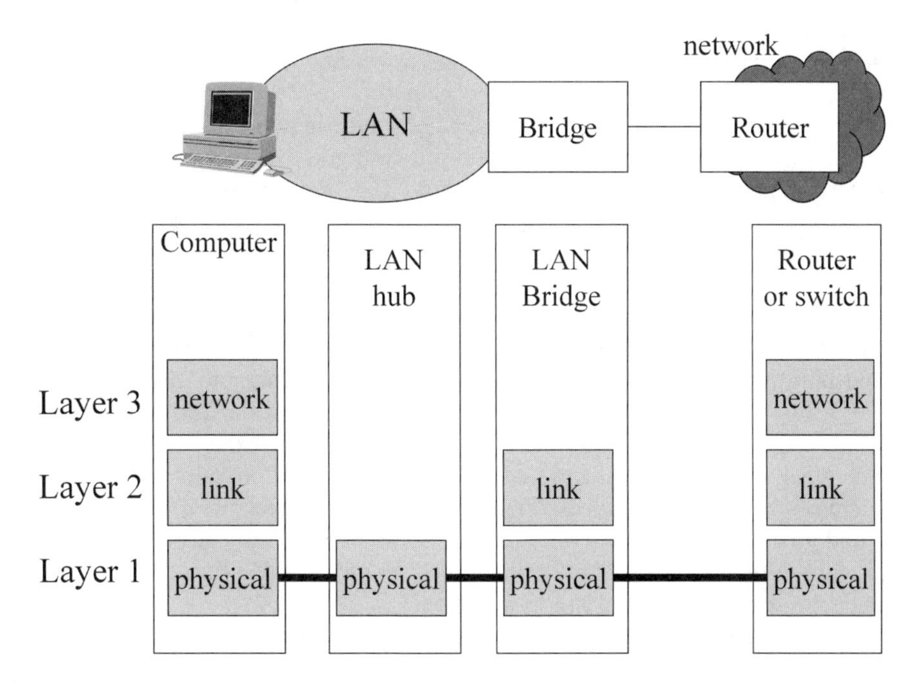

Figure 10.13 The various layers of communications interface

while the other devices in between simply carry this information without interpretation or alteration.

By standardising interfaces in layers as we describe above, the job of equipment design and manufacture is simplified, since similar components can be used in each of the for the realisation of the individual layers. The same transmitters and receivers used at the physical layer, for example, can be used. In some cases, these devices are in effect connected back-to-back, purely for the purpose of relaying the signal (as at the LAN Hub), while in other cases (at the router) the signal must be decoded for a decision about which route the information should be sent on when leaving the router (hence the term *router*: the network protocol and network address is interpreted to decide upon the relevant route (next hop) in the connection).

The problem with layered interfaces is that in many cases, a string of various interface definitions with references like X.21, V.24, V.28, G.703, PRI, E1 are required to define the interface fully. This confuses most people, including many experienced telecommunications engineers.

To try to ease the confusion (or maybe add to it!), Figures 10.14–10.16 illustrate the most common types of interfaces which may occur in radio access networks. It will be clear from the charts that defining an interface as 'RJ-45' (this is just a connector-type) will not help to determine whether the protocols running on the interface are intended to support an 'E1 leaseline', a BRI (Basic Rate ISDN) line, or something else. Similarly, defining the interface as 'frame relay' does not tell us what electrical interface or connector interface we will need. Even for STM-1, we have to specify whether the interface is 'electrical' or 'optical'. The optical interface must be specified in terms of the particular type of optical fibre and wavelength being used (there are several: G.652, G.653, G.654, etc.), and there are various different alternative optical cable connectors (e.g. SC-connector or ST-connector).

Leaseline interfaces (Figure 10.14) generally carry information 'transparently' from one point to another. Leaselines are generally pre-configured lines, which permanently interconnect to specific devices or network ports. The 'pipe' provided by a leaseline is permanently available and cannot be switched on demand to an alternative address. As a result, no network protocol or signalling is required to indicate call set-up information (including the destination address) to the network.

There are various types of analog and digital voice telephone port interfaces (Figure 10.15) used to connect telephone handsets to public telephone networks. The form of the socket and plug connectors used usually varies from one country to another, as can the number of wires actually used in the socket (so carrying a phone from one country and trying to use it in another does not always work — the bell, for example, does not always ring!). The V5.1 and V5.2 interfaces, as we discussed in earlier chapters, are interfaces used for *concentration* of interfaces. This interface is only used as a network interface between and an *Access Network* (*AN*) (e.g. a radio base station) and a *Local Exchange* (*LE*) (or central office switch). The V.x interfaces are only used in Europe. The equivalent north American interface is the TR-303 interface.

Data network interfaces are the most complicated of all, since several different 'leads' in the connection socket are needed to allow the communicating computer and the network to coordinate and control one another, so that they can communicate in an orderly manner.

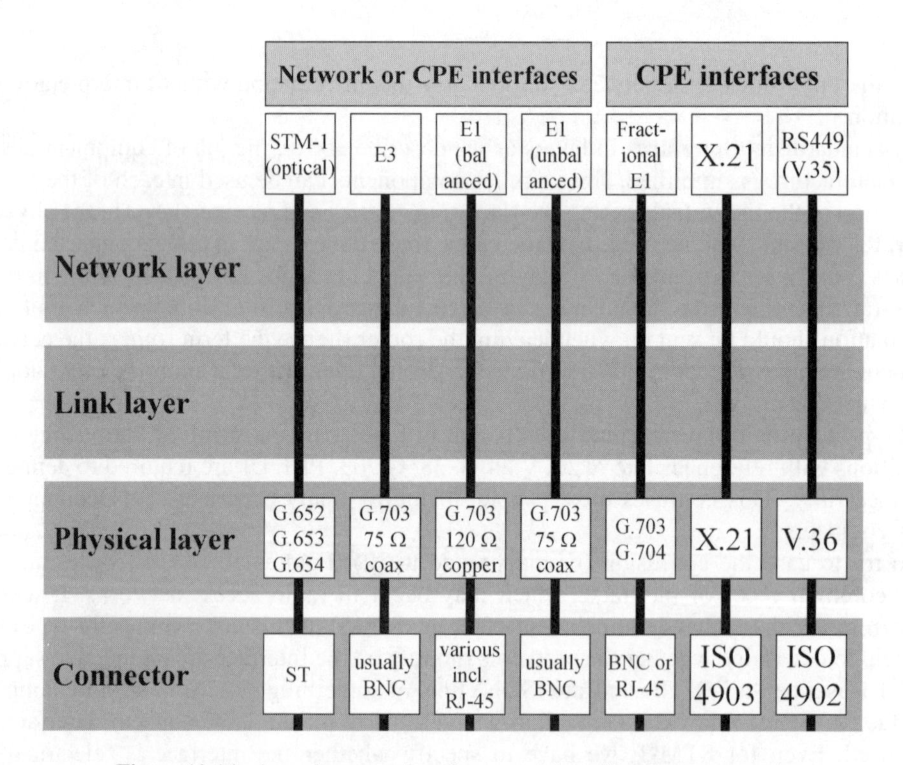

Figure 10.14 Leaseline-type interfaces used in radio acccess networks

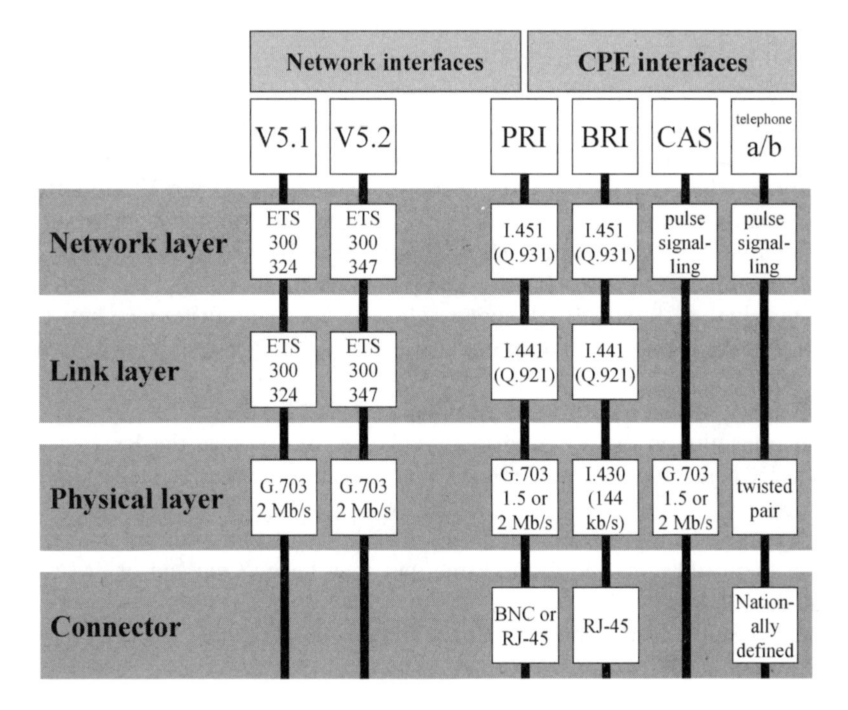

Figure 10.15 Voice telephone interfaces used in radio access networks

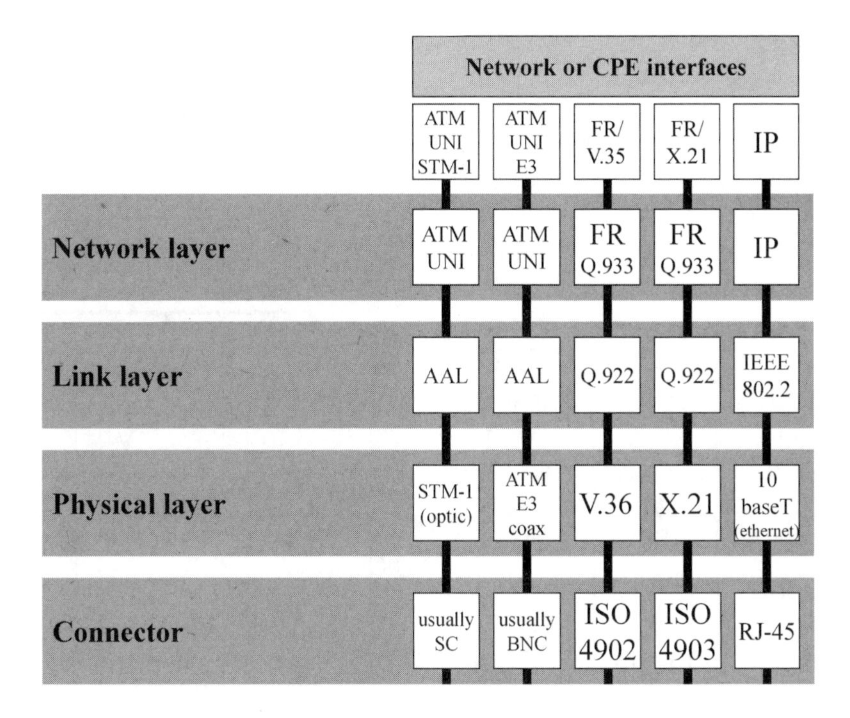

Figure 10.16 Data networking interfaces used in radio access networks

These leads include leads like *DTR* (*Data Terminal Ready*), busy, reset, acknowledge, etc. This requires a multiple pin socket, and there are a number of further variations here, as Figure 10.17 shows. Thus, for example, we see from Figure 10.17 that the common use of the term 'V.24-interface' is not very clear . . . is this intended to mean V.24/V.28 with a 25 pin D-plug (ISO 2110) or a V.35 or V.36 interface? Worse still, many people refer to a 'V.35 interface' when they actually mean the more common 37-pin V.36 interface.

In the case of data network interfaces, it is normal to distinguish between the *Data Terminal Equipment* (*DTE*) and the *Data Circuit Termination Equipment* (*DCE*). These correspond to the customer premises equipment (e.g. his computer) and the radio subscriber terminal, as Figure 10.18 shows. There is not much difference between a DTE and a DCE, but the difference is important to establish for the following reasons:

- the DTE synchronises itself to the clocking provided by the DCE;
- the DTE transmits on the Tx (transmit) leads and receives on the Rx (receive) leads, while the DCE receives on the Tx leads and transmits on the Rx leads.

If you wrongly configure the DTE and DCE, or if you use the wrong cable to interconnect them (they may look the same, but may be *null modem* or *crossed* cables), then things do not work!

With this, we move to the subject of network synchronisation.

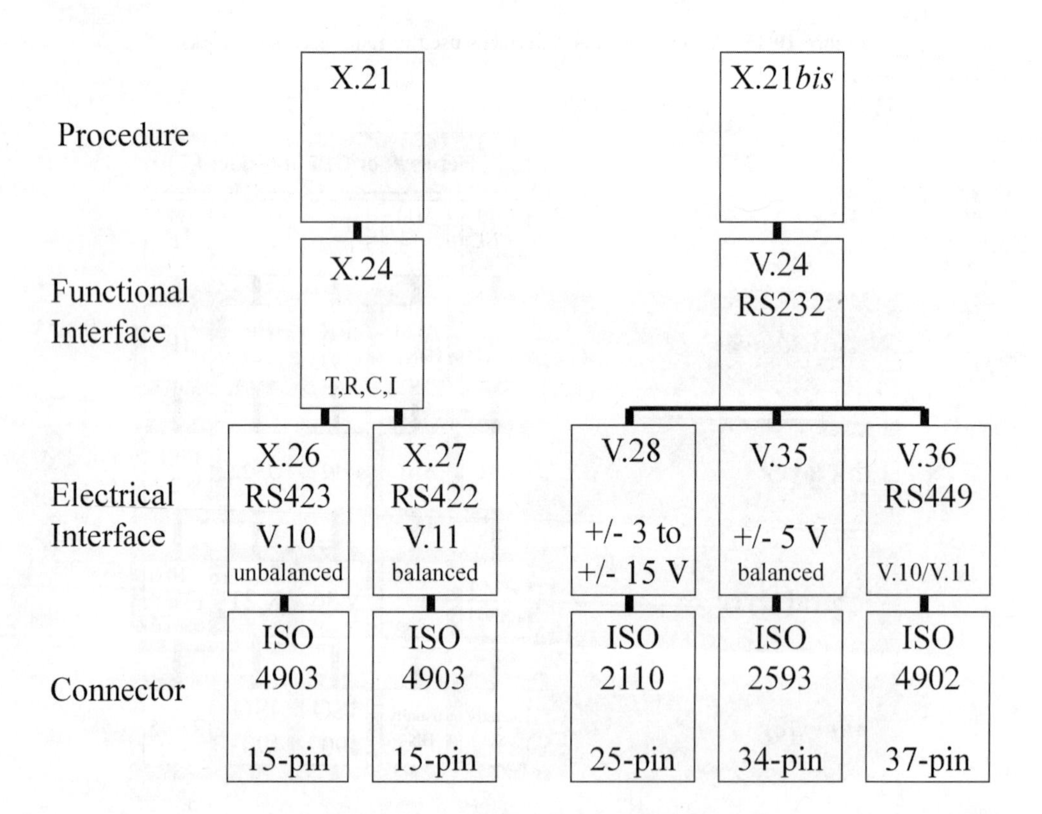

Figure 10.17 Physical data network interfaces

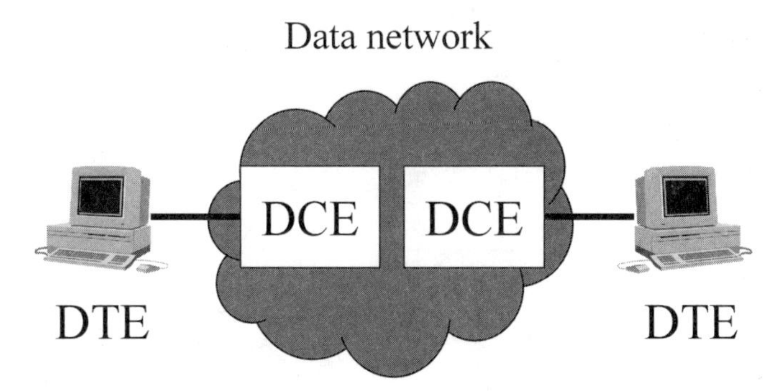

Figure 10.18 Data Terminal Equipment (DTE) and Data Circuit Termination Equipment (DCE)

10.5 Network Synchronisation

Without proper network *synchronisation* of all the components in a network (including backbone and access network components, as well as end-user equipment) the communications carried by a digital network are subjected to quality degradation by bit errors. This we learned in Figure 10.9. As a result, it is normal to design and build networks according to a strict *network synchronisation plan.*

The purpose of synchronisation is to remove all short, medium and long term digital signal degradation effects caused by differences in bitrates of transmitters and receivers in the network by keeping them all in step (synchronised). By so doing, we eliminate the *jitter* or *wander* which migh otherwise occur. In the very short term, synchronisation between transmitter and receiver takes place at a bit level, by *bit synchronisation*, which keeps the transmitting and receiving clocks in step, so that bits start and stop at the expected moments. In the medium term timeframe it is also necessary to ensure *character* or *word synchronisation*, which prevents any confusion between the last few bits of one character and the first few bits of the next. If we interpret the bits wrongly, we end up with the wrong characters. Finally there is *frame synchronisation*, which ensures data reliability (or *integrity*) over even longer time periods.

Figure 10.19 illustrates a strictly hierarchical *network synchronisation* plan. The plan shows the use of a highly accurate *master clock* to provide for accurate synchronisation of the entire network. As a back-up (to protect against failure of the master clock or isolation of the switch to which it is connected), a *back-up clock* is also in use. The clocking signal is distributed to all the switches and other equipment in the network by means of the normal trunks which interconnect them.

Each device in Figure 10.19 has a 'first choice' and 'second choice' of clock source. The first choice source (denoted by 1) is used as long as it is available. This derives primarily from the external *master clock*. Should the 'first choice' trunk fail, then the device is instead synchronised to the 'second choice' trunk clock source (2). Overall, the scheme is arranged so that the master clock source will tend to prevail, so long as it does not become isolated from the rest of the network. Should it become isolated, the back-up clock source will take over. In our example, the back-up clock source is provided by the internal clock of another switch.

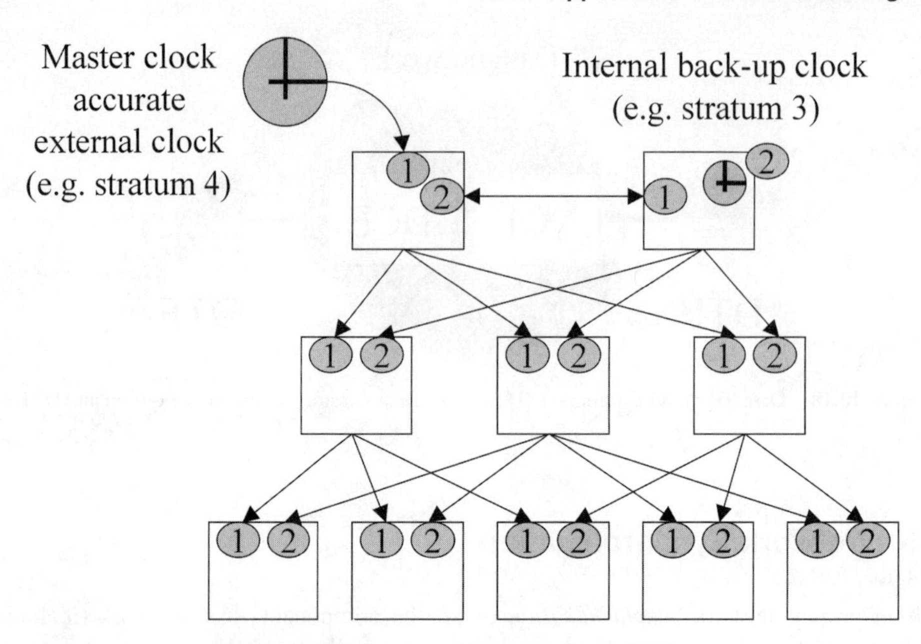

Figure 10.19 Typical network synchronisation plan

The accuracy of a clock is defined in terms of its *stratum level*. The most accurate clocks are *stratum 4* clocks. Such clocks are either caesium clocks, or are derived from national public agencies, from the *GPS* (*Global Positioning System*) satellite system, or from another public operator's network. (Indeed, it is sometimes wise to use the clock of the dominant operator in a particular country, so that problems do not arise for traffic passing between the two networks due to lack of synchronisation.) Most internal equipment clocks (when provided) are of *stratum 3* quality. A clock of stratum 1 or 2 quality is not generally adequate for providing network-wide synchronisation.

As with most network synchronisation plans, the network of Figure 10.19 is designed in a strictly hierarchical fashion. Devices at each 'layer' of the network have roughly equivalent quality of clocking. No single trunk, switch or device failure will cause the loss of network synchronisation.

In contrast to Figure 10.19, in which all the switches partake in a *synchronous* network (such as that provided by an *SDH* [*Synchronous Digital Hierarchy*] network) a modern *ATM* (*Asynchronous Transfer Mode*) network may be synchronised in a number of ways:

- by distribution of the SDH (trunkline) clock directly from one trunk to another;
- by means of *SRTS* (*Synchronous Residual Time Stamp*), or
- by *adaptive synchronisation.*

Generally, synchronisation in *asynchronous* networks is more difficult than in *synchronous* networks, but equally important, particularly when attempting to use such networks for carriage of *Constant Bit Rate* (*CBR*) type signals (i.e. synchronous signals) such as voice and video. The best advice is to stick to a single synchronisation method throughout the entire ATM network.

In Radio networks it is also advisable to clock customer end-user devices from the network, with the radio terminal acting as DCE (Figure 10.18) and the CPE (customer premises equipment) as DTE. Lack of synchronisation manifests as *bit slips* and *bit errors*.

10.6 Network Services, Signalling and Protocols

The most efficient networks are optimised for the carriage of the particular services, signalling systems and data protocols which they are designed to carry. The interfacing considerations we considered earlier in the chapter. Without the right interfaces, the services will not work at all. Another consideration in the network design needs to be the efficiency of the network as a whole. Considerable cost savings in the *backhaul network* (that connecting the radio base station or *Access Network* (*AN*) to the backbone switch) are possible as the result of a *concentration* function and interface, as we discussed in Figure 1.5 of Chapter 1. Such an interface for voice telephone traffic is provided by the V5.2 interface (see Figure 10.15). For data and *multimedia* networks, an equivalent concentration could be provided by means of an ATM, IP or other data trunk interface (e.g. ATM UNI of Figure 10.16).

10.7 Network Propagation Delays

Unduly long propagation delays encountered by a signal as it traverses a network can lead to problems of echo and to *bit errors* as we discussed earlier. Such propagation delays are caused either by very long connections (signals travel through a telecommunications network at about one-third the speed of light, or 10 ms for each 1000 km), by signal processing (for example, due to speech compression or low rate encoding) or due to protocol delays.

As far as possible, the network should be designed to minimise propagation delays, but there is always a residual unavoidable delay. This has to be taken into account when designing applications to run over the network, and extra equipment, such as echo cancellors, may be necessary to overcome possible signal quality degradation.

Most network transmission plans allocate a potion of overall allowed end-to-end propagation delay to each sub-network making up a possible end-to-end connection. However, even though a given sub-network may conform with its own allocated target value, this does not necessarily relieve that sub-network of a role in compensating end-to-end signal degradations.

Where the one-way propagation delay in a voice telephone network exceeds about 30 ms, it is advisable to use echo cancellors to eliminate the echo which distracts talkers. Such echo cancellers, when used, are best located at the nearest possible locations nearest to the two ends of the connection (i.e. in the access network). So, even though a radio access network might have kept within the recommended ETSI propagation delay for radio access networks of 10 ms one-way propagation, it might still be necessary to use echo cancellers in that radio access network to overcome the effects of an overall end-to-end propagation delay of more than 30 ms for the connection as a whole.

In the case of data communications, long propagation delays can have detrimental effects which may not immediately be obvious. I have myself experienced computer software applications which like to have each command acknowledged before the next command is issued. This is all very well where the propagation delay is quite short, but very time consuming. Imagine a sentence in which you wait for the listener to confirm he heard each word in turn . . .

> ''**Hello**'' . . . (wait for message to reach listener) . . . [listener acknowledges] . . . (wait for the acknowledgement) . . . ''**how**'' . . . (wait for message to reach listener) . . . [listener acknowledges] . . . (wait for the acknowledgement) . . . ''**are**'' . . . (wait for message to reach listener) . . . [listener acknowledges] . . . (wait for the acknowledgement) . . .'' **you**'' . . . (wait for message to reach listener) . . . [listener acknowledges] . . . (wait for the acknowledgement).

Well if that wasn't long enough and painful enough, imagine now that the 'talking' software has an extra 'feature' which repeats words if it decides, after a 'timeout' period, that the message must have got lost on the way. Consider what chaos results if the 'timeout' is set to a length of time less than that required to receive the acknowledgement. The 'talking' software keeps on saying ''hello'', and thus never completes the message. In data communications, the timeout values and other configuration parameters are crucial to good network performance!

10.8 Network Management and Monitoring

The capability to monitor and control the network, and in particular the subscriber terminals remotely is very important to good service. This relies upon the availability of good network management software, and an adequate control and monitoring capability of the equipment. An extensive *Management Information Base* (*MIB*) and standardised network management protocols (such as SNMP or CMIP) are valuable assets in this regard.

As well as requiring monitoring capabilities for performance monitoring of the network, for fault diagnosis and repair, it may also be necessary to provide monitoring capabilities for state security agencies. In Germany, as in some other European countries, it is a legal requirement of all licensed public telecommunications network operators that they should provide facilities which enable the national security agencies (e.g. police) to 'tap' telephone calls or other types of communications without either the network operator or the user in question being aware of the 'tapping' activity. This may require special provisions in either the radio access network equipment or in the telephone or data switching equipment.

10.9 Service Accounting

Naturally, most public network operators are keen to be paid by their customers for the communications services they provide. The ability to charge for services on a usage basis (either per minute of 'conversation time' or per megabyte of data information carried) relies upon the collection of accounting information recording the amount of use. Usually, such *accounting records* (or *Call Data Records*, *CDRs*) are collected in voice or data switches. Where the radio base station part of an access network performs only a multiplexing or concentration function, it is not necessary to collect the accounting records in the radio base station. However, as soon as local switching between one radio network

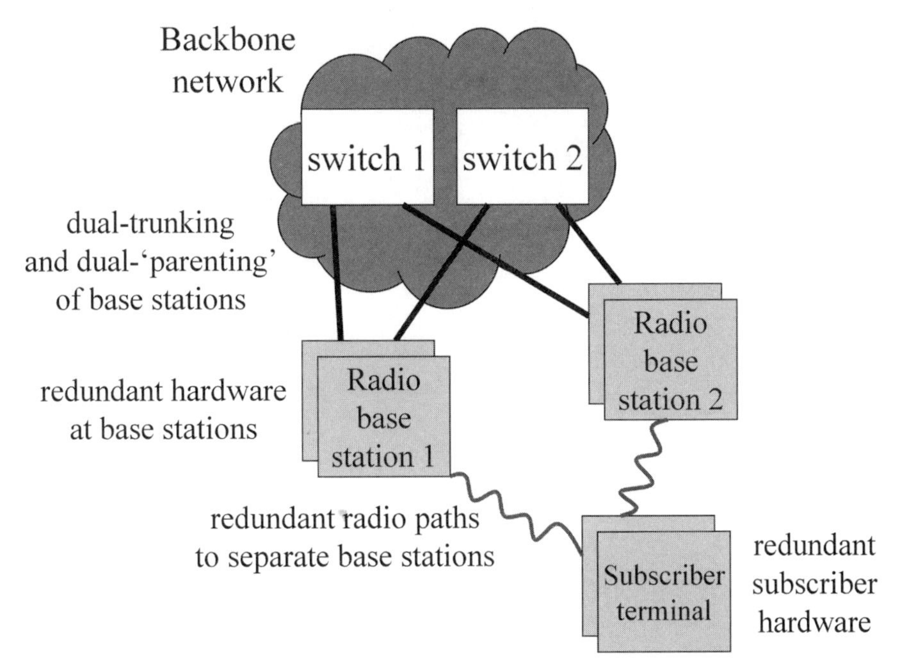

Figure 10.20 Redundant network configuration

subscriber and another is possible without having to leave the base station, then the base station too will have to collect such accounting records.

10.10 Building Networks Robust against Failure

Many network operators like to try to improve the overall resilience and availability of their networks by designing redundant hardware and a redundant topology to connect the base station to the backbone. As in fixed-wired networks, such redundancy greatly reduces the likelihood of the base station going 'off-air', or of becoming isolated from the rest of the network (Figure 10.20). Unfortunately, however, mere redundant hardware at the subscriber terminal is not an effective safeguard against the inevitable unavailability caused by rain fading (Chapter 7). For really good connection availability there is no alternative but to use 'double-homing' of subscriber terminals onto two or more physically diverse base stations (Figure 10.20).

It may only be possible to reach a second 'back-up' base station by means of a much longer radio link than that used for the normal 'home' base station. In this case, as we discussed in Chapter 7, an antenna with a higher gain may need to be used.

11

Wireless Local Loop (WLL)

The term *Wireless Local Loop* (WLL) (as opposed to broadband wireless access) has come to be synonomous with radio access networks designed for narrowband (i.e. low bit rate or narrow bandwidth) applications, in particular as a means of *Plain Old Telephone Service (POTS)* network access, or for basic rate *ISDN (Integrated Services Digital Network)*. In this chapter, we present the main characteristics of WLL systems, and review some of the commercially available products. We also review the architectural structure of a typical standardised system (using DECT as an example), and discuss the use of mobile telephone network technology for the provision of fixed wireless access.

11.1 The Use and Characteristics of Wireless Local Loop Systems

Wireless Local Loop (WLL) systems tend to be used for one of two reasons:

- By 'monopoly' or dominant public telephone network operators in regions with poor cable infrastructure as the most economic means of providing a basic telephone service.
- By new competing operators in 'deregulated' markets who are keen to build their own access network infrastructure rather than have to rely on the resources and services of their main competitor (the ex-monopoly carrier).

WLL systems for low speed telephony applications typically are designed to operate in the radio range 1–3 GHz. In particular, there are a number of point-to-multipoint systems which work in the 2.1 GHz, 2.2 GHz and 3.5 GHz (3.4–4.2 GHz) bands. Some use has also been made of the cordless telephone technology (including telepoint technology and DECT — Digital European Cordless Telephony) for providing a 'limited mobility' form of local telephone network access. In addition, there are a range of other point-to-point type systems in unlicensed bands which have been used by some operators.

The point-to-multipoint systems can be classified into one of two broad types:

- Systems including concentration functions, which connect multiple remote telephones by means of a radio link and a base station to a single V5.2 trunk (Figure 11.1a).
- Systems with no concentration function (it might be more appropriate to call such systems *Multiple-Point-to-Point (MPP)* systems (Figure 11.1b)).

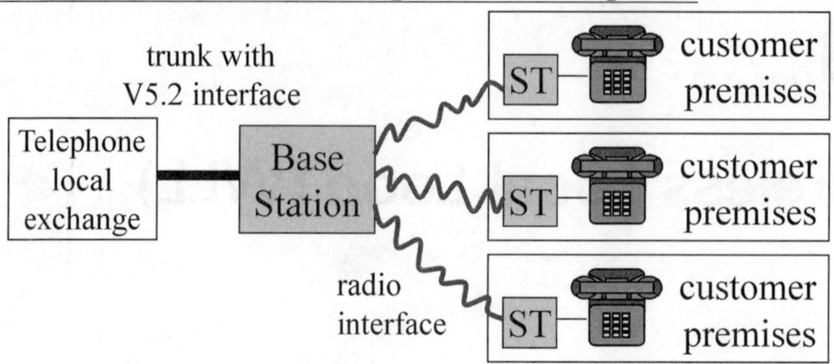

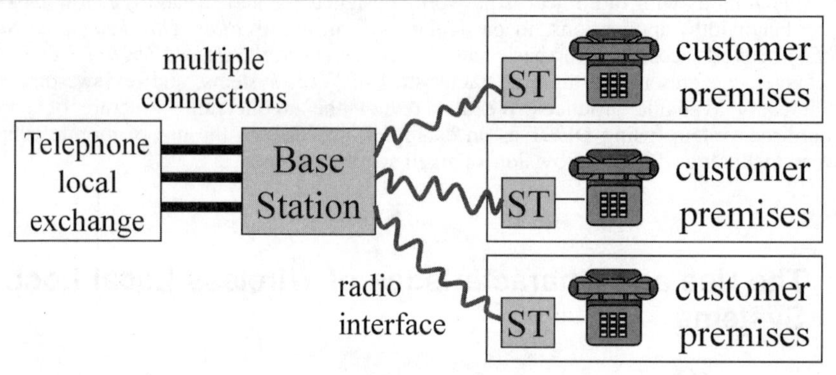

Figure 11.1 Point-to-multipoint and multiple-point-to-point architectures of telephone WLL systems

The advantage of using a concentrating interface such as V5.2 is that fewer ports are required to connect the base station to the remote switch, and less *backhaul* capacity is required on the trunk interconnecting the two. This can save a significant amount of cost. These interfaces are defined by ETSI (European Telecommunications Standards Institute) as *access network interfaces*. The equivalent north American interface is TR-303.

11.2 Access Network Interfaces

The emergence of new technology (including fibre as well as radio) in the access network between customer premises and the exchange site has brought with it new problems and opportunities. The problems arise from the need to devote effort to standardisation of new interfaces, the opportunity is the new service functionality thereby made possible, together with the scope for network restructuring and cost optimisation.

Two types of interface have been defined by standardisation work on transmission technology for the access network. These are *Local Exchange* (*LE*)-to-*Access Network* (*AN*) interfaces (designated *V5-interfaces* by ETSI), and the *Subscriber-Network Interface* (*SNI*). Figure 11.2 illustrates these interfaces.

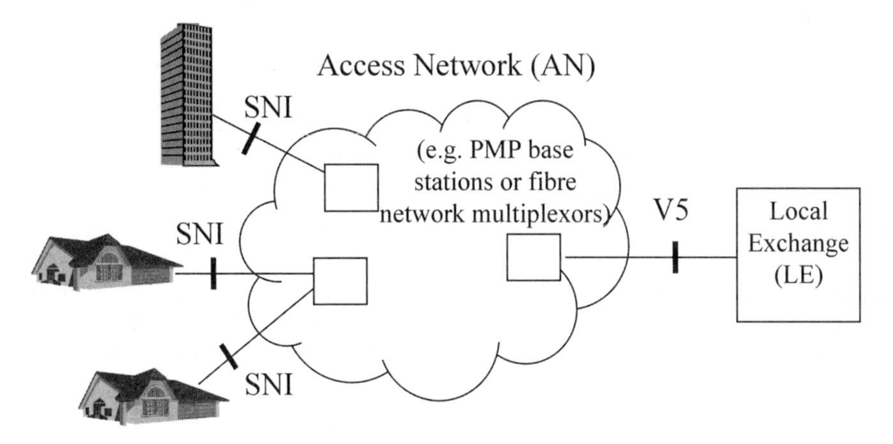

Figure 11.2 Access network interfaces

11.3 ETSI V5-Interfaces

In conjunction with the modernisation of the East German telephone network and the introduction of its *OPAL* (*optical access line*) technology, Deutsche Telekom recognised the potential for savings in access network lineplant, in the number of customer ports needed on telephone exchanges and in the number of telephone exchanges needed to supply a given region. This could be done by inclusion of *concentration* functions within the OPAL network. This lead to the development of the ETSI V5-interfaces.

As Figure 11.3a illustrates, the access network need only support sufficient connections across itself for the actual number of telephone calls in progress. Historically, copper access networks had provided a permanent connection line for each end user (Figure 11.3b). This configuration requires many more connections within the access network and many more local exchange ports.

In the example of Figure 11.3, ten end-user terminals are connected to the local exchange. It is assumed that only a maximum of two of these terminals are in use at any

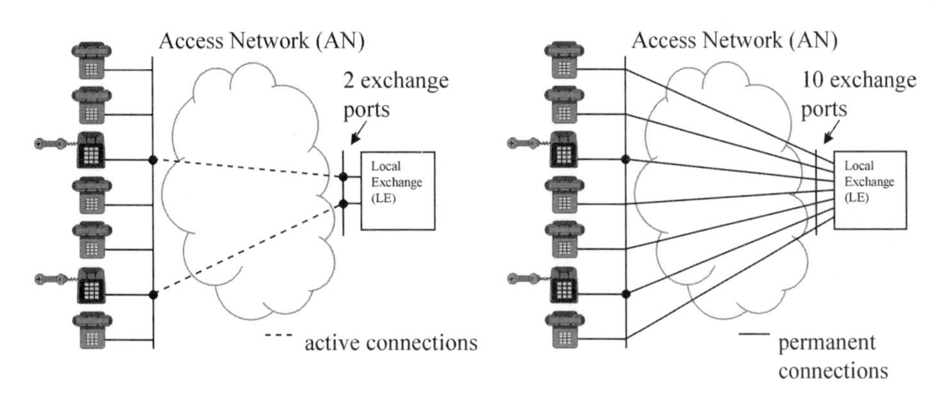

Figure 11.3 The effect of a concentration function in the access network

one time. In the case of Figure 11.3a, a concentrating function (i.e. simple switching function) within the access network ensures that only two through connections are required to be carried and only two ports are required at the exchange. In Figure 11.3b, no *concentration* is undertaken by the access network, so that ten connections and ten exchange ports are necessary.

Before the access network can undertake the concentration function, a new signalling procedure must first be defined, since it would otherwise no longer be possible for the exchange to know (merely by port of origin) which customer was wishing to make a call. The local exchange requires this information so that the correct customer is billed for the call. Similarly, for incoming calls, the local exchange must be able to signal to the access network which destination customer is to be connected. This signalling is defined in the ETSI specifications for its V5.1 and V5.2 interfaces.

V5.2 Interface

The V5.2-interface is defined in ETSI standard ETS 300 347. It defines a method for connecting up to 480 customer lines of 64 kbit/s capacity (480 simple telephone lines, 240 ISDN basic rate access lines or 16 ISDN primary rate access lines, or an appropriate mix thereof) via an access network to a telephone or ISDN local exchange. The access network may be connected using up to 2 Mbit/s lines to the local exchange. Figure 11.4 illustrates the V5.2 interface.

Since the V5.2 interface provides for a concentration function (like Figure 11.3a) to be undertaken by the access network, the number of traffic carrying channels at the V5.2 interface (between AN and LE) may be less than the number of customer connections required from the AN to customer premises. The protocol of V5.2 is complex, and not covered in detail here. It bears some resemblance to ISDN signalling (in the ISDN D-channel as defined by ITU-T recommendation Q.931). The main elements and terminology of the interface are as follows:

- *bearer channel*—this is a channel with a bitrate of 64 kbit/s (or an integral multiple thereof) which is used to carry customer telephone signals or ISDN data services;

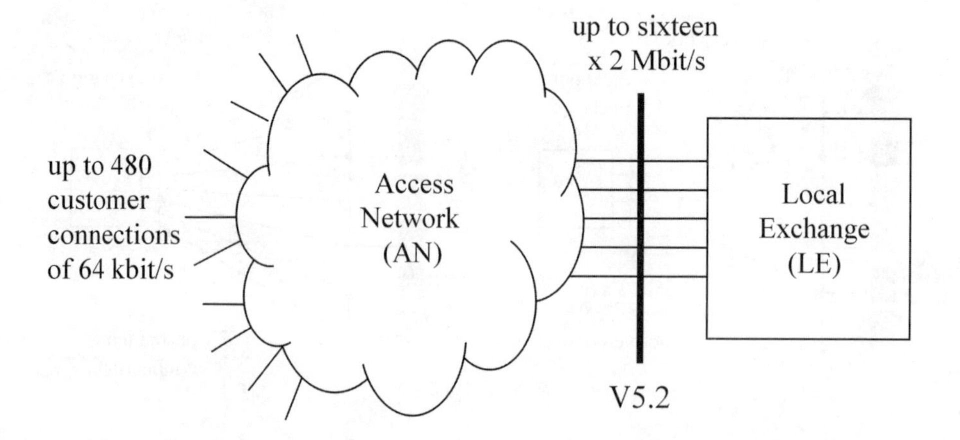

Figure 11.4 V5.2 Interface between local exchange and access network

- *Bearer Channel Connection* (*BCC*) *protocol*—this is a protocol which allows the LE to control the AN in the allocation of *bearer channels*. It is one of the types of information which may be carried by an *information path*;
- *communication path* (*c-path*)—this is the path needed to carry signalling or data-type information across the V5.2 interface. Apart from the *BCC protocol*, a *c-path* is also used for carriage of the ISDN D-channel signalling and packet or frame data originated by the various customer ISDN connections;
- *communication channel* (*c-channel*)—this is a 64 kbit/s allocation at the V5.2 interface configured to carry a *communication path*;
- *logical communication channel* (*logical c-channel*)—this is a group of one or more *c-paths*;
- *physical communication channel* (*physical c-channel*)—this is an actual 64 kbit/s timeslot allocated at the V5.2 interface for carrying logical *c-channels*. A *physical c-channel* is configured for communication and signalling, and may not be used to carry *bearer channels*;
- *active c-channel*—this is a *physical c-channel* which is currently carrying a *logical c-channel*. When not carrying a logical channel, the same physical c-channel becomes a *standby c-channel*;
- *standby c-channel*—this is a physical c-channel which is not currently carrying a logical c-channel.

Thus the 64 kbit/s *timeslots* traversing the V5.2 interface are subdivided into *bearer channels* and *c-channels* by assignment (i.e. when the network is configured). The bearer channels serve to carry user telephone and ISDN or data connections. The *c-channels* serve (on an as-needed basis) to carry the *BCC protocol* for allocation of bearer channels to individual calls, and to carry the ISDN D-channel signalling and data information between the end user terminal and the local telephone or ISDN exchange.

V5.1 Interface

The V5.1 interface is a simpler version of the V5.2 interface in which the concentration feature (Figure 11.3a) is not included. As a result, there are always as many 64 kbit/s channels permanently connected to the *Local Exchange* (*LE*) as there are 64 kbit/s telephone channels presented to end customers. While there is no possibility for traffic concentration as there is with V5.2, at least V5.1 allows individual 64 kbit/s channels or *BRI* (*Basic Rate ISDN* access line) ports to be consolidated into groups of 30 lines, each 30 presented on a single E1 (2 Mbit/s) port interface at the local exchange.

V5.1 is usually regarded by operators as the first step to V5.2. It allowed the adoption of new generation access network technology while the full specification and development of the concentration function (V5.2) took place. However, V5.1 is beginning to fall out-of-use now that many manufacturers offer the V5.2 interface.

11.4 Commercially Available Telephone and Basic Rate ISDN (BRI) Wireless Local Loop Systems

A number of systems have been developed by various manufacturers for telephone and ISDN (Integrated Services Digital Network) wireless local loop application. These systems aim to provide for telephone and full 64 kbit/s connection service. These systems use a variety of different and mostly unstandardised radio interfaces and modulation techniques, but they share a number of common attributes:

- they typically use low bit rate or restricted bandwidth channels to carry simple voice channels (e.g. 8–10 kbit/s or 4 kHz per voice channel);
- they are relatively long range systems, and thus suited to provide low density coverage over large (typically rural) areas;
- they use modulation schemes which allow for adequate telephone and fax transmission quality but also permit low subscriber terminal costs to be achieved through mass production.

One of the increasingly popular *multiple access* techniques used for this type of low bit rate system is *CDMA* (*Code Division Multiple Access*). CDMA is well suited to the application since it is relatively immune to signal fading and interference, and thus relatively robust. As a result, it can be cheap and easy to install. Importantly for CDMA, the relatively low bit rate of the individual user channels relative to the overall radio bandwidth used by the PMP system, enables a large *spreading factor* to be used. As we discussed in Chapter 5, a large spreading factor is critical to good operation of CDMA systems.

Example technologies include Lucent Technologies' *Airloop* system, the *Proximity i* system of Northern Telecom (NORTEL), *Airspan* (a system developed by DSC in cooperation with British Telecom), and a number of others. Which (if any) of these systems survives in the long term remains to be seen. Critical will be the cost per user, as well as the technical system performance.

In addition to these proprietary WLL systems, there have also been a number of attempts by public operators to use DECT (Digital European Cordless Telephony) technology as an alternative means of WLL. As we shall see in the remaining discussion, the DECT system is not eminently well-suited to WLL because of the poor radio range of the system (resulting from its low power), and because of the fact that all users share the same radio spectrum (so making traffic capacity planning of the spectrum almost impossible). Nonetheless, the overall architecture of the system, the authorisation methods and modulation schemes used, the clever way in which *repeaters* are incorporated and the protocols which are used are likely to become a model for the design of new equipment. For this reason, we study the system in a little more detail.

DECT (Digital European Cordless Telephony)

The DECT standards, developed by ETSI, have grown to be a sophisticated set of point-to-multipoint radio system specifications. Their complexity rivals even those of fully mobile telephone network standards. The initiative for their development grew from the desire to develop a *Common Air Interface* (*CAI*) for digital wide area cordless telephones. Along the way, a number of other features have been built in:

- security measures against unauthorised use of the handset and overhearing of conversation;
- mobile station tracking — so that incoming calls can be forwarded to the DECT telephone user, no matter where he is;

- full handover of mobile stations from one cell to the next;
- 64 kbit/s traffic carrying capability, to provide for correct functioning of ISDN data terminals over DECT;
- *OSI* (*Open Systems Interconnection*) compatibility of the DECT protocols (to enable systems from all different types of manufacturers to be used with one another).

Since the architecture and the suite of standards are in many ways a good model for the standardisation of fixed wireless access and Wireless Local Loop specifications, we review them in detail. Specific WLL standards (as well as proprietary equipment designs in the interim) are likely to develop using all or part of the DECT standards.

Figure 11.5 illustrates the reference model of the DECT system.

The most important part of DECT is the radio *common air interface*, D_3. This allows for the connection of *portable radio terminals* (*PT*) (the *portable part*, *PP*, of the system) to *fixed radio terminations* (*FT*) (being the *fixed part*, *FP*, of the system) using a *cordless* (i.e. radiolink) connection. The fact that the interface is standardised allows handsets produced by one manufacturer to be used in conjunction with another manufacturer's base stations. In a WLL application, this gives the possibility for the operator only to have to install the base stations, and then to be able to sell the telephone handsets through normal retail channels.

The advantage of the DECT interface over predecessing cordless telephone technologies is the high speech quality afforded by a digital radio connection, and the extra security measures added to guard against overhearing and unauthorised use, be it malicious or unintended. The extra security is afforded by means of data *encryption* and by *smart card* (so-called *DECT Authorisation Module*, *DAM*) user identification.

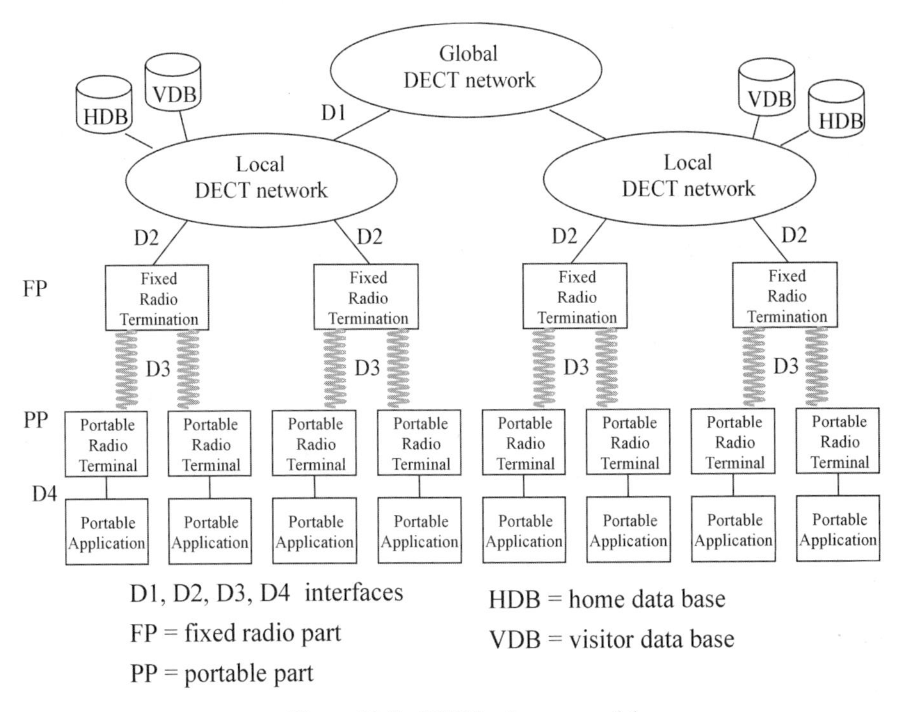

Figure 11.5 DECT reference model

DECT Handover

The interfaces D1 and D2 and the functions *HDB* (*Home Data Base*) and *VDB* (*Visitor Data Base*) support the ability to receive incoming calls in a wide area DECT network, and also support *roaming* between cells. Each *fixed radio termination* (*FT*) controls a cell within a DECT radio network. Roaming between cells is controlled by a local network function comprising HDB and VDB, which perform similar functions to the *Home Location Register* (*HLR*) and *Visitor Location Register* (*VLR*). The handover in DECT is by means of *mobile controlled handover* (*MCHO*), in which the mobile station alone decides from which base station it is receiving the strongest signal, and thus connects to this base station. When instead the signal from a neighbouring base station is stronger, the handset initiates handover to this base station and controls the transfer process. This is claimed to lead to faster and more reliable handover even than that achieved in mobile telephone networks. This method compares with the mobile assisted handover of GSM in which the *mobile switching centre* and *base stations* control the handover based on information provided by the mobile. The decision to initiate handover in the DECT system is based upon the mobile unit's measurement of the *RSSI* (*Received Signal Strength Indicator*), *C/I* (*Carrier to Interference*) and *BER* (*Bit Error Rates*) of alternative signals.

In a WLL application, the ability to handover or roam to neighbouring base stations could be a useful asset for overcoming heavy rain fading in a particular base station coverage area, or for bridging base station outage periods caused by hardware failures.

The Radio Relay Station Concept in DECT

The main problem which makes DECT economically unviable as a technology for WLL is its limited range. The range of a single hop within the DECT system is relatively limited (typically 200 m), although under ideal conditions with very high gain (and thus very directional) base station antennas, several kilometres have been achieved. There has thus been effort made in the design of the system to find a means of extending the range. The radio relay station concept allows for relaying of connections (i.e. concatenation of several radio links) to allow the *portable radio termination* (*PT*) to stray somewhat further away from the *fixed radio termination* (or *Radio Fixed Part*, *RFP*). Relay stations may be either *Fixed Relay Stations* (*FRS*) or *Mobile Relay Stations* (*MRS*), as Figure 11.6 illustrates.

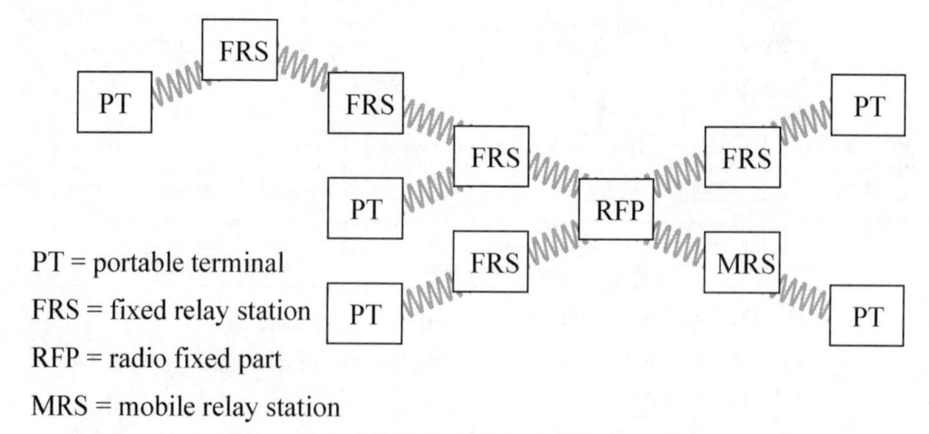

Figure 11.6 DECT fixed and mobile relay stations

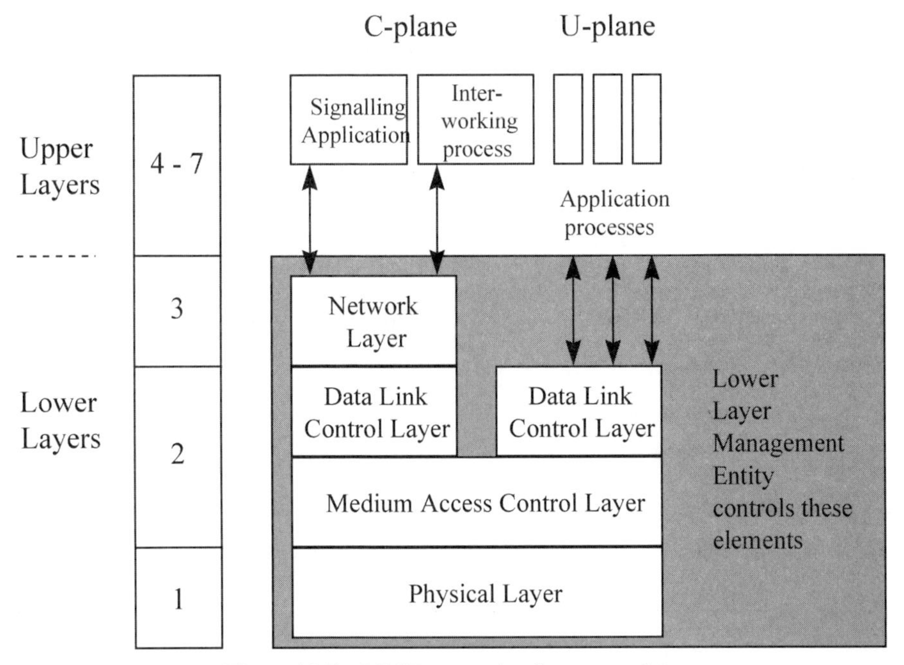

Figure 11.7 DECT protocol reference model

Up to three relay stations may be traversed, but the topology must be a star centred on the RFP.

The drawback of DECT relaying is that multiple radio channels are used to connect a single connection or call, making it impracticable for high traffic volume networks (such as those found in WLL applications). In addition, the connection quality is likely to be degraded.

The DECT Air Interface (D_3-interface)

The DECT air interface is designed to be OSI compliant. It therefore comprises layered protocols for *physical layer*, *medium access control* and *data link control* for both the *control-plane* (*c-plane*) and *user plane* (*u-plane*), as Figure 11.7 illustrates.

The *c-plane* (*control-plane*) protocol stack is used for setting up and controlling connections (like telephone signalling). The *u-plane* (*user-plane*) protocol stack is that used during the *conversation phase* of a call or connection—to convey the users' speech or data. The lower layer management entity is the set of network management functions provided to monitor and reconfigure the protocols as necessary for network operation.

The characteristics of the *physical layer* of the radio interface are listed in Table 11.1.

The multiple access scheme is based on TDMA, as illustrated in Figure 11.8.

A single slot may comprise either a *basic physical packet P32* (a full slot), a *short physical packet P00* (for a short signalling burst) or two half slots (*low capacity physical packet P08*).

The basic physical packet P32 is of 424 bytes length, subdivided into the *S-field* (for synchronisation) and the *D-field* (for carriage of data). The D-field is further subdivided

Table 11.1 DECT air interface—physical layer

Radio band	1880–1900 MHz
Number of radio channels	10
Radio channel separation	1.728 MHz
Transmitter power (max)	250 mW
Channel multiplexing	TDMA (time division multiple access)
Duplex modulation	TDD (time division duplexing)
TDMA frame duration	10 ms
Timeslots per TDMA frame	24
Modulation	GFSK (Gaussian Frequency Shift Keying)
Total bit rate	1152 kbit/s per cell
User channels	B-channel: 32 kbit/s (user) A-channel: 6.4 kbit/s (signalling)

into *A*- and *B-fields*, whereby the *A-field* is a permanent signalling channel (for *c-plane* protocol) and the *B-field* is the user data information filed (*u-plane* protocol). The various parameters within the *A-field* have the functions listed in Table 11.2.

Full 64 kbit/s ISDN *bearer channels* (i.e. user information channels for ISDN 64 kbit/s data) may be transmitted over DECT networks by the occupation of two B (bearer) channels.

Advantages and Problems of DECT when used for WLL

The DECT system has a number of capabilities which are attractive to prospective operators of WLL (Wireless Local Loop) networks for providing access to public telephone or ISDN networks:

Table 11.2 DECT signalling parameters (A-field)

Parameter	Purpose
TA	*A-field* information type
Q1, Q2	Quality control bits used as handover criteria
BA	*B-field* information type
A-field information	The field used for carriage of MAC and higher layer c-plane protocol information
R-CRC	Cyclic redundancy check (for error detection)

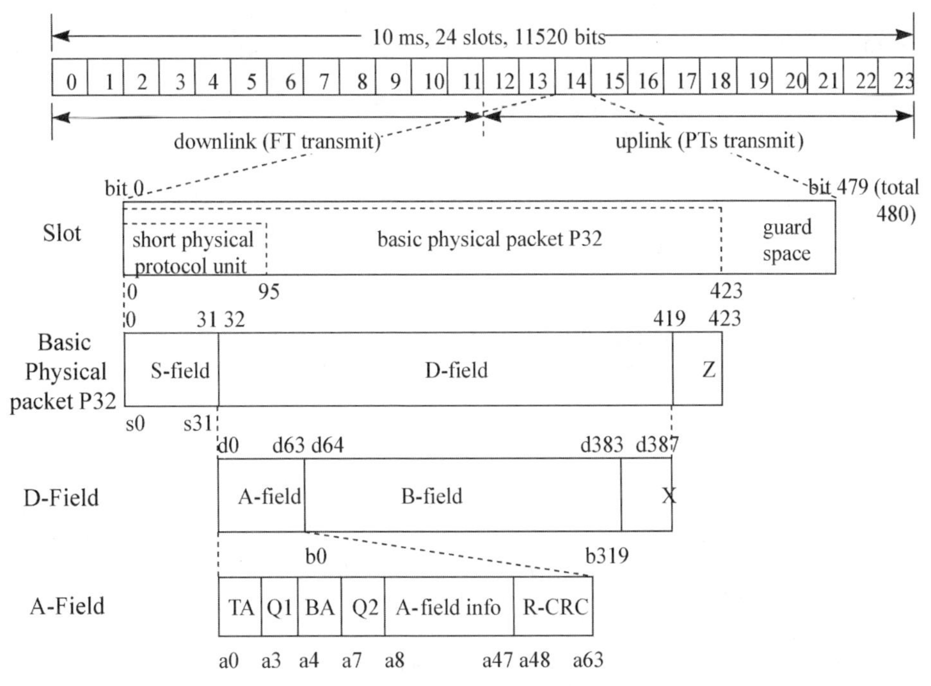

Figure 11.8 TDMA frame structure in DECT

- Standardised radio interface, allowing subscriber terminal equipment from different manufacturers to be employed;
- Easy installation of subscriber terminals enables them to be sold through normal retail channels;
- Full authentication of the handset identity protects against fraudulent usage;
- High quality digital transmission;
- Powerful and flexible protocols allow in principle for support of multimedia services;
- Repeater concept allows basic base station range to be increased;
- Handover concept could be used to overcome connection outages due to unavailability of a 'home' base station due to heavy rain fading or hardware failure;
- Mass production potential of handsets allows for affordable *Customer Premises Equipment* (*CPE*).

Unfortunately, however, there are a number of problems associated with the use of DECT technology as a solution for WLL in public telephone networks:

- The DECT radio spectrum cannot be sub-divided and allocated to individual users. This means that public network operators using the DECT system for WLL would have to share the spectrum with private users using DECT cordless phones. This is impractical for public operators, who are thereby denied the control over the usage of the spectrum, and thus stand to suffer uncontrollable capacity problems at times of high demand.
- The very restricted range of the system means that a huge number of base stations would be required to create a nationwide coverage. This makes DECT in its current form uneconomic as a WLL technology.

- The restricted amount of spectrum (20 MHz in total) limits the scope for a 'fair' allocation of reasonable bandwidth to a number of competing operators in a deregulated market. The restricted spectrum availability also precludes the use of the repeater concept in areas where high connection density is required, since it is inefficient to allocate more than one radio channel to any single connection.

Despite the problems, I am certain that much of the technology and many of the ideas developed initially for DECT will find their way into other technology.

11.5 Mobile Telephone Networks used as a Means of Fixed Wireless Access

In a number of countries where the density of fixed telephone network penetration remained very low until the mid or even late 1990s, there have been attempts to use the mobile telephone network as a means of fixed wireless access. We therefore include here a brief overview of this technology, and a review of the difficulties (mainly economic) associated with its use as a *universal service* access technology.

Since the early 1980s there has been rapid deployment of nationwide mobile telephone networks virtually across the globe. The 'mobile' generation has demanded mobile communications. Initially, the systems were analogue and relatively low capacity (Table 11.3). The early analogue mobile telephone networks were popular with businessmen as 'carphones', and their popularity lead quickly to network capacity problems. As a result, the the World Administrative Radio Council (WARC) allocated a new band in the 900 MHz range. Subsequently, there was pressure to digitalise the radio speech channels, as well as the control channel, to improve radio spectrum usage. Digital channels allow closer channel spacing, due to reduced interference between channels. The adoption of digital technology enabled, first, greater efficiency in the use of available radio bandwidth, and therefore higher traffic volumes; secondly dramatic price reductions resulting from miniaturisation and large scale production of components. Further, the introduction of competing cellular radio telephone network operators as the first stage of deregulation and competition for the traditional monopoly telephone companies simultaneously resulted in much reduced prices for mobile telephone services. The result was the worldwide boom in

Table 11.3 Comparison of analogue cellular radio network types

	AMPS (USA)	C 450 (Germany)	NMT 450 (Scandinavia)	NMT 900 (Scandinavia)	TACS (UK)
Uplink band	824–849 MHz	450–455 MHz	453–458 MHz	890–915 MHz	890–915 MHz
Downlink band	869–894 MHz	461–466 MHz	463–468 MHz	935–960 MHz	935–960 MHz
Channel spacing	30 kHz	20 kHz	25 kHz	25 kHz	25 kHz
Multiplexing	FDMA	FDMA	FDMA	FDMA	FDMA
Modulation	PSK	FSK	FSK	FSK	PSK
Number of channels	833	222	180	1000	1000

Table 11.4 Comparison of DCS-1800 (PCN) with GSM, USDC and PDC

Parameter	GSM	DCS-1800 (GSM-1800)	USDC (ADC) (US or American digital cellular system)	PDC (JDC) (Japanese personal digital cellular system)
Uplink band	890–915 MHz	1710–1785 MHz	824–849 MHz	940–960 MHz
Downlink band	935–960 MHz	1805–1855 MHz	869–894 MHz	810–830 MHz
Channel spacing	200 kHz	200 kHz	30 kHz	25 kHz
Duplex spacing	45 MHz	95 MHz	45 MHz	130 MHz
Multiplexing	TDMA/FDD	TDMA/FDD	TDMA/FDD	TDMA/FDD
Modulation	GMSK	GMSK	DQPSK	DQPSK
Speech data rate	13 kbit/s (6.5 kbit/s)	13 kbit/s	<13 kbit/s	<11 kbit/s
Frequencies	124	374	832	800
Time slots per radio frequency	8 (16)	8	3	3
Total available speech channels	992 (1984)	2992	2496	2400
Data service	9.6 kbit/s	9.6 kbit/s	4.8 kbit/s	4.8 kbit/s
Maximum speed of mobile station	250 km/h	250 km/h	100 km/h	100 km/h
Output of handheld unit	2, 5, 8 or 20 Watt	0.25 or 1 Watt		

mobile telephony. The predominant technical standards (Table 11.4) are those of *GSM* (*global system for mobile communication*) and *PCN* (*Personal Communications Network*, also known as *DCS-1800 — Digital Cellular System/1800 MHz*. This is also increasingly being called GSM-1800, as the basic network architecture and components are the same as GSM — only the band of radio operation is different).

Mobile telephone networks were the first type of large scale networks to employ a *cellular* construction of base stations to give full coverage, the scope to handover or roam from one base station (or even network) to another. It was for mobile telephone networks that much of the early work on frequency re-use planning (Chapter 9) was undertaken. The basic components of each of the different types of network are similar. Figure 11.9 illustrates the architecture of a GSM or DCS network.

The *Mobile Switching Centres* (*MSCs*) are the main controlling elements of a GSM network (a national network typically comprises 10–25 MSCs). Each controls a given geographic area over which a number of *Base Transmitter Stations* (*BTS*) are spread (typically 5–8 km radius cells). The *BSC* (*Base Station Switching Centre*) is the control element for the base transmitter stations, but need not be collocated with the BTS. Thus, in a dense metropolitan area, several antenna sites may be used, but requiring only one small BSC switching site. *Home Location Registers* (*HLR*) and *Visitor Location Registers* (*VLR*) share the function of tracking the location of the mobile user when in his *home* or in a foreign network, respectively.

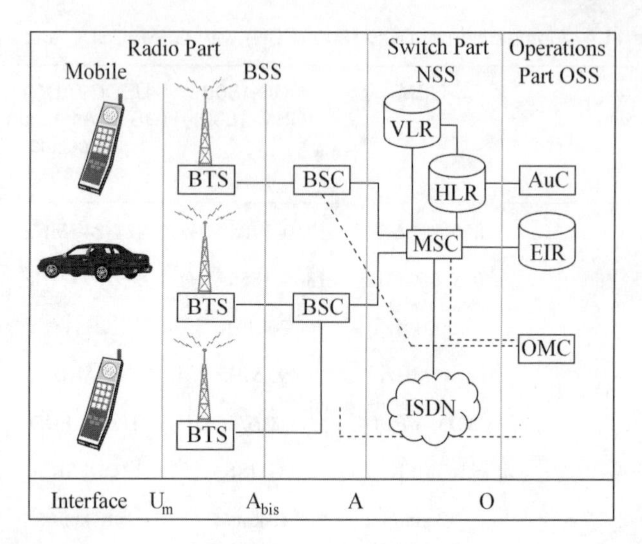

AuC	Authentication Centre	HLR	Home Location Register
BSC	Base Station Switching Centre	MSC	Mobile Switching Centre
BSS	Base Station Sub-system	NSS	Network & Switching Sub-system
BTS	Base Transmitter Station	OMC	Operations & Maintenance Centre
EIR	Equipment Identity Register	OSS	Operations & Support Sub-system

Figure 11.9 Main components of a GSM or DCS network

The MSC provides for call control and switching, and for gateway functions to other mobile or fixed networks. The GSM system has a number of additional security features compared with its predecessors. These aim to reduce the problem of stolen handsets. A *SIM card*, containing a small user identification chip, and information concerning the user's configuration must be inserted into the user's handset before it will operate. This chip enables an authorisation procedure to be carried out at each call set up between mobile station, MSC and the *Equipment Identification Register* (*EIR*). Should a handset or a card be lost or stolen, then the EIR may be updated to record this information. The EIR contains a *blacklist* of barred equipment and a *grey list* of equipment not functioning correctly or for which no services are registered. A *white list* contains all registered users and relevant subscription services.

The radio part of the GSM system uses a 25 MHz radio band. The *uplink* channels (mobile to base station) occupy the band 890–915 MHz. The *downlink* (base station to mobile) channels are between 935–960 MHz, whereby uplink and downlink channels of a particular channel pair are separated by 45 MHz.

The radio band is subdivided into 124 carrier frequency pairs, each of 200 kHz uplink and 200 kHz downlink bandwidth. Each carrier is coded digitally using *TDMA* (*Time Division Multiple Access*) as explained in Chapter 5. The normal TDMA frame used in GSM is illustrated in Figure 11.10.

The usable individual circuit bitrate is around 24 kbit/s (114 bits per circuit, every 4.615 ms). Frequency hopping (Chapter 5) provides for protection against radio fading caused by *multipath* effects, and also gives some protection against criminal overhearing of the radio signal. A further GSM measure against overhearing is a key-coded data

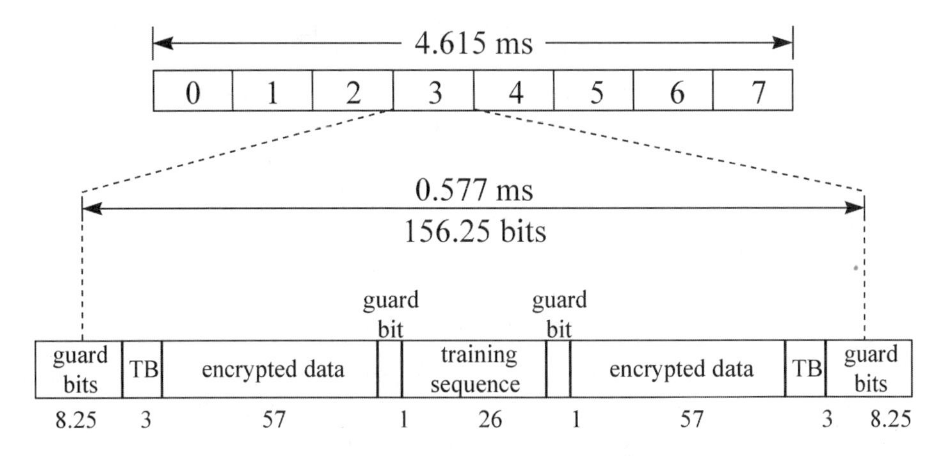

Figure 11.10 TDMA frame used in GSM

encryption, whereby the key is held on the SIM card in the mobile station and the *Authorisation Centre* (*AuC*) provides the encryption algorithm.

The radio modulation is *GMSK* (*Gaussian Minimum Shift Keying*), a form of phase shift keying (PSK—binary *phase modulation*). The *PCM* (*Pulse Code Modulation*) coding of the speech signal uses an *ADPCM* (*Adaptive Differential PCM*) algorithm (see Chapter 3) with a bit rate around 10 kbit/s. The remaining bitrate is needed for signalling, error correction, encryption and other protocol functions (e.g. *handoff*, etc.).

The U_m-interface is the mobile telephone network equivalent of the ISDN U-interface. It is the standard radio interface and protocol allowing mobile handsets from many different manufacturers access to the mobile network. Figure 11.11 illustrates the classification of the various ISDN-like end user termination tapes.

New developments of the GSM system will increase the intelligence and functionality of the system. Already, sophisticated mailbox services, short message services (displayed on

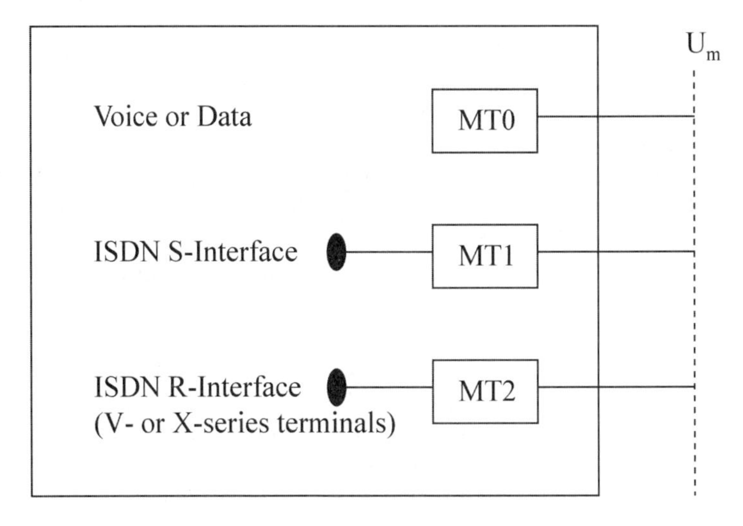

Figure 11.11 Mobile station termination types for GSM

the display of the mobile station as a short text), fax and data services are becoming common network offerings.

One of the major lessons learned from both DCS-1800 and GSM has been how quickly viable alternative networks to the traditional, terrestrial-based public telephone can be established. GSM operators have become very adept at finding and buying or leasing small footprint sites where they may erect radio towers which serve simultaneously as base stations for the radio links to mobile stations and masts for the installation of point-to-point microwave radio links as backhaul links (Chapter 12) to *mobile switching centres.*

The Difficulties of using Mobile Telephone Networks for Fixed Wireless Access

Clearly, the advent of mobile telephone networks has been a great success. Indeed, in some countries (for example, in eastern Europe) the number of mobile telephones is beginning to rival the total number of fixed telephones. The geographical coverage of the networks has reached 90–95% of land coverage even in countries where perhaps less than 50% of houses in rural areas have a telephone. The range of services of mobile networks has quickly come to include not only basic telephony, but also ISDN, fax and even Internet access. In parallel, the monthly rental of a mobile telephone service and the cost of the calls has fallen to affordable levels. Meanwhile, the mobile network operators have become rich. All this has not escaped the attention of regulators and government economic planners, some of whom have laboured with the idea of adopting mobile telephony as an easy and quick means to provide for *universal telephone service*—providing telephone service to everyone at an affordable price, almost as if this was a basic right of the national constitution.

Trials using mobile telephone networks as a basis for rural fixed telephone service have taken place in a number of countries, but while these trials have proven that the technology can provide a quality of service comparable with a 'fixed' telephone line, a number of economic questions have remained. The trials typically use an 'outdoor' high gain antenna mounted on the roof of the customer's premises. The high gain antenna has three beneficial effects in the context of a rural *fixed* wireless access:

- The higher gain of the subscriber terminal antenna serves to increase the effective coverage area and range of each base station cell, thereby minimising the investment necessary in further base stations to give full coverage.
- The higher fade margin arising from the higher antenna gain improves the availability of the service, enabling it to rival the quality expectation from a *fixed* service (and to dispel a widespread belief that radio is poor quality when it rains).
- The fixing of the antenna to the wall prevents the customer from using the mobile telephone handset as a *mobile* service at a subsidised *fixed* network tariff.

In particular, attention has focussed on the use of some of the old analogue mobile telephone networks (e.g. NMT-450) which are no longer at full capacity because of the migration of large numbers of customers to the more modern digital telephone systems (GSM). Such networks are mostly 'written off' (i.e. fully depreciated) financially, and can therefore be operated at 'marginal cost'. In addition, the relatively low radio frequency band (450 MHz) means that the base stations have relatively large coverage areas and range. As a result, comparatively little investment is necessary to set up the base stations necessary to provide 100% geographical coverage.

Some government, economic and development planners have contemplated making direct grant subsidies to support the establishment of the necessary base stations to give full geographic coverage. However, this alone has not helped them solve the remaining dilemmas:

- What is the remaining useful 'lifetime' of obsolescent (e.g. analogue NMT-450) equipment? A 'marginal' investment may not be justified if the likely 'lifetime' is only five years or so. It might be economically more efficient to make a much larger investment in civil works, conduits and glass fibre or other wireless infrastructure if this has a much longer lifetime.
- How can the mobile network operator be obliged to continue offering a universal service in remote geographical areas when the overall running costs of the network are no longer economically self-supporting?
- What about the possible re-use of the radio frequency band in conjunction with more modern and efficient technology? Most NMT-450 network operators, for example, are considering whether to scrap their old technology in favour of a digital GSM-450 system. Using a single integrated GSM network across all three frequencies (450 MHz, 900 MHz and 1800 MHz) and a common pool of switches and other hardware makes for considerable cost savings through the economies of scale.

Does it make sense? Can it be made to work? I guess we will have to wait and see what develops. Maybe the most likely outcome is that the continuing growth in demand for mobile telephone service will simply continue to drive down the cost of service provision, so making *mobile* networks the cheapest technology for 'basic' telephone service. By this time, full geographical coverage will be inevitable — first, because only with full coverage can an operator get full access to the maximum potential number of customers, and secondly, because of the competitive 'differentiation' made possible by a full roaming capability.

11.6 Iridium, Globalstar and the Evolution Towards the Universal Mobile Telephone Service (UMTS)

Our review of mobile telephone network systems with a capability for providing *fixed* wireless access would not be complete without a brief mention of the emerging *Universal Mobile Telephone Service* (*UMTS*).

With the mobile telephone boom of the early 1990s in full swing in the developed countries, attention swung towards extending the possibilities of mobile roaming to worldwide coverage, and to the increased penetration of mobile telephone services in less developed parts of the world. The result was a number of new satellite technology initiatives. In these, the *Base Transmitter Station* (*BTS*) and *Base Switching Centre* (*BSC*) functions of the mobile network (Figure 11.9) are combined into one unit which is now carried by a low orbiting satellite. The low orbit is necessary to reduce the transmit power needed in the mobile stations, and also to reduce the undesirable long propagation time of signals transmitted to and from *geostationary* orbits at 37,000 km altitude (Chapter 10).

The low orbit creates a significant technological challenge, since each individual satellite now moves at considerable velocity relative to the earth's surface, being at best, 'visible' from a particular point on the ground for a maximum of 15 minutes per orbit.

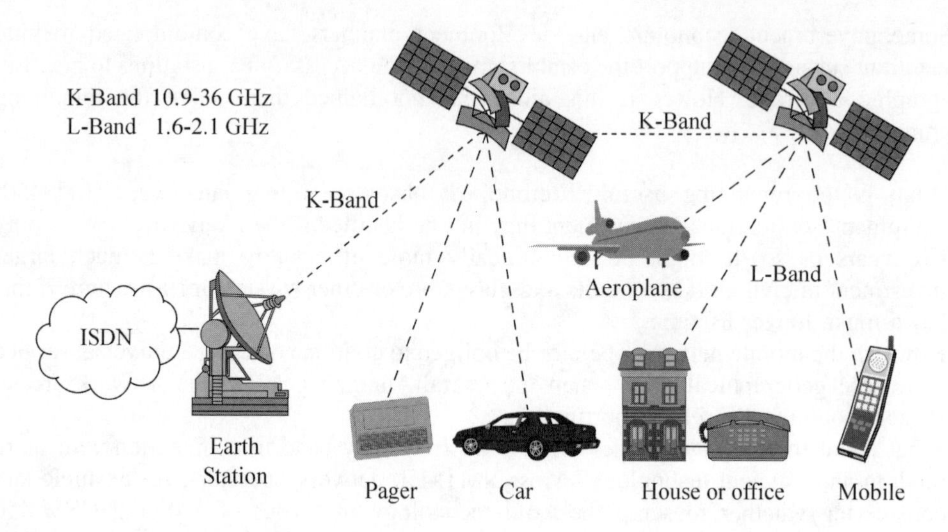

Figure 11.12 The Universal Mobile Telephone Service (UMTS)

Table 11.5 Comparison of satellite mobile telephone systems

	Iridium	Globalstar	Odyssey	Teledesic
Consortium leader	Motorola	Loran, Qualcom	TRW-Matra	Bill Gates
Number of satellites	66	48	10	840
Orbit type and altitude	LEO (low earth, circular orbit), 780 km, six orbit paths	LEO (low earth, circular orbit), 1414 km, 14 orbit paths	MEO (medium earth, circular orbit) 10,355 km, three orbit paths	LEO (low earth orbit) 840 km
Satellite inclination	86.4°	47°/65°	55°	
Minimum elevation	8°	10°	18°	40°
	2.6 ms	4.7 ms	34.5 ms	
No. of spot beams (cells/satellite)	48	6	163	
Multiplexing	FDMA/TDMA	FDMA/CDMA	FDMA/TDMA	TDMA/SDMA
Modulation	QPSK	QPSK	DPSK	
Bitrate	2.4/4.8 kbit/s	4.8/9.6 kbit/s	4.8 kbit/s	
Satellite switching	On-board	Transparent	Transparent	
Satellite weight	689 kg	262 kg	1130 kg	

Like GSM, each of the proposed satellite systems divides the coverage area (the earth's surface) into a number of *cells*, within each of which a number of radio channels permit individual users to make telephone calls. The difference is that the cells are now typically the size of a European country, so that the capacity in terms of traffic density is much less than for a terrestrial-based GSM system. The advantage of the satellite systems, of course, is the global coverage and roaming capability. Figure 11.12 illustrates the concept in simple terms and Table 11.5 compares four of the systems originally proposed to go into service over the 1998–2000 timeframe.

Resulting from these initiatives, new interest was generated in the standards fora to create a UMTS by enabling roaming between cells of a land-based GSM-type mobile network and a satellite-based cell. The idea is that a single, but perhaps *dual-mode* handset could register for terrestrial GSM network service in those countries where it was available, and where not, could be reached by a satellite service. In this way, the coverage of the satellite-based system could be complemented by the higher traffic capacity of land-based systems in more telephone traffic-dense areas.

Trials of the systems have commenced, but the economic problems of the operators (for example, Iridium filing for bankruptcy) indicate that the system is not yet affordable. Over time this type of system is bound to appear, but more time, it seems, is necessary to improve the technology to a more affordable price.

11.7 Shorthaul Microwave Radio for WLL

As the demand from public network operators for wireless local loop technology is growing, so radio spectrum regulators and manufacturers alike are increasingly looking to the possibility of using previously unused parts of the radio spectrum. Ever high frequency bands are being licensed for point-to-point and point-to-multipoint radio systems. Already frequencies in some countries are earmarked at 40–42 GHz and 54 GHz.

At such very high frequencies, the range of the system is very limited. On the other hand, the shorter range means more frequent frequency re-use (see Chapter 9), and thus higher densities of connections from the same radio bandwidth. As long as the market demands it, new systems are bound to keep appearing!

12

Backbone, Backhaul and High Capacity Access Radio Systems

Microwave radio was first used on a large-scale basis in public telecommunications networks for providing high-capacity long-haul point-to-point trunk links or 'backbone' networks between major cities across a country. Microwave towers and radio systems in the range 1–10 GHz appeared. Later, the advent of the mobile telephone networks in the 1980s created a need to establish rapidly a 'backhaul' infrastructure to connect remote *Base Transmitter Stations* (*BTSs*) of these networks to their corresponding *Mobile Switching Centres* (*MSCs*). The age of 'medium haul' and 'shorthaul' point-to-point microwave communications dawned, using new frequencies at 13–38 GHz and smaller sized antennas (30 cm and 60 cm diameter). Meanwhile a new era of 'backhaul' communications is about to emerge, as the demand for even higher density of connections grows (to enable the backhaul connection of the new micro-BTSs (lampost-top mobile base stations). In this era we can expect the use of point-to-multipoint systems and even higher frequency (very-short-haul) point-to-point systems. The same technology also provides an ideal basis for high-capacity network access links. In this chapter, we describe the typical architecture of backbone and backhaul networks and the considerations which should go into their design. In particular, we also describe the recent migration in telecommunications transmission technology from *PDH* (*Plesiochronous Digital Hierarchy*) to *SDH* (*Synchronous Digital Hierarchy*) and *SONET* (*Synchronous Optical Network*) and the reasons for the migration.

12.1 Backbone Networks

When microwave radio was first used in telecommunications networks, it was because no further radio spectrum was available in the MHz range (sub-1 GHz). New technology was necessary to adapt to the higher frequencies, which were initially in the range from about 1 GHz to 10 GHz. The frequencies were used for long-hop backbone transmission systems and satellite *uplinks* and *downlinks*.

Using large antennas (240 cm, or up to 18 m in diameter in the case of satellite *earth stations*) at these (relatively low) microwave frequencies very long distances can be spanned using a single hop (typically up to about 70 km terrestrially but 37,000 km in a satellite uplink or downlink). Indeed, one of the biggest impediments to longer terrestrial links is the curvature of the earth, which demands very tall towers to maintain the line-of-sight path needed (Figure 12.1).

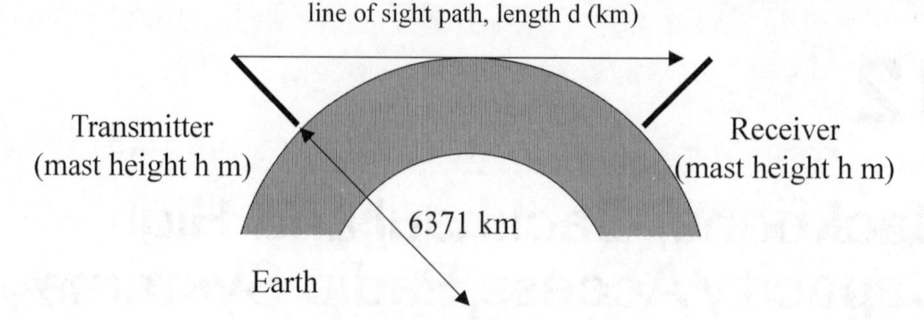

Figure 12.1 At low frequencies the curvature of the earth and the height of the antenna mast limits the length of a line-of-sight path

Since the radius of the earth is 6371 km, then simple Pythagoras geometry leads us to the relationship between the maximum possible path length (the one which 'brushes' the earth's surface at the mid-point of the path as shown in Figure 12.1):

Maximum line-of-sight path length, d (in km) $\simeq \sqrt{(51\,\mathrm{h})}$ where h is the height of transmitter and receiver masts in metres.

Thus with quite tall masts of 50 metres (equivalent to a 20-storey building) at each end of the link, we can only expect a link range of the order of 50 km! A slightly longer range is possible by positioning even taller masts than this on carefully selected hills.

In the longest-established point-to-point radio bands within the spectrum range from 1 GHz to 10 GHz, the bit rate capacity of a single radio channel is very large. These bands and radio systems were specifically used for high capacity trunk connections of capacities typically around 90 Mbit/s or 140 Mbit/s. Because the radio channels were thus quite high bandwidth (and thus quite a high percentage of the available bandwidth within the band), it was necessary to be very efficient in the re-use of individual frequencies. Thus, it became common, as we already discussed in Chapter 9, to realise longhaul links (i.e. longer than 70 km) by means of *multiple-hop* links, zig-zagging between tall microwave towers on hilltops (Figure 12.2).

The basic methodology for calculating the availability of the end-to-end link is as we discussed in detail in Chapter 7. It is normal for such links to be planned for very high availability (e.g. 99.999%), because of the importance of the trunks in a telecommunications backbone network. However, as we also learned in Chapter 8, a modification factor needs to be applied to 'scale down' the cumulative unavailability of multiple-hop links (Figure 12.3).

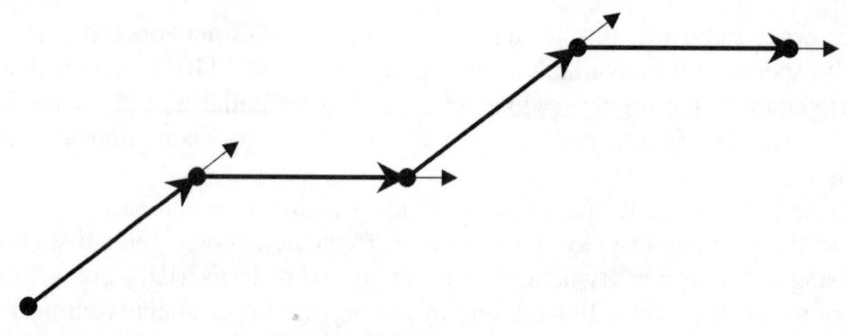

Figure 12.2 Zig-zag layout of a long haul multiple-hop backbone transmission link, using the same radio channel on each hop

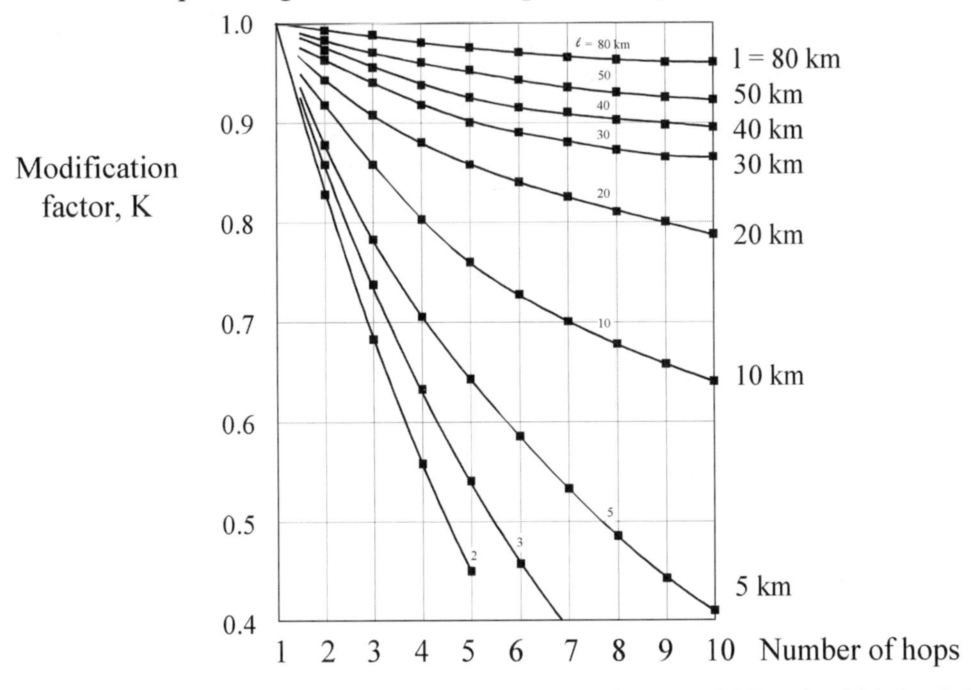

Figure 12.3 Modification factor for 'scaling down' the cumulative *unavailability* of multiple-hop links (Courtesy of ITU-R recommendation F.530)

The radio equipment typically used for longhaul high capacity backbone links is very sophisticated and expensive. Typically the equipment includes every possible capability for the provision of *path diverstiy* (i.e. *space diversity* and *frequency diversity*), as well as providing for hardware *redundancy* or *protection.* We discussed all of these capabilities in detail in Chapter 8. All of these capabilities are designed to maximise the link availability.

Because it was often the case that the radio links themselves were the best or only means of telecommunication from a remote microwave tower, it became common practice for network operators to demand the availability of an *engineering order line* across the link. The engineering order line was an extra, low bandwidth telephone channel carried by the radio equipment as a 'parallel' connection to the main high bandwidth or *payload* connection. The engineering order line enables two technicians at either end of the link to communicate easily with one another by phone, without requiring a separate telephone line or expensive multiplexing equipment (to break a single telephone channel out of the normal payload channel) to be installed in the remote location.

Initially, most backbone microwave radio systems were analogue, but the clear advantage of digital radio (which is capable of removing nearly all of the signal distortions caused by fading — as we saw in Chapter 10) has meant that most systems nowadays are *Digital Radio-relay Systems* (*DRS*). Initially, most digital radio systems were based on the *Plesiochronous Digital Hierarchy* (*PDH*), but there is a rapid migration to the *Synchronous Digital Hierarchy* (*SDH*), for the reasons we explain below.

12.2 History of the Synchronous Digital Hierarchy (SDH)

The *Synchronous Digital Hierarchy* (*SDH*) was developed based upon its North American forerunner *SONET* (*Synchronous Optical Network*). SDH is the most modern type of transmission technology, and as its name suggests, based upon a synchronous multiplexing technology. The fact that SDH is synchronous adds greatly to the efficiency of a telecommunications transmission network (independent of whether that network is radio, optical-fibre or wire-based), and makes the network much easier to manage.

The Problems of PDH Transmission

Historically, digital telephone networks, modern data networks and the transmission infrastructures serving them have been based on a technology called *PDH* (*Plesiochronous Digital Hierarchy*). Three different PDH hierarchies evolved, as we summarised in Figure 12.1. They share three common attributes:

- they are all based on the needs of telephone networks, i.e. offering integral multiples of 64 kbit/s channels, synchronised to some extent at the first multiplexing level (1.5 Mbit/s or 2 Mbit/s);
- they require multiple multiplexing stages to reach the higher bit rates, and are therefore difficult to manage and to measure and monitor performance, and relatively expensive to operate;
- they are basically incompatible with one another.

Each individual transmission line within a PDH network runs *plesiochronously*. This means that it runs on a clock speed which is nominally identical to all the other line systems in the same operator's network, but is not locked *synchronously* in step (it is *free-running*). This results in certain practical problems (as we discussed in Chapter 10). Over a relatively long period of time (say one day), one line system may deliver two or three bits more or less than another. If the system running slightly faster is delivering bits for the second (slightly slower) system, then a problem arises with the accumulating extra bits.

Eventually, the number of accumulated bits become too great for the storage available for them, and some must be thrown away. The occurrence is termed *slip* and can result in *bit errors*. To keep this problem in hand, *framing* and *stuffing* (or *justification*) bits are added within the normal multiplexing process, and are used to compensate. These bits help the two end systems to communicate with one another, slowing up or slowing down as necessary to keep better in step with one another. The extra framing bits account for the difference, for example, between 4×2048 (E1 bitrate)=8192 kbit/s, and the actual E2 bitrate (8448 kbit/s — see Figure 12.4).

Extra framing bits are added at each stage of the PDH multiplexing process. Unfortunately, this means that the efficiency of the higher order line systems (e.g. 139 264 kbit/s — usually termed 140 Mbit/s systems) are relatively low (91%). More critically still, the framing bits added at each stage make it very difficult to break out a single 2 Mbit/s *tributary* from a 140 Mbit/s line system at an intermediate point without complete demultiplexing (Figure 12.5). This makes PDH networks expensive, rather inflexible and difficult to manage.

(a) Europe

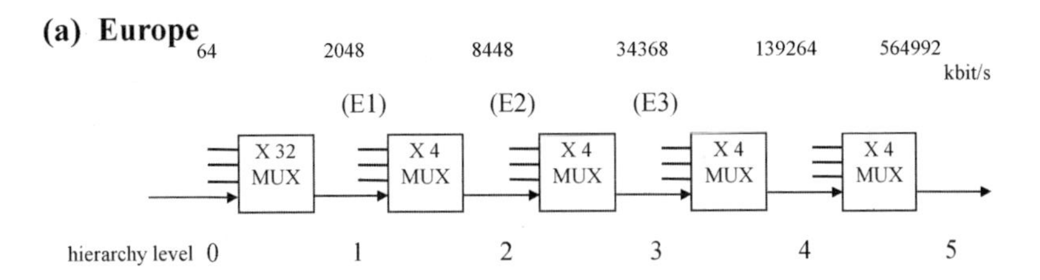

(b) North America

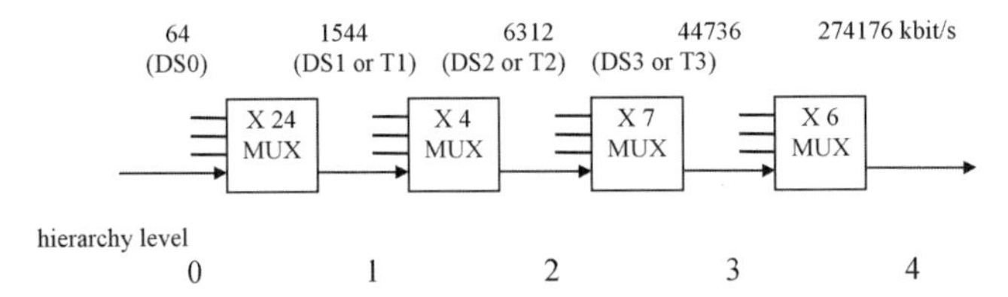

(c) Japan

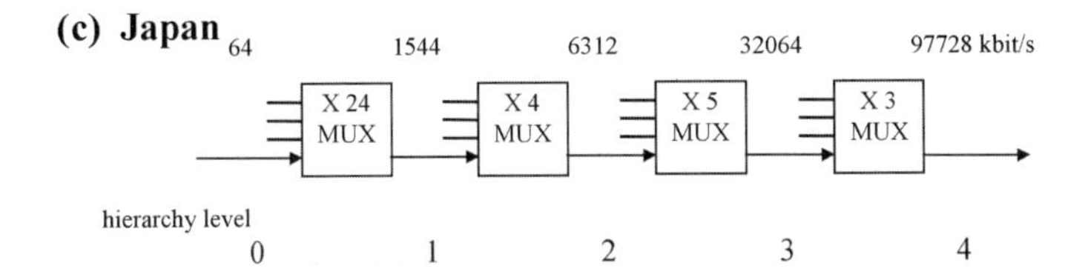

Figure 12.4 The various Plesiochronous Digital Hierarchies (PDH; ITU-T/G.571)

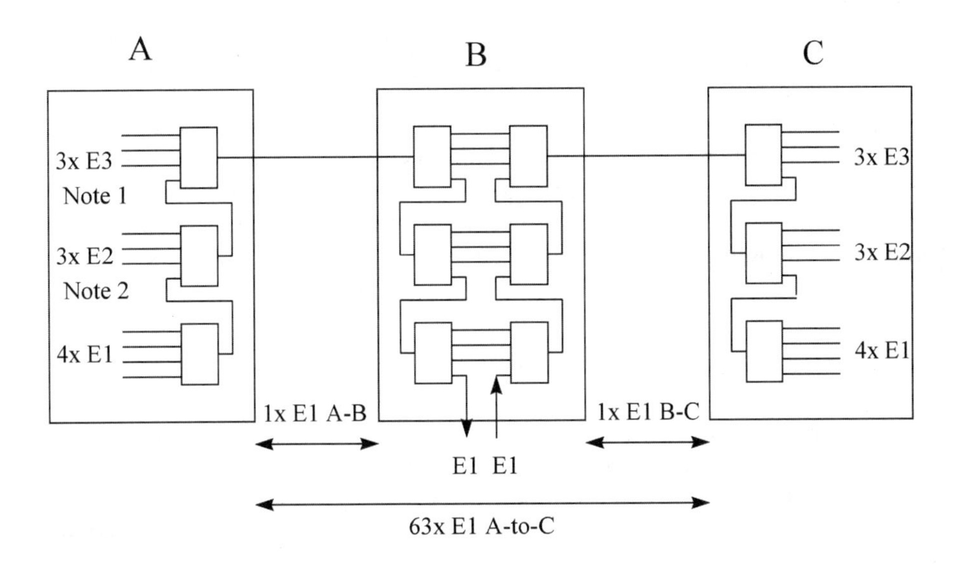

Figure 12.5 Breaking-out 2 Mbit/s (E1) from a 140 Mbit/s line system at an intermediate exchange

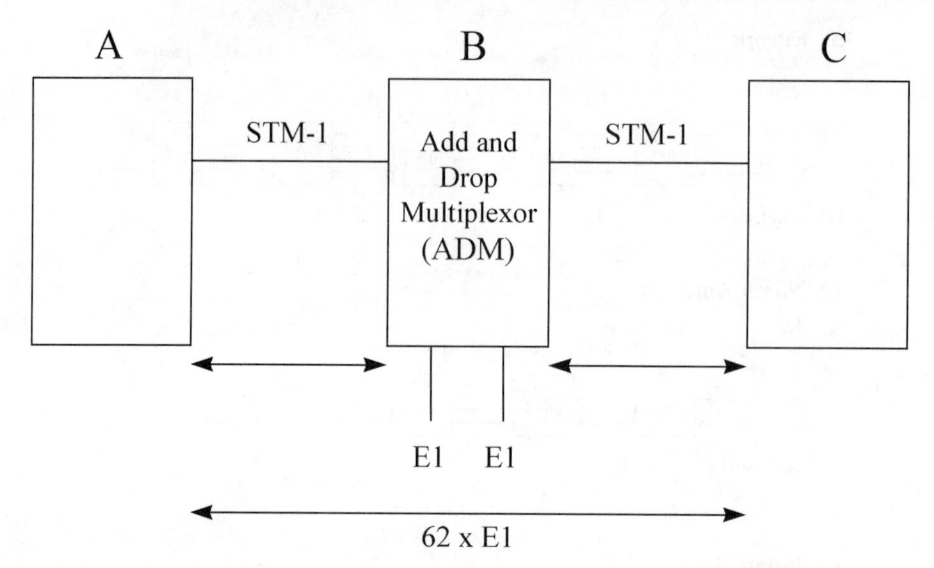

Figure 12.6 Add/drop multiplexor used to break-out 2 Mbit/s (E1) from a 155 Mbit/s (STM-1) line at an intermediate exchange

SDH, in contrast to PDH, requires the synchronisation of all the links within a network. It uses a multiplexing technique which has been specifically designed to allow for the *adding* and/or *dropping* of the individual *tributaries* within a highspeed bit rate. Thus, for example, a single *Add and Drop Multiplexor* (*ADM*) is required to break out a single 2 Mbit/s *tributary* from an *STM-1* (*Synchronous Transport Module*) of 155 520 kbit/s (Figure 12.6).

Other major problems of the PDH are the lack of tools for network performance management and measurement now expected by most public and corporate network managers, the relatively poor availability and range of high speed bit rates and the inflexibility of options for line system back. Before SDH, link back-up tended to be on a *1 main+1 standby* protection basis, making back-up schemes costly and difficult to manage. These problems have been eliminated in the design of SDH through in-built flexibility of the bitrate hierarchy, integration of the optical units into the multiplexors, ring structure topologies and in-built performance management and diagnostic functions.

The Multiplexing Structure of SDH

This section reviews, for the benefit of backbone network transmission planners, the detailed multiplexing structure of SDH.

As is shown in Figure 12.7, the *containers* (i.e. available bit rates) of the synchronous digital hierarchy have been designed to correspond to the bit rates of the various PDH hierarchies. These containers are multiplexed together by means of *Virtual Containers* (abbreviated to *VCs*, but not to be confused with *Virtual Channels*, which are also so abbreviated), *Tributary Units* (*TU*), *Tributary Unit Groups* (*TUG*), *Administrative Units* (*AU*), and finally, *Administrative Unit Groups* (*AUG*) into *Synchronous Transport Modules* (*STM*).

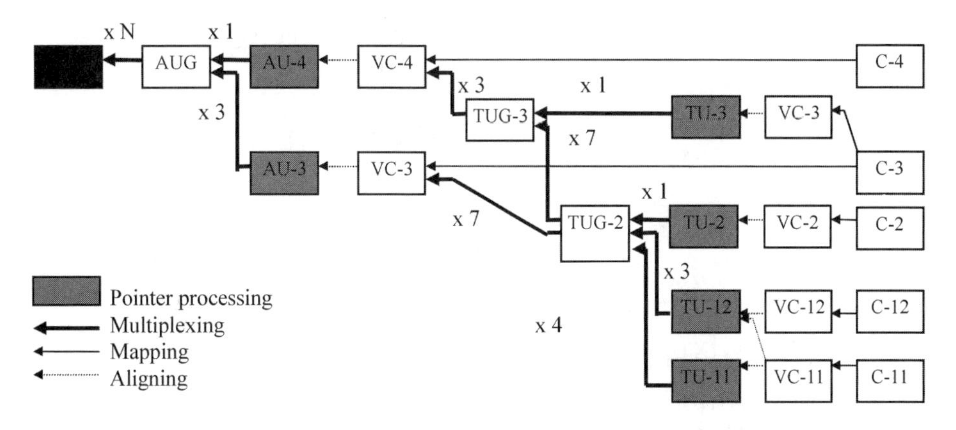

Figure 12.7 Synchronous Digital Hierarchy (SDH) multiplexing structure (ITU-T /G.709)

The basic building block of the SDH hierarchy is the *Administrative Unit Group (AUG)*. An AUG comprises one AU-4 or three AU-3s. The AU-4 is the simplest form of AUG, and for this reason, we use it to explain the various terminology of SDH (containers, virtual containers, mapping, aligning, tributary units, multiplexing, tributary unit groups).

The *container* comprises sufficient bits to carry a full *frame* (i.e. one cycle) of user information of a given bit rate. In the case of *container 4* (*C-4*), this is a field of 260 × 9 bytes (i.e. 18,720 bits). In common with PDH, the *frame repetition rate* (i.e. number of cycles per second) is 8000 Hz. Thus, a *C4-container* can carry a maximum user throughput rate (*information payload*) of 149.76 Mbit/s (18720 × 8000). This can either be used as a raw bandwidth or, say, could be used to transport a PDH link of 139,264 Mbit/s.

To the container is added a *path overhead* (*POH*) of 9 bytes (72 bits). This makes a *Virtual Container* (*VC*). The process of adding the POH is called *mapping*. The POH information is communicated between the point of *assembly* (i.e. entry to the SDH network). It enables the management of the SDH system and the monitoring of its performance.

The virtual container is *aligned* within an *Administrative Unit* (*AU*) (this is the key to synchronisation). Any spare bits within the AU are filled with a defined filler pattern called *fixed stuff*. In addition, a *pointer* field of 9 bytes (72 bits) is added. The pointers (three bytes for each VC — up to three VCs in total (9 bytes maximum)) indicate the exact position of the virtual container(s) within the AU frame. Thus in our example case, the AU-4 contains one 3 byte pointer indicating the position of the VC-4. The remaining 6 bytes of pointers are filled with an idle pattern. One AU-4 (or three AU-3s containing three pointers for the three VC-3s) are *multiplexed* to form an AUG.

To a single AUG is added 9 × 8 bytes (576 bits) of *section overhead* (*SOH*). This makes a single *STM-1 frame* (of 19,440 bits). The SOH is added to provide for *block framing* and for the maintenance and performance information carried on a transmission line *section* basis. (A section is an administratively defined point-to-point connection in the network, typically an SDH-system between two major exchange sites, between two intermediate multiplexors or simply between two regenerators.) The SOH is split into 3 bytes of *RSOH* (*regenerator section overhead*) and 5 bytes of *MSOH* (*multiplex section overhead*). The RSOH is carried between, and interpreted by, SDH line system *regenerators* (devices appearing in the line to *regenerate* laser light or other signal,

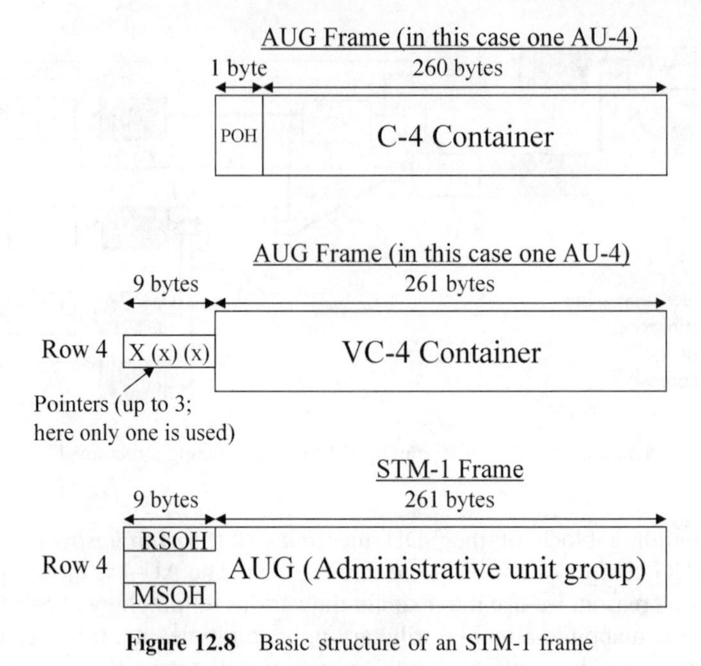

Figure 12.8 Basic structure of an STM-1 frame

thereby avoiding signal degeneration). The MSOH is carried between, and interpreted by the devices assembling and disassembling the AUGs. The MOH ensures integrity of the AUG.

Since the *frame repetition rate* of an STM-1 frame is 8000 Hz, the total line speed is 155.52 Mbit/s (19440 × 8000). Alternatively, power of four (1, 4, 16, etc.) multiples of AUGs may be multiplexed together with a proportionately increased section overhead, to make larger STM frames. Thus, an STM-4 frame (4 AUGs) has a frame size of 77,760 bits, and a line rate of 622.08 Mbit/s. An STM-16 frame (16 AUGs) has a frame size of 311,040 bits, and a line rate of 2488.32 Mbit/s.

Tributary Unit Groups (*TUGs*) and *Tributary Units* (*TUs*) provide for further breakdown of the VC-4 or VC-3 payload into lower speed tributaries, suitable for carriage of today's T1, T3, E1 or E3 line rates (1.544 Mbit/s, 44.736 Mbit/s, 2.048 Mbit/s or 34.368 Mbit/s).

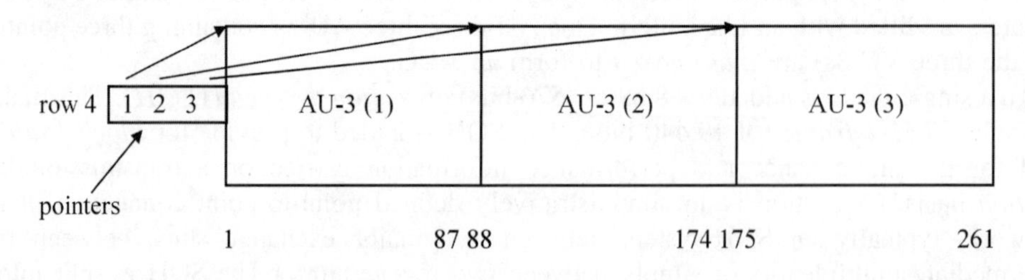

Figure 12.9 AUG frame arranged as 3 × AU-3

Table 12.1 Payload rates of SDH containers

Container type	Container frame size	Frame repetition rate	Capable of carrying PDH line type (kbit/s)
C-11	193 bits	8000	T1 (1544)
C-12	256 bits	8000	E1 (2048)
C-21	789 bits	8000	T2 (6312)
C-22	1056 bits	8000	E2 (8448)
C-31	4296 bits	8000	E3 (34368)
C-32	5592 bits	8000	T3 (44736)
C-4	260 × 9 bytes	8000	139,264

Figure 12.8 shows the gradual build up of a C-4 container into an STM-1 frame. The diagram conforms with the conventional diagrammatic representation of the STM-1 frame as a matrix of 270 columns by nine rows of bytes. The transmission of bytes, as defined by ITU-T standards is starting at the top left hand corner, working along each row from left to right in turn, from top to bottom row. The structure is defined in ITU-T recommendations G.707, G.708 and G.709.

The Tributaries of SDH

The structure of an AUG comprising three AU-3s is similar to that for an AUG of one AU-4 except that the area used in Figure 12.6 for VC-4 is instead broken into three separate areas of 87 columns, each area carrying one VC-3 (Figure 12.9). In this case, all three pointers are required to indicate the start positions within the frame of the three separate VCs. The various other TU and VC formats follow similar patterns to the AUs and VCs presented (TUs also include pointers like AUs). Table 12.1 presents the various container rates available within SDH. Note that the terminology C-12 is intended to signify the hierarchical structure and should not therefore be called C-twelve, but instead C-one-two. The relevant VC is VC-one-two, etc.

Figure 12.10 shows an alternative demultiplexing scheme, based upon the sub-multiplexing of a VC-4 container into three *tributary unit groups-3* (*TUG-3s*). In this case, the first three columns are used as path overhead, and each TUG occupies a total of 86 columns, but the individual TUGs are *byte interleaved.* This sub-multiplexing scheme lends itself better to the carriage of PDH signals.

The sub-multiplexing of the TUG-3s themselves may be continued as shown in Figure 12.11, where each TUG-3 is sub-multiplexed into 7 × TUG-2, also using *byte interleaving*. Finally, as Figure 12.11 also shows, the TUG-2s may be subdivided into *byte interleaved* TU-11 tributaries (for T1 rate of 1.544 Mbit/s) or TU-12 tributaries (for E1 rate of 2.048 Mbit/s).

The individual *containers* (C-11 or C-12) may be packed into the TU-11 (synonymous with VC-11) or TU-12 (synonymous with VC-12) in one of three manners, using either:

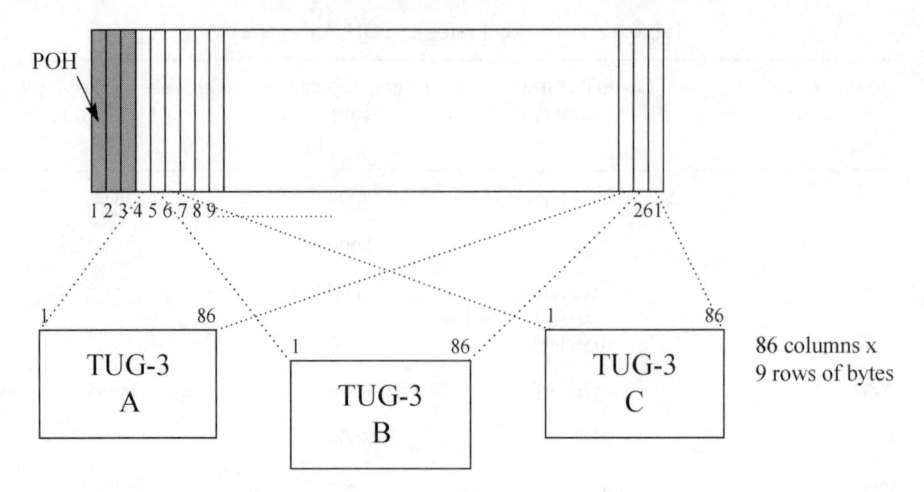

Figure 12.10 VC-4 submultiplexing scheme as 3 × TUG-3 using byte interleaving

- no framing (i.e. *asynchronously*);
- *bit synchronous framing*;
- *byte synchronous framing*.

The *asynchronous* and *bit synchronous framing* methods allow a certain number of bits for *justification*. This enables 1.5 Mbit/s or 2 Mbit/s tributaries of an SDH transmission network to operate in conjunction with PDH or other networks running on separate clocks (i.e. not running synchronously with the SDH). *Byte synchronous framing*, in contrast,

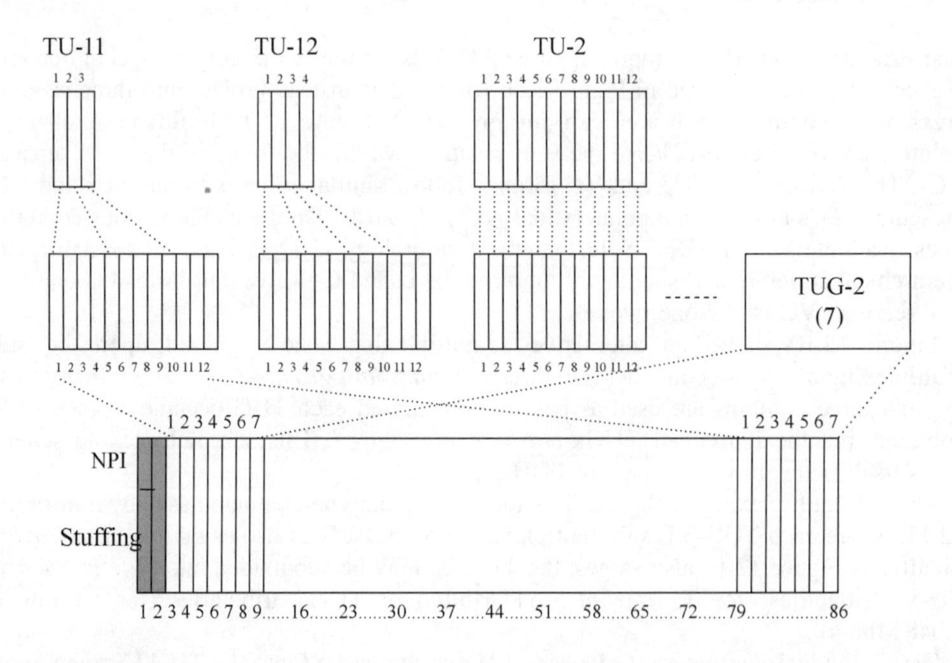

Figure 12.11 TUG-3 submultiplexing into 7 × TUG-2; TUG-2 submultiplexing

Table 12.2 Meaning and function of the fields in the SDH path overhead (POH)

Field	Name	Function
BIP-2	bit inserted parity	error check function
FEBE	far end block error	indication of received BIP error
L1, L2, L3	signal label	indication of VC payload type
remote alarm	remote alarm	indication of receiving failure to transmitting end
J1	path trace	verification of VC-n connection
B3	BIP-8 parity code	error check function
C2	signal label	indication of VC payload type and composition
G1	path status	indication of received signal status to transmitting end
F2	path user channel	provides communication channel for network operating staff
H4	multiframe indicator	multiframe indication
Z3, Z4, Z5	bytes reserved for national network operator use	reserved

demands common clocking. The advantage is the ability to directly access 64 kbit/s subchannels within the 1.5 Mbit/s or 2 Mbit/s tributary using *add and drop* methods (Figure 12.5). In addition, *byte synchronous* streams are simpler for the equipment to process.

Path Overhead (POH)

Figure 12.12 illustrates the *path overhead* (*POH*) formats used for creating VC-1, VC-2, VC-3 and VC-4 containers. This information is added to the corresponding *container*.

The meanings and functions of the various bits and fields are given in Table 12.2.

Section Overhead (SOH)

The diagram and table of Figure 12.13 illustrate the constitution of the section overhead.

Network Topology of SDH Networks

SDH equipment is designed to be used in the construction of transmission networks in redundant ring topologies from all different types of transmission media. In particular, SDH microwave radio has become popular even amongst network operators previously 'committed' to fibre. In microwave radio they see the opportunity to extend the SDH topology even to remote outlying areas, and to close otherwise open-ended fibre spurs with radio. Figure 12.14 illustrates such a mixed fibre/radio network.

A number of specific equipment types are foreseen by the standards as the building blocks of such networks. These are also illustrated in Figure 12.14.

BIP-2		FEBE	unused	signal label L1	L2	L3	remote alarm
1	2	3	4	5	6	7	8

(A) VC-1 / VC-2 POH byte

1	2	3	4	5	6	7	8
J1							
B3							
C2							
G1							
F2							
H4							
Z3							
Z4							
Z5							

(B) VC-3 / VC-4 POH (9 rows x 1 byte[8 bits])

Figure 12.12 Path overhead (POH) formats for VC-1, VC-2, VC-3 and VC-4

Columns (bytes)	1	2	3	4	5	6	7	8	9
Rows 1	A1	A1	A1	A2	A2	A2	C1		
2	B1			E1			F1		
3	D1			D2			D3		
* 4									
5	B2	B2	B2	K1			K2		
6	D4			D5			D6		
7	D7			D8			D9		
8	D10			D11			D12		
9	Z1	Z1	Z1	Z2	Z2	Z2	E2		

* Row 4 is used for the AUG frame pointers

Field	Function
A1, A2	framing
B1, B2	parity check for error detection
C1	identifies STM-1 in STM-n frame
D1-D12	data communications channel (DCC - for network management use)
E1, E2	orderwire channels (voice channels for technicians)
F1	user channel
K1, K2	automatic protection switching (APS) channel
Z1, Z2	reserved

Figure 12.13 The SDH section overhead (SOH)

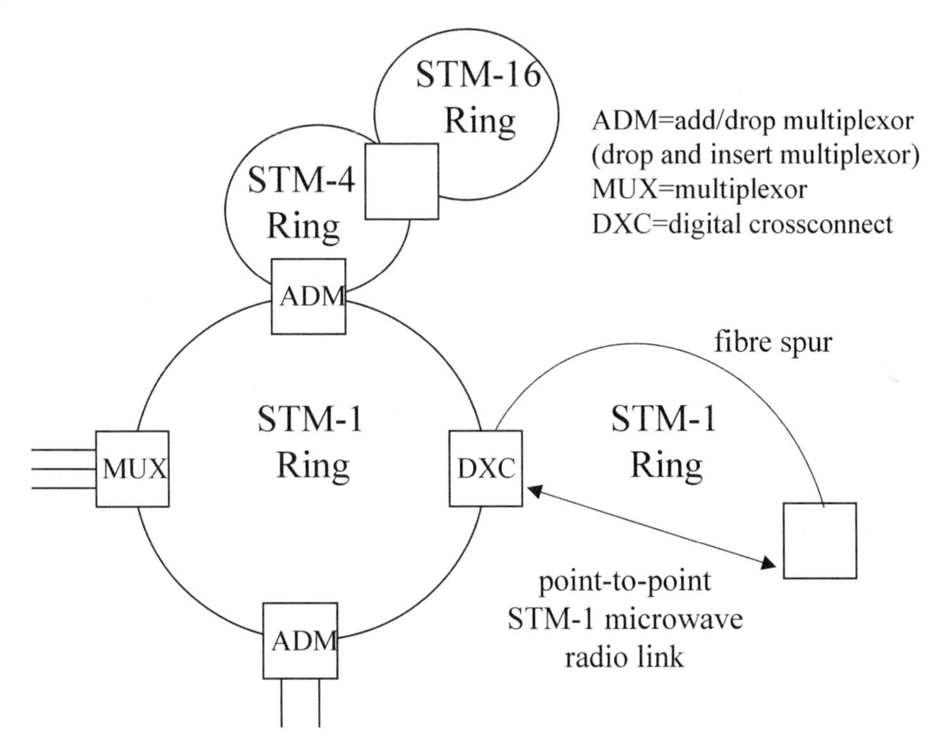

Figure 12.14 Mixed technology, ring topology and generic equipment types used in SDH networks

SDH multiplexors allow 2 Mbit/s and other sub STM-1 rate tributaries to be multiplexed for carriage by an SDH network. *Add/Drop Multiplexors* (*ADM*) allow tributaries to be removed from the line at an intermediate station without complete demultiplexing (as we discussed in Figures 12.5 and 12.6). *Cross connectors* (or *DXC*, *digital cross connectors*) allow for the flexible interconnection and reconfiguration of tributaries between separate sub-networks or rings. STM-4 and STM-16 multiplexors allow concentration of STM-1 signals onto high speed 622 Mbit/s (STM-4) or 2.5 Gbit/s (STM-16) backbone networks.

Structuring of the network in interconnected rings allows for easy back-up (*restoration*) of failed connections in the network. In a highly meshed network *n*:1 (as opposed to 1:1) restoration is possible by choosing any alternative route to the destination. In simpler networks and single rings a 1:1 restoration may be possible, but leaves at least 50% of the capacity unused for most of the time. *N*:1 restoration is useful in reducing the amount of normally unused plant (and thus costs) in cases where failures are rare.

Management of SDH Networks

Compared with PDH networks, SDH networks are more efficient and easier to administrate (due to the availability of add/drop multiplexors). Using a C-4 container at its full capacity (i.e. 149.76 Mbit/s), we achieve a system efficiency using SDH of 96% (c.f. 91% with PDH). However, apart from these benefits there is one other significant advantage — SDH networks are much easier to manage in operation. Partly this is due to the fact that SDH was conceived as a technology for a whole network (rather than a set of individual links);

Table 12.3 Comparison of SDH and SONET hierarchies

North American SONET	Carried Bitrate / Mbit/s	SDH
VT 1.5	1.544	VC-11
VT 2.0	2.048	VC-12
VT 3.0	3.152	—
VT 6.0	6.312	VC-21
—	8.448	VC-22
—	34.368	VC-31
—	44.736	VC-32
—	149.76	VC-4
STS-1 (OC-1)	51.84	—
STS-3 (OC-3)	155.52	STM-1
STS-6 (OC-6)	311.04	—
STS-9 (OC-9)	466.56	—
STS-12 (OC-12)	622.08	STM-4
STS-18 (OC-18)	933.12	—
STS-24 (OC-24)	1244.16	—
STS-36 (OC-36)	1866.24	—
STS-48 (OC-48)	2488.32	STM-16
STS-96 (OC-96)	4976.64	—
STS-192 (OC-192)	9953.28	STM-64

partly it is due to the fact that SDH is simply more modern, and therefore the available network management tools are more advanced.

The SDH standards, in contrast to PDH standards, set out a set of functions for monitoring and reconfiguration of remote equipment. This is achieved by the dedication of a defined management channel within the section overhead. This is referred to as the *Data Communications Channel* (*DCC*), or sometimes the *Embedded Communications Channel* (*ECC*). A number of ITU-T recommendations are in the course of preparation to define functions compatible with the telecommunications management network which can be supported by the DCC. These will include facilities for performance monitoring (by sending a continuous given bit pattern and measuring the received signal), remote loopback and testing facilities, as well as remote configuration capability.

12.3 SONET (Synchronous Optical NETwork)

SONET is the name of the North American variant of SDH. It is the forerunning technology which led to the ITU's development of SDH. The principles of SONET are very similar to those of SDH, but the terminology differs. The SONET equivalent of an SDH synchronous transfer module (STM) has one of two names, either *Optical Carrier (OC)* or *Synchronous Transport System (STS)*. The SONET equivalent of an SDH Virtual Container (VC) is called a *Virtual Tributary (VT)*. Some SDH STMs and VCs correspond exactly with SONET STS and VT equivalents. Some do not. Table 12.3 presents a comparison of the two hierarchies.

12.4 Backhaul in Mobile Telephone Networks

In the 1980s the advent of large scale mobile telephone networks created demand for *backhaul* networks for connecting the *Base Transmitter Stations (BTSs)* of the mobile networks to remotely located *Base Station Controllers (BSCs)* and *Mobile Switching Centres (MSCs)*. The need was met by deployment of large scale short and medium-haul point-to-point radio networks which operated in newly opened frequency bands between 10 GHz and 40 GHz (in Europe at 13 GHz, 15 GHz, 18 GHz, 23 GHz, 26 GHz, 38 GHz).

The new radio bands (in Europe) were structured to have relatively narrow radio channels of 3.5 MHz bandwidth, which was to correspond with a transmission bitrate of 2 Mbit/s (E1). (In the USA the corresponding radio channel bandwidth was 2.5 MHz, and the basic bit rate block was 1.5 Mbit/s (T1), as we saw in Figure 12.4.) The bit rate of 2 Mbit/s corresponds to the smallest multiplexing transmission level. This was the basic transmission system bit rate used to interconnect BTS, BSC and MSCs (Figure 12.15). (As an aside: as technology has moved on since the 1980s, all manufacturers of shorthaul PTP equipment are now able to carry 2×2 Mbit/s (so-called 2E1) of capacity by means of a single 3.5 MHz channel.)

Shorthaul and medium haul *point-to-point* radio links were deployed in their thousands during the 1980s, and extensive nationwide networks appeared as mobile operators' *backhaul* networks. These networks typically comprise links of multiple E1 capacity (typically $2 \times$ E1, $4 \times$ E1, $8 \times$ E1, $16 \times$ E1) arranged in a star-topology (Figure 12.15).

Links nearest to the MSC tend to have higher capacity (e.g. 8E1 or 16E1). Since many outlying BTSs depend upon a single link, it is also normal to use redundant hardware (or *protected links*) at these locations. Further towards the periphery of the network, lower capacity links are used (4E1), until either 2E1 or 1E1 links are used to connect individual outlying BTSs.

There are two main advantages in using short- and medium-haul point-to-point microwave systems for mobile backhaul:

- because of the highly directional nature of the antennas of point-to-point systems, very high capacity networks can be achieved by re-using frequency;
- the small size of the antennas needed at higher frequencies (typically 30 cm and 60 cm diameter parabolic antennas) are not too visually obtrusive, and do not need very strong mountings to put up with the wind forces on them.

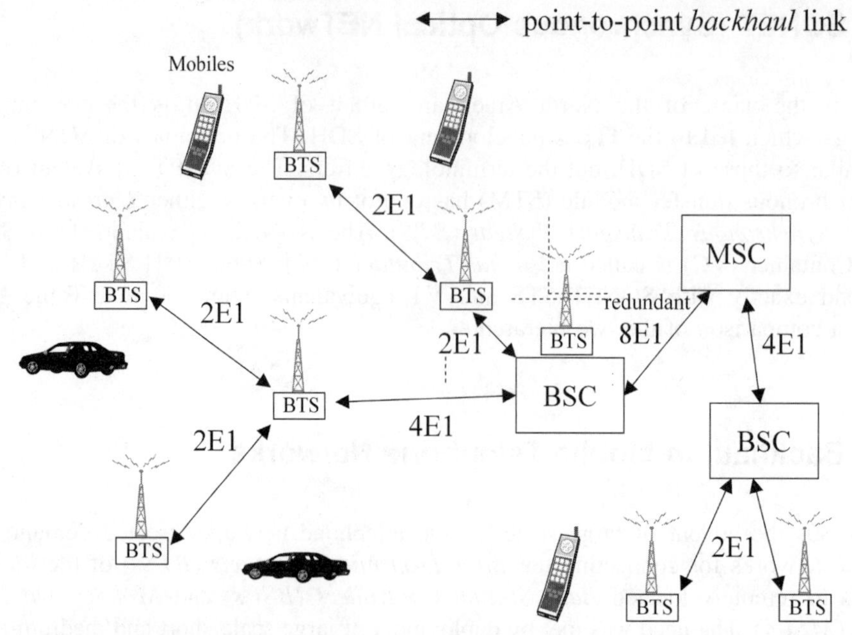

Figure 12.15 Typical star-topology of mobile backhaul networks

Because most mobile backhaul networks are based on multiples of E1 connections, it is also nowadays common to build cross connects into the network at the nodal points to enable redundant topologies to be developed, and to allow for easy remote configuration and re-configuration of lines (Figure 12.16). One of the most popular cross connect devices is the DXX from Ericsson, for example.

Progressively, point-to-point links have been used to connect ever larger numbers of ever-smaller capacity *Base Transmitter Stations* (*BTS*). It has now reached the point where many BTSs only require a backhaul link of *n**64 kbit/s (typically 384 kbit/s). Meanwhile, as we have seen, the capacity of the links themselves has grown to a typical minimum bitrate per link of 2 E1 (i.e. 4 Mbit/s in total). As a result, much of the link capacity is wasted. This is a pity, since it obviously impacts on the overall frequency re-use plan, and thus the capacity of the backhaul network as a whole. To counter this problem, the cross connects of Figure 12.16 help to some extent. However, to have to install a cross connect at every BTS is not always economically viable, as they are relatively expensive devices. For this reason, it is also sometimes possible to 'chain' BTSs using a special *ABIS-protocol*, which is specifically designed progressively to consolidate multiple trunks of *n**64 kbit/s bit rate at individual BTSs along the chain, thus enabling all the BTSs to share a single E1 between each BTS site in the chain (Figure 12.17). This not only saves the capacity required in the backhaul PTP network, but also the number of E1 ports required at the switch (either BSC or MSC).

Shorthaul and medium haul PTP systems are quickly installed on simple steelwork. Typically, the antennas of these systems are also quickly aligned using a voltmeter or other measurement device (e.g. laptop computer) as we discussed in Chapter 6.

Just like their bigger brothers, the longhaul PTP systems, shorthaul and medium-haul PTP systems usually offer a number of *wayside channels* for the use of the network

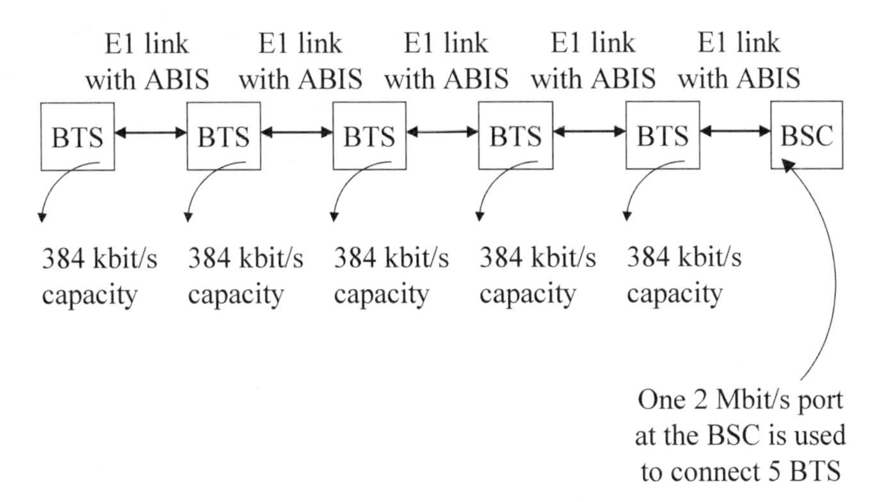

Figure 12.16 Typical use of cross connects in a backhaul network

operator. Wayside channels are low-bit rate data, *alarm extensions* or low bandwidth voice (*engineering order wire*) connections carried in parallel to the main *payload* channels across the link.

Low bit rate data wayside channels (typically one or two of 9.6 kbit/s or 64 kbit/s bit rates) are most commonly used to connect other devices at the remote location for the purpose of network management. (Either the network management port of the remote BTS or other equipment can be directly connected to the data wayside channel or a dedicated router network could be used as the basis of an Internet-protocol-based network management network. Such a router network allows a single data wayside channel to be used to connect to multiple remote devices for network mangement.)

Engineering order wires (when provided) are intended to provide a telephone connection between technicians at both ends of the link. This connection can be a useful means of

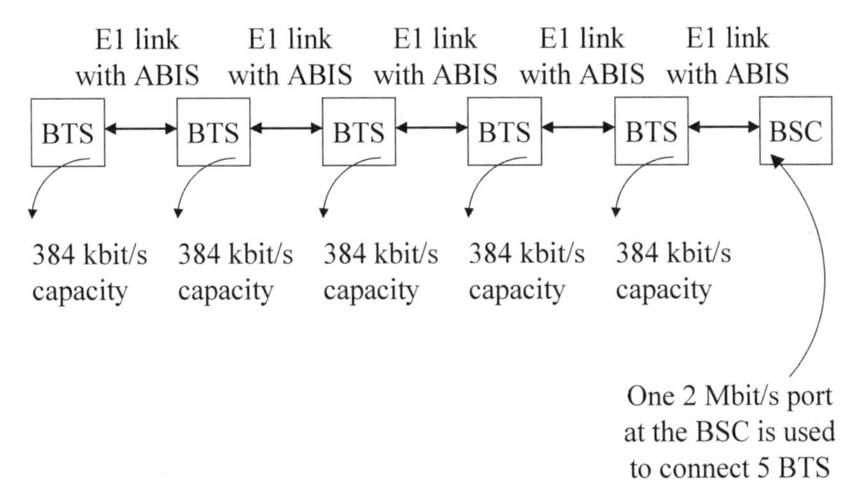

Figure 12.17 The ABIS-protocol allows timeslot and port consolidation on chained PTP links between mobile network BTSs and the BSC

communication between the two technicians during, for example, the installation and antenna alignment period. (During good weather at the time of installation, an approximate alignment of the two antennas based upon a compass reading may be sufficient to establish communication, whereafter the engineering order wire can be used for the final accurate alignment of the antennas and configuration of the equipment.)

12.5 Point-to-Multipoint Backhaul

As the demand for mobile telephone connections continues to grow in metropolitan areas, there has come to be a need for even higher densities of base stations, each using only a few 64 kbit/s (e.g. 2*64 kbits) channels as its backhaul connection. No longer is the range of the PTP system a critical factor in determining which radio band to use. Now maybe a number of *micro-base-stations* (on lamp-post tops, etc.) are within only a few kilometres range.

To establish point-to-point links to each micro-base-station may be impractical. Ever more landlords and local authorities, for example, are not willing to allow more than 4–7 antennas to be installed on a single rooftop. There is thus growing interest amongst mobile telephone network operators for the use of *point-to-multipoint* radio systems as a means of BTS backhaul (Figure 12.18).

By using point-to-multipoint radio for the backhaul links, a number of remote BTSs can be connected to a central BSC collocated with the PMP base station. Only a single sector antenna is required to connect multiple remote BTSs. Perhaps the sector illustrated in Figure 12.18 has a total capacity of 4 Mbit/s. This can be split between a number of remote BTSs in a number of different ways (e.g. 0.5 Mbit/s, 0.5 Mbit/s, 1 Mbit/s, 1 Mbit/s and 1 Mbit/s). However, not only this; the allocation could be changed from one minute to the

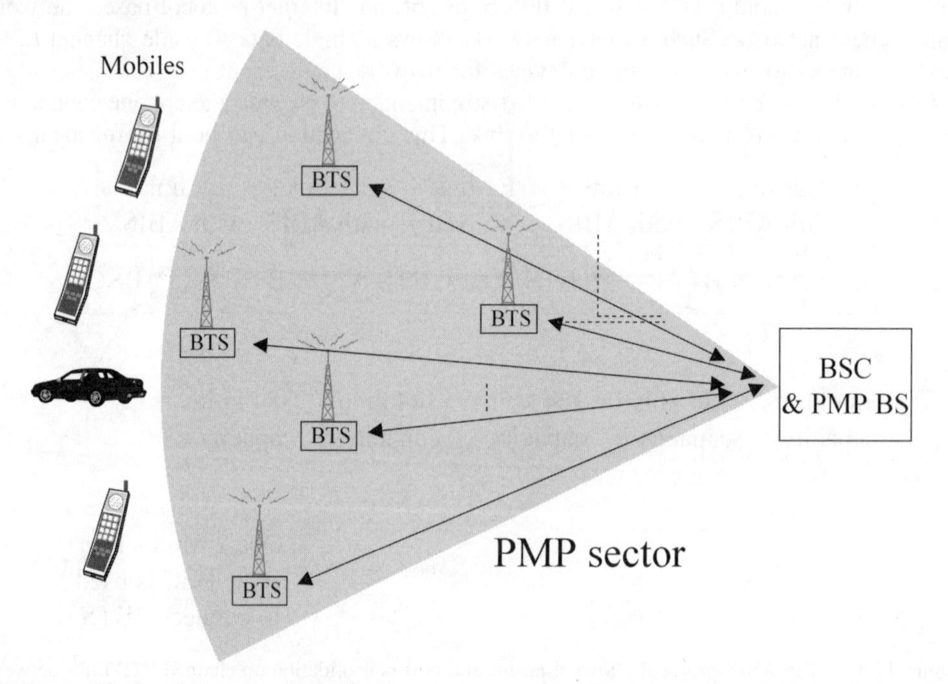

Figure 12.18 Using point-to-multipoint as a means of backhaul in mobile telephone networks

next according to the actual capacity needs of the BTSs, according to the prevailing number of mobile telephone connections. (This capability for *bandwidth-on-demand* might re-allocate the 4 Mbit/s of our example to 0.25 Mbit/s, 1 Mbit/s, 0.25 Mbit/s, 1.25 Mbit/s and 1.25 Mbit/s only a moment later.) Such bandwidth-on-demand is not possible in point-to-point backhaul networks. It might have practical use in a mobile telephone network by enabling the same PMP base station to be used as a backhaul mechanism for both metropolitan BTSs used mainly during the day and suburban BTSs used more during the evening. The BTS can hereby share not only the backhaul network, but also the switch ports. The most modern types of PMP system provide for bandwidth allocation on demand by means of ATM (Asynchronous Transfer Mode). We shall discuss such systems in a little more depth in Chapter 14. Meanwhile, it is worth noting that the use of an ATM PMP backhaul system seems also likely to be well-suited to the new generation of ATM-based mobile telephone networks which are about to emerge (*GSM3, global system for mobile communication phase 3 and UMTS, universal mobile telephone service*).

A further advantage of the star topology of Figure 12.18 (in comparison with the 'cascaded-star' topology of Figure 12.15) is that each BTS is connected by a single radio link to the associated BSC. This makes for better availability of the backhaul connection (i.e. less sensitivity to rain fading), and also reduces the need for so much redundant radio system hardware. In the PTP topology of Figure 12.15, the high capacity links near the BSC and from the BSC to the MSC were made redundant, since otherwise the failure of this link might lead to the loss of a large number of BTS and consequent severe loss of coverage. In the case of Figure 12.18, redundant hardware at the PMP base station may be considered sufficient, meanwhile avoiding redundant radio hardware at the remote BTS. After all, the loss of a single BTS is maybe not so catastrophic to the overall mobile network coverage.

Before leaving the subject of backhaul networks, it is worth mentioning the role of the different types of system (PTP and PMP) and the different types of modulation within PMP (4-QAM, 16-QAM and other *higher modulation* schemes, as we discussed in earlier

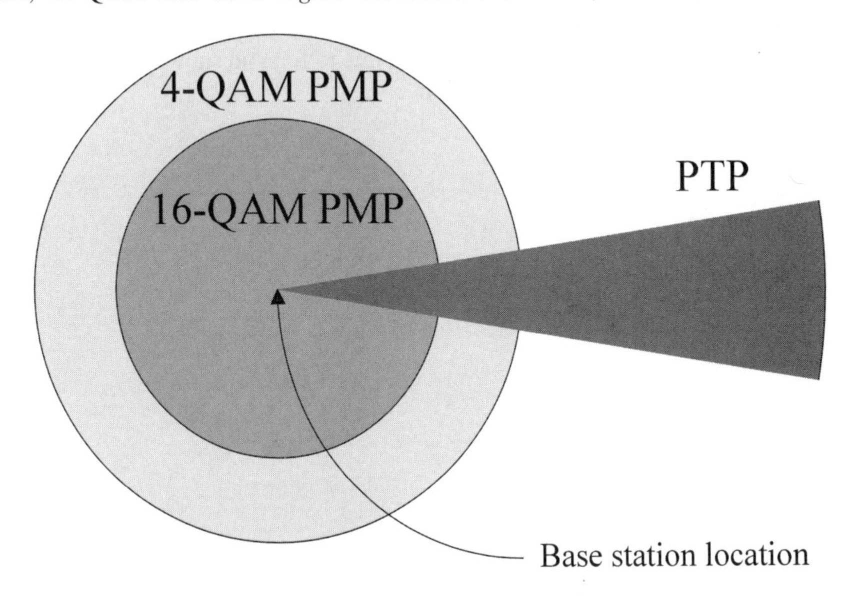

Figure 12.19 Different range and coverage of PTP and PMP technologies used for backhaul

chapters). Figure 12.19 shows the different relative coverage areas. PTP gives the greatest range in a single hop. 16-QAM PMP gives the greatest density of connections for a restricted area in the immediate vicinity of a PMP base station. 4-QAM PMP provides a compromise of range, reliability and medium density of remote stations.

12.6 Point-to-Point for High Capacity Access

The demand for radio network access and backhaul capacity continues to grow — driven not only by an increased demand for more connections, but also by a simultaneous demand for ever-higher bitrates on each of the connections. This is causing both radio regulators and equipment manufacturers alike to seek new portions of usable radio spectrum for such applications, and new more efficient technological solutions to make better use of the spectrum already available. Three different measures may help to meet this demand:

- The allocation of wide-bandwidth radio bands at very high frequency (40–42 GHz and 54 GHz) will provide for radio bandwidth. In addition, because the radio range of such high frequencies is very limited (so-called *very shorthaul* point-to-point [1–3 km]), it means that there is significant extra capacity available from the increased frequency re-use potential of these new bands.
- Higher modulation schemes will need to be used to enable high bit rate links to be realised within practical radio channel sizes. Thus, for example, it is now normal on high capacity PTP links to use at least 16-QAM to allow 155 Mbit/s to be carried in a single 56 MHz channel.
- New technologies in as yet hardly used parts of the spectrum. An example is the increasing use of laser point-to-point communications link systems for providing links between nearby buildings (up to about 1 km). Such laser systems have even more limited range than very shorthaul PTP microwave radio. This is because of the poor propagation characteristics of the even higher frequency of the electromagnetic waves. Laser systems are particularly prone to fading due to fog and snow, as well as to the rain fading with which we are already familiar.

13

Data Networks and Radio

Data communications have revolutionised not only office life, but also everyday life over the last ten years or so. The new data applications made possible by the Internet and the World Wide Web have dramatically changed office life — extending to every part of the company's business — including not only the bookkeeping department, but also the product design, the manufacturing, the mobile sales force, the haulage fleet and the travelling executive. Data communications are also becoming a vital part of everyday life — home banking and home shopping, 'teleworking' and 'distant' or 'interactive' education. For this reason, there is a growing focus amongst radio equipment manufacturers to develop wireless access solutions optimised for carriage of data and Internet communications. In this chapter, we consider the particular demands which the carriage of data places upon a radio system, and we review some of the basic types of data communications 'access' networks.

13.1 The Nature and Demands of Data Communication

Data communications differentiate themselves from telephone, television and video communications in two important respects:

- Data communications are highly *bursty* in nature. The required bit rate of information carriage fluctuates wildly between high speed transmission (with no reception) to high speed reception (with no transmission) with maybe long *idle* periods in between when no data are sent in either direction (there is no need for a constant bit rate *circuit* to exist between the two ends of a data communication for the duration of the 'call' as is the case for telephone communication).
- Data communications are highly sensitive to errors and other signal corruptions caused during transmission of the signal.

Mainly for these two reasons, data communications technology diverged from telephone network (circuit-switching) technology. Specialist data communications equipment (including *packet switches, bridges, gateways* and *routers*) is used to *switch* data traffic. In a similar manner, we can expect radio access network equipment to emerge which is optimised for carriage of data traffic. This equipment will:

- be able to support multiple *bursty* data communications devices sharing radio bandwidth on a *statistical multiplexing* basis (we discuss the benefits of this next);

- be resilient to bit error corruptions during transmission; and
- use some kind of standardised radio *air interface protocol* and/or a *Medium Access Control* (*MAC*) protocol for controlling the data transfer.

13.2 Real Communication is 'Asymmetric' and 'Bursty'

Most 'classical' telecommunications transport systems (including radio ones) make the assumption of *symmetric* two-way (*duplex*) communications; but most communications, particularly data communications are usually *asymmetric* and *half-duplex*. Usually, only one person talks at a time (or one computer is mainly transmitting or mainly receiving). Often one person (or one computer) talks much more (or sends more data) during the 'conversation' than the other.

Take the example of a person 'surfing' the Internet. The 'surfer' receives large *web page* downloads, but all he sends are a few commands to say which web pages should be downloaded.

Given the asymmetrical nature of most communications, it is inefficient (particularly in radio, where spectrum resources are limited) to 'permanently' allocate two-way (i.e. full duplex) channels to communications channels, since this wastes 50% of the available spectrum. At all points in time, one or other of the two directions is idle!

Some modern data radio systems have been specifically designed for the carriage of asymmetric communication, and are thus well-suited to providing for the carriage of high bit rates which are only required on a 'burst'-type basis. What do we mean by 'bursty'? Let us use an example to explain.

An *Internet* or other data network user, for example, typically may request a file download or *web page* download only about once every two minutes. For a few seconds, he would like to use the maximum possible bit rate available (e.g. 8 Mbit/s) so that the download is completed in the minimum possible 'response' time. Then for the next two minutes he may need no transmission capacity (and thus no radio spectrum) at all.

Let us now consider how a radio system based upon a highly sophisticated and modern form of data communications called *ATM* (*Asynchronous Transfer Mode*) can be used to provide carriage and statistical multiplexing to support this type of *bursty* service while simultaneously maximising the efficiency in the use of radio spectrum.

13.3 Statistical Multiplexing by Means of a Combination of TDMA and ATM

One methodology which has appeared for efficient *statistical multiplexing* of bursty data connections is a combination of *TDMA* (*Time Division Multiple Access*) and ATM. The technique was pioneered by Netro Corporation, one of the current leaders in the field of broadband wireless access. It is illustrated in Figure 13.1.

In the *downstream* direction (from the *Base Station* (*BS*) to the *Subscriber Terminals* (*STs*)), there is a continuous transmission of an 8 Mbit/s digital signal on each 7 MHz radio channel. The transmission, however, is segmented into *slots* of a fixed duration. These slots are directed individually at the various subscriber terminals.

In 'classical' TDMA schemes (as we saw in Chapter 5), the allocation of the downstream slots takes place on an entirely *synchronous basis*—slot one to ST1, slot two to ST2, slot three to ST3, slot 4 to ST4, then start again at ST1, etc. Meanwhile, in the *upstream* direction, a similar and *symmetrical slot* scheme is used, also with *synchronous* allocation of the transmission slots between each of the remote transmitters. In this way, both the downstream and the upstream bit rates of 8 Mbit/s is subdivided into strictly incremental and symmetrical duplex multiples of 64 kbit/s.

By contrast, the Netro *AirStar* system (Figure 13.1) does not allocate either the upstream or the downstream slots on a synchronous basis. Instead, a more flexible allocation scheme is used, made possible by 'addressing' each of the slots to its intended destination individually. This allows for *asymmetric* communication channels, leading to the possibility of *statistical multiplexing* of the *virtual connections* which are the basis of modern data communications.

The protocol used by Netro to control the whole procedure, ensuring that upstream slots are not sent by multiple remote terminals at once, is called *CellMAC*. The whole procedure and its associated *air interface protocol* enables far greater efficiency in the use of the radio spectrum, as we shall see.

We consider in Figure 13.1 the allocation of a contiguous series of nine slots in both upstream and downstream directions. In the downstream direction, five of the slots have been addressed (labelled) for subscriber terminal 1 (ST1), two for ST2 and one each for ST3 and ST4. The instantaneous bit rates received downstream by each of the devices is therefore 8 Mbit/s * 5/9; 8 Mbit/s * 2/9; 8/9 Mbit/s and 8/9 Mbit/s, respectively. Meanwhile, the allocation of the upstream slots during the same time period is different, as we see in Table 13.1.

Each of the TDMA slots in the scheme is arranged to correspond with a *single ATM* cell. The ATM methodology applied to the TDMA slots allows each slot to be individually *addressed* and sent or received from the individual subscriber terminals on an *asynchronous* (i.e. non-regular) basis. Thus, a given subscriber terminal is able to receive less slots than it sends (*asymmetric downstream*), or to send more slots than it receives (*asymmetric upstream*), or to use duplex capacity of *symmetrical* upstream and downstream bit rates.

13.4 Re-use of Upstream and Downstream Duplex Channels

By using an ATM methodology, Netro pioneered the possibility for statistical multiplexing of data communications on the radio interface. The possibility arises because in reality no communications are actually symmetric, requiring equal bandwidth (or bit rate) in both directions of communication for the entire period of connection. Rarely is one talking when one is listening, and rarely is a data device sending much when it is receiving a large file. In 'classical' wireless systems which allocate *n**64 kbit/s duplex channels, at least half of the radio spectrum is wasted for most of the time (since only one out of each pair of duplex radio channels is actually used at any one time). However, by using ATM, this is not the case. A single 8 Mbit/s duplex channel could be used to connect *Internet Service Provider* (*ISP*) and Internet 'surfer' at the same time. The ISP delivers at the full 8 Mbit/s, using the entire upstream path, while the 'surfer' receives at full 8 Mbit/s on the 8 Mbit/s downstream path of the same duplex radio channel (Figure 13.2).

- BS to ST (Downstream:) continuous ATM cells sent by BS
- STs select the cells relevant to themselves and can de-crypt

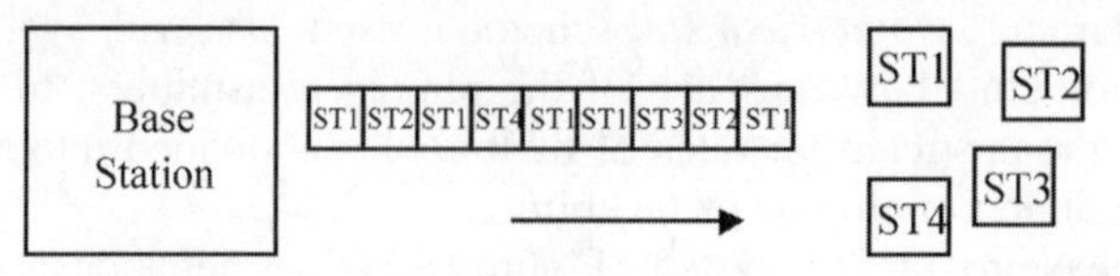

Each time slot = ATM cell plus system header

- ST to BS (Upstream): 'slotted' ATM cells sent by STs
- only STs with 'grants' may send slots

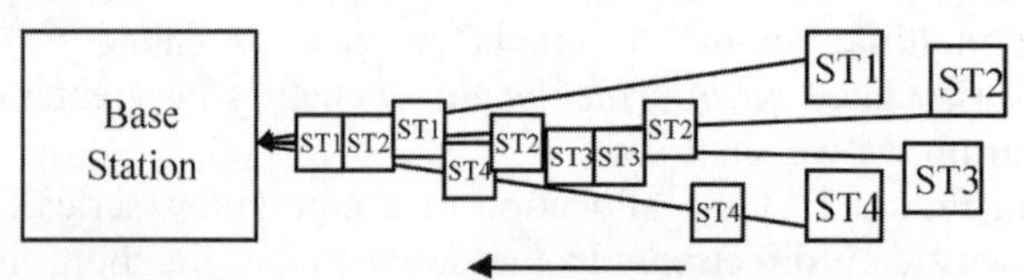

Figure 13.1 Netro's ATM-based TDMA technique

The separate use of upstream and downstream channels, as we have seen in Figure 13.2, offers the potential for a 2:1 better use of the spectrum. However, in addition, the fact that, even during a communication, the participants are neither 'talking' nor 'listening' for a good proportion of the time, leads to a possibility for further *statistical multiplexing*. This makes for very efficient carriage of 'bursty' data, as we explain next by means of an example.

More Customers can be Connected Simultaneously

Imagine a WLL system with 8 Mbit/s of capacity in a given sector, allocated to individual customers on a 'permanent' or 'bandwidth-on-demand' basis of n*64 kbit/s allocations. At

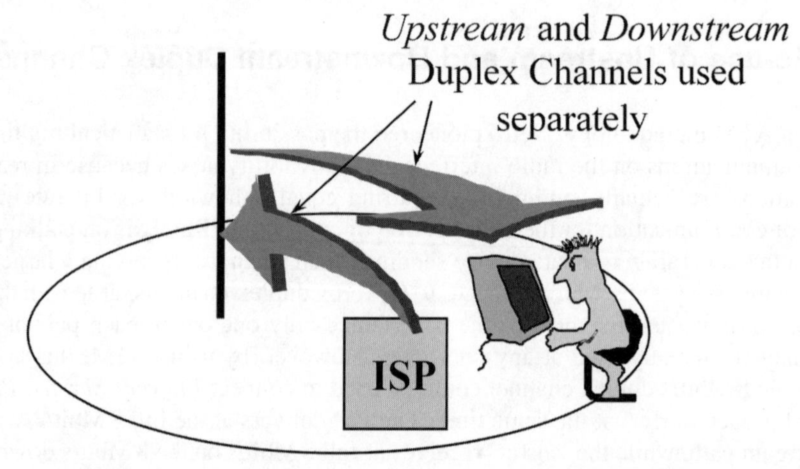

Figure 13.2 Re-use of the upstream and downstream duplex channels

Table 13.1 Asymmetric allocation of TDMA slots in upstream and downstream directions

Subscriber terminal	Instantaneous *upstream* bit rate (Mbit/s)	Instantaneous *downstream* bit rate (Mbit/s)
ST1	4.44	1.78
ST2	1.78	2.66
ST3	0.89	1.78
ST4	0.89	1.78

maximum, 120 simultaneous Internet 'surfers' can be connected to the network for active communication (with one permanently allocated 64 kbit/s channel each). Perhaps these users remain connected for many hours each evening, blocking use by any further users. Meanwhile, they only actually 'click' to receive a web page about once every two minutes, so wasting much of the transmission capacity. And having clicked, the download takes perhaps 15 seconds, because the linespeed is only 64 kbit/s. So the actual line loading is 15 seconds/120 seconds, or 12.5% usage of the downstream channel, and virtually 0% usage of the upstream channel (6.25% efficiency in use of total spectrum employed).

By using instead a TDMA/ATM scheme as illustrated in Figure 13.1, we can allocate each user only a *virtual* connection (by this we mean that the end device can react as if the connection were available all the time, but by means of statistical multiplexing we only actually consume the valuable radio spectrum when information or data actually needs to be sent). Thus, many more users can be connected than the 120 of the 'classical' WLL system. Based on the above example in which 15 seconds of actual usage is made of each available 120 seconds, at least eight times as many more customers can be connected simultaneously (and this before we consider the further capacity still left idle in the upstream direction). This is not the only benefit, though, for ATM/TDMA also gives far better quality of service to the end customer, for now the webpage download occurs at the full 8 Mbit/s — so taking only about one eighth of a second for the same download that previously needed 15 seconds. The customer is much happier!

Error Correction Ensures the Quality of Radio Data Communications!

It is commonplace in modern digital radio systems to use *Forward Error Correction (FEC)* to minimise signal corruptions of digital signals transmitted over radio paths. We discussed this in Chapter 6. FEC provides for considerable improved quality and resistance to signal corruption and errors. However, for very sensitive data applications, even the use of straightforward FEC may not yield high enough connection quality. It is therefore sometimes also further complemented by a technique known as bit interleaving.

Bit Interleaving

Imagine that our normal forward error correction scheme uses a four-bit code to protect an eight bit payload. For the sake of our example, let us assume that this code is able to detect up to 1 errored bit within the complete frame of payload and FEC bits (12 bits in total).

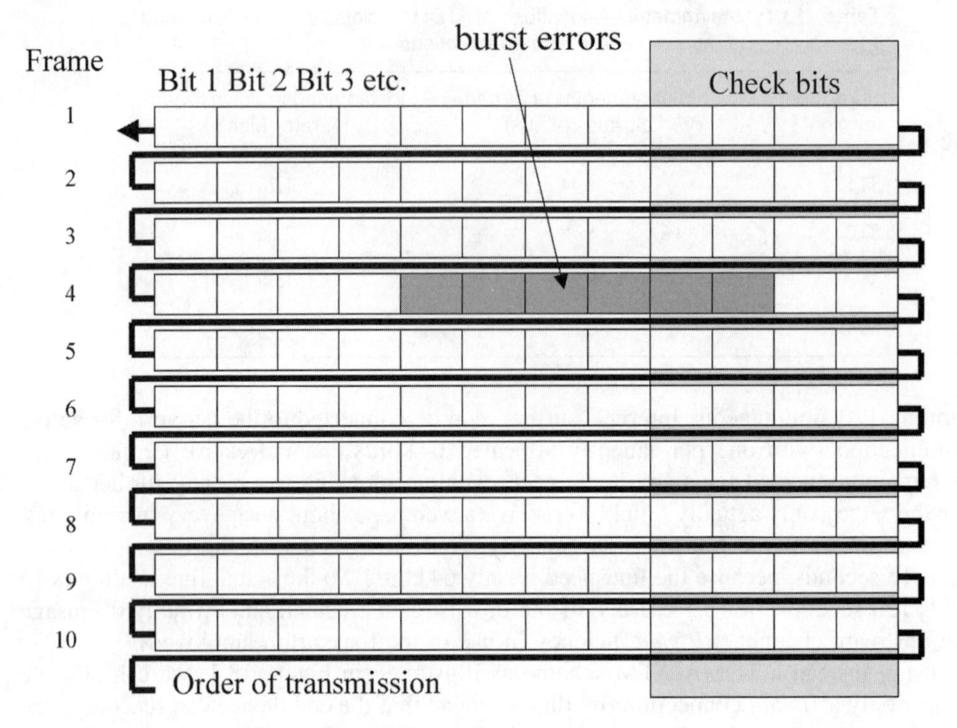

Figure 13.3 The effect of burst errors when sending frames sequentially

Now let us consider the reality of radio. Radio fades tend to lead to *burst errors* (i.e. strings of many bit errors, one after the other), though over longer periods of time quite high quality average BER (Bit Error Ratio) is achievable. For the sake of a simple example, let us assume that a burst of errors is typically six consecutive bit errors, but that the long term average BER is six bit errors in 120 bits sent. (In reality, many more than six consecutive errors may constitute a *burst error*, but much better BER is typically achieved than 6/120=0.05. Typical values of radio systems are less than than one error in 10^6 bits sent; BER better than 10^{-6}.)

Let us consider the consequence of the burst error of six consecutive bit errors, should we send the ten frames of 12 bits each sequentially, one after the other. In Figure 13.3 we have assumed that the burst error occurs on the 41st–46th bits of 120 (10 frames in total). The greyed out bits are those which get corrupted as a result of the *burst error*.

Because the entire burst error occurs within frame 4 of Figure 13.3, the FEC is unable to cope with the errors, since the maximum correctable number of errors, as we have said, is only one per FEC frame.

This problem can be overcome by means of *frame* or *bit interleaving*, as shown in Figure 13.4. Figure 13.4 illustrates an alternative order of transmission of the bits of the 10 consecutive frames. It is important that the FEC codes of each frame are coded before the interleaving. The bits are then sent bit 1 of frame 1, bit 1 of frame 2, bit 1 of frame 3 . . . up to bit 1 of frame 10, then bit 2 of frame 1, etc. When the bits are sent like this, a burst error of the bits 41–46 leads to only one bit error per frame in frames 1–6 (as shown in Figure 13.4). However, one bit error per frame is correctable by the FEC, so we are able to eliminate the effects of burst errors by *bit interleaving*.

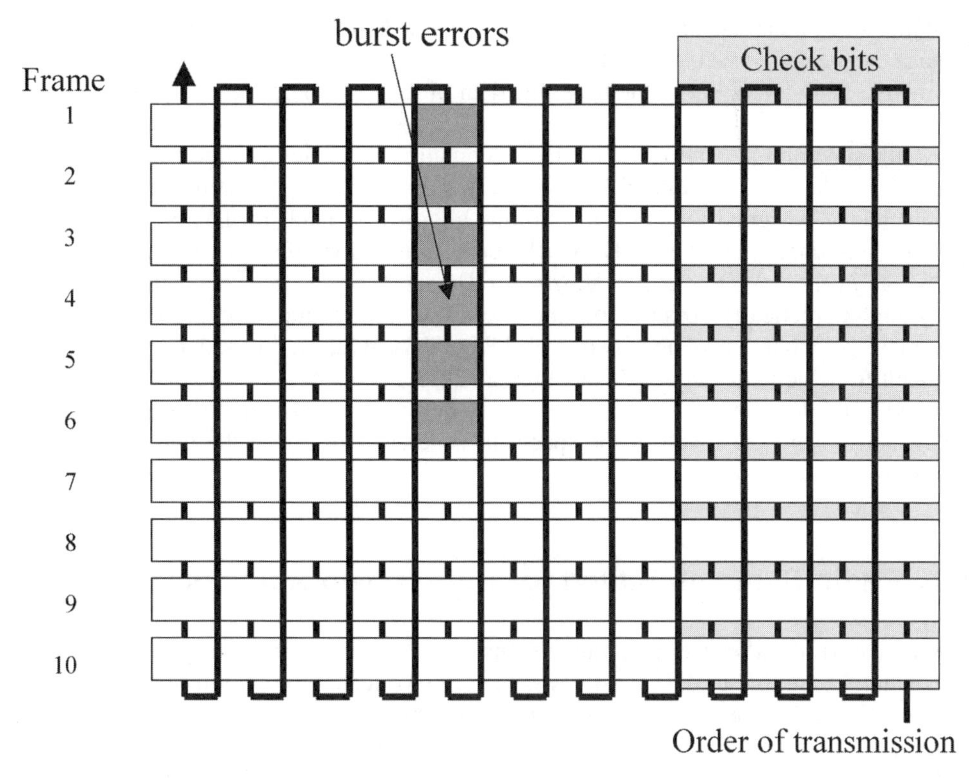

Figure 13.4 The elimination of the effects of burst errors by using bit interleaving to complement forward error correction

The disadvantage of bit interleaving is the extra delay inflicted upon the propagation time of the signal. The first frame cannot be sent until at least 10 frames are ready to be transmitted. At 125 μs per frame, this equates to an extra one-way propagation delay of at least 1.25 ms! Not significant, you might think, but as we discussed in Chapter 10, cumulative propagation delays can mount up to cause other signal quality problems.

13.5 Time Division Duplex (TDD)

Before we move on to describe some of the basic types of radio systems used for data communications, we should also note and explain the increasing trend towards consideration of *Time Division Duplexing* (*TDD*) for data radio communications systems.

Time division duplex as opposed to *Frequency Division Duplex (FDD)* creates two separate channels for *transmit* and *receive* directions of communication (or *downstream* and *upstream*) by sub-dividing the entire available radio spectrum over time rather than pre-allocating exactly one half portion of the frequency range to transmit and the other half to receive. The advantage of TDD is the ability to adjust the relative amount of spectrum used for the transmit and receive directions over time, so enabling asymmetrical connections to be *statistically multiplexed* over a period of time. Thus, the entire available spectrum could be dedicated to the downstream direction of Figure 13.2 (if the ISP were connected to another base station or by means of fibre to the network). This would double again the rate at which a web page can be delivered over the Internet.

However, TDD too is not without its practical problems. There are two principal ones:

- First, the spectrum allocations of the radio regulations agencies do not always suit TDD. Thus while the *DECT* (*Digital European Cordless Telephone*) system we discussed in Chapter 11 is designed for TDD, most of the microwave bands for PTP and PMP usage at higher frequencies are not. Even where the regulations theoretically might permit TDD (as in the *LMDS* (*Local Multipoint Distribution Service*) band in the USA, it may not be practically possible to build a radio with the required tuning range to make the most potential out of the theory of TDD. Very wide tuning ranges of radios are difficult to achieve at frequencies as high as 28 GHz!
- Secondly, the frequency re-use planning may be made much more complicated by the uncertainty over how much of the spectrum is actually in use in the two directions of transmission.

13.6 Radio Systems Suited for Data Network Access

Since the world of radio data communications is undergoing rapid change as the leading manufacturers scramble to develop a new generation of technology optimised for *Internet* and *World Wide Web* access, it is difficult to anticipate the full capabilities of radio data communications which might already be under development. We shall therefore have to make do with a quick review of some of the more widely-used standardised systems as an introduction to the design considerations and problems of practical radio data networking. We shall discuss:

- radiopaging;
- mobile data networks and TETRA;
- Wireless LANs (Local Area Networks), including the ETSI-standardised system, HIPERLAN in particular.

Radio Paging

Radio Paging was the first major type of network enabling transfer of short data messages to mobile recipients. Initially, it was a method of alerting an individual in a remote or unknown location (typically by 'bleeping' him) to the fact that someone wishes to converse with them by phone. Subsequently, the possibility to send a short text message to the mobile recipient has become commonplace.

To be paged, an individual needs to carry a special radio receiver, called a *radio pager*. The unit is about the size of a cigarette box, and is designed to be worn on a belt, or clipped inside a pocket. The person carrying the pager may roam freely and can be *paged* provided they are within the radiopaging *service area*. The service may provide a full nationwide coverage.

Paging is carried out by the caller dialling a specialised telephone number, as if making a normal telephone call. Text messages (consisting of alphanumeric characters) are either relayed by the radiopaging operator, or sometimes it is also possible to input the message using videotext, email or a similar data network service. The most advanced receivers when used in a suitably equipped radiopaging network, are capable of messages up to 80 characters long.

The key elements of a radio paging system are the *Paging Access Control Equipment* (*PACE*), the paging *transmitter* and the paging *receiver*. The PACE contains the electronics necessary for the overall control of the radio paging network. It is the PACE which codes up the necessary signal to alert only the appropriate receiver. This signal is distributed to all the radio transmitters serving the whole of the geographic area covered by the radio paging service. On receiving its individual alerting signal, the receiver bleeps.

A special digital code is used between the PACE and all the paging receivers. It enables each receiver to be distinguished and alerted. A number of digital codes were developed in the late 1970s and early 1980s, amongst them the Swedish MBS code (1978), the American GSC code (1973) and the Japanese NTT code (1978). The most important code, now common throughout the world, is that stimulated by the British Post Office. Known as the *POCSAG* code, after the advisory group that developed it (the *Post Office Code Standardisation Advisory Group*), it was developed over the period 1975–1981 and was accepted by the *CCIR* (*Consultative Committee for International Radio* — the forerunner to *ITU-R*) as the first international radio paging standard. It has a capacity of two million pagers (per zone) and a paging rate of up to 15 calls per second. Further, it has the capability for transmitting short alphanumeric messages to the paging receiver. It works by transmitting a constant digital bit pattern of 512 bits per second. The bit pattern is segregated into *batches*, with each batch sub-divided into eight *frames*. A particular pager will be identified by a 21-bit *radio identity code*, transmitted within one (and always the same one) of the eight *frames*. It is this code, when recognised by the paging receiver, that results in the alerting bleep.

The pager itself is a small, cheap and reliable device. Most are battery operated, but if the pager were to be on all of the time, the battery life would be very short, so a technique of battery conservation has become standard. We have already described how the radio identity code is always transmitted in the same frame of an eight frame batch to a particular receiver. This means that receivers need 'look' only for their own identity code in one particular frame, and can be 'switched off' for seven-eighths of the time; this prolongs battery life.

Paging receivers include a small wire loop aerial, and because of the low battery power can only detect strong radio signals. This fact needs to be taken into account by the radio paging system operator when establishing transmitter locations and determining transmitted power requirements, and by the user when expecting important calls. The *radio fade* near large buildings can be a major contributor to low probability of paging success.

The paging access control equipment stores the database of information for determining which zones the customer has paid for, and for converting the telephone numbers dialled by callers into the code necessary to alert the pagers, and in addition, it performs the coding of textual alphanumeric messages. The PACE also has the capability to queue up calls if the incoming calling rate is greater than that possible for alerting receivers over the radio link. Further, the PACE prepares records of customer usage, for later billing and overall network monitoring.

The most advanced modern paging radio paging systems are *satellite paging systems*. These work in exactly the same way as terrestrial radio paging systems, except that the transmitted signal is relayed via a satellite in order to achieve a global coverage area. This enables the roaming individual to receive his messages wherever he is in the world.

Mobile Data Networks

Mobile radio is an awkward medium for carrying data. Interference, fading, screening by obstacles, and the hand-off procedure between cells all conspire to increase errors, and the error rate over mobile radio can be as high as 1 in 50 bits.

Very basic systems with slow transmission speeds (say 300 bit/s) have been used. At these rates little data is lost, and connections that are lost can be re-established manually. However, for more ambitious applications, error correcting procedures must be used — normally a combination of forward error correction and automatic *re-request retransmission.* In this technique, sufficient redundant information is sent for data errors to be detected, and the original data reconstructed even if individual bits are corrupted during transmission. Typical speeds achieved are 2.4–4.8 kbit/s.

The appearance of mobile data networks was largely stimulated by the taxi industry. *Press to speak* private mobile radio systems first appeared in taxis as a means for controlling taxi fleet movements. A taxi customer calls a telephone number, where a number of operators act to accept orders and *despatch* available taxis to pick clients up. The *despatching* process occurs by radio. After each 'drop-off', a taxi driver registers his position and receives instructions about where he can 'pick-up' his next client.

By the mid-1980s, the *press to speak* despatch systems had become unable to cope with the size of some of the large metropolitan taxi despatch consortia. It was getting difficult to be able reliably to contact all the drivers, and wearing on the drivers always to have to listen out for calls. Computer despatch systems were being introduced for the automation of taxi route planning, and the natural extension was direct computer readout to the individual drivers of their planned activities. By computer automation, it became possible to ensure despatch of a client order to a particular taxi driver, who could be automatically prompted to acknowledge its receipt and his acceptance of the order. Simple confirmation by the driver ensures precise computer tracking of pick-up time and a successfully completed fare. Subsequent computer analysis of journey time statistics could further help future journey planning.

Now there was a need for data networking via radio. Most of the systems devloped to answer this need evolved from the previous press-to-speak *Private Trunk Mobile Radio* (*PTMR*) systems used in the taxi and regional haulage business beforehand. As a result they tend to use a similar radio frequency range for operation, and a similar 12.5 kHz or 25 kHz channel spacing. The derived user data bit rates achievable are typically around 7200 bit/s per connection, but once the overheads necessary to ensure the reliable and bit error free transport of the user data are removed, the effective data rate of some systems does not exceed 2400 bit/s. Miserable, you might think, when compared to fixed network data applications running at 64 kbit/s or even higher rates, but quite adequate for the short packet (i.e. around 2000 byte packet messages (approximately 2000 characters)) for which the systems were developed.

The three best known manufacturers of low speed mobile data networks are Motorola (its *Modacom* system), Ericsson (Eritel's *Mobitex* system) and *ARDIS*. The systems find their main application in private network applications within metropolitan or regional operations (for haulage or taxi companies), or on campus sites, essentially providing radio-based X.25 packet networks, as Figure 13.5 shows. There have been a number of attempts at providing commercial nationwide and even international public service networks, but these have not been a great success.

TETRA (Trans-European Trunked Radio System)

Despite the relatively low interest in low speed mobile data networks, and the emergence of the GSM and DECT systems as overpowering competitors both for voice service via *trunk mobile radio* and data carriage via *Modacom*-like low speed mobile data networks, there has been continued effort applied by ETSI to agree the *TETRA* (*Trans-European Trunked Radio*) series of standards. These are intended to provide for harmonisation of trunk mobile radio networks across Europe, opening the way for pan-European services and the use of identical equipment.

Work on the TETRA standards started in ETSI in 1988, when a system to be called *Mobile Digital Trunk Radio System* (*MDTRS*) was foreseen. This was renamed TETRA in 1991. A series of standards have now been published, which can be classified into two different broad system categories:

- TETRA V+D is a system for integrated *voice and data*;
- TETRA PDO is a system for *packet data only*.

The first of these systems is intended as an ISDN-like replacement for analogue trunk mobile radio systems. The second system is a standardised version of the *Modacom*-like systems, but with higher data throughput capabilities. Table 13.2 lists the bearer and teleservices planned to be made available.

Figure 13.6 illustrates the basic architecture of the TETRA system. The concept foresees a normal connection between a *Line Station* (*LS*) and a *Mobile Station* (*MS*) via a *base station* and *Switching and Management Infrastructure* (*SwMI*). Thus, a typical example would be a taxi computer despatch centre as a line station connected to the fixed ISDN network, accessing one or more (typically many) mobile stations. Similar to the DECT system, the data base is conceived to take over *home data base* and *visitor data base* functions, in order to allow roaming of mobile stations between different base stations and

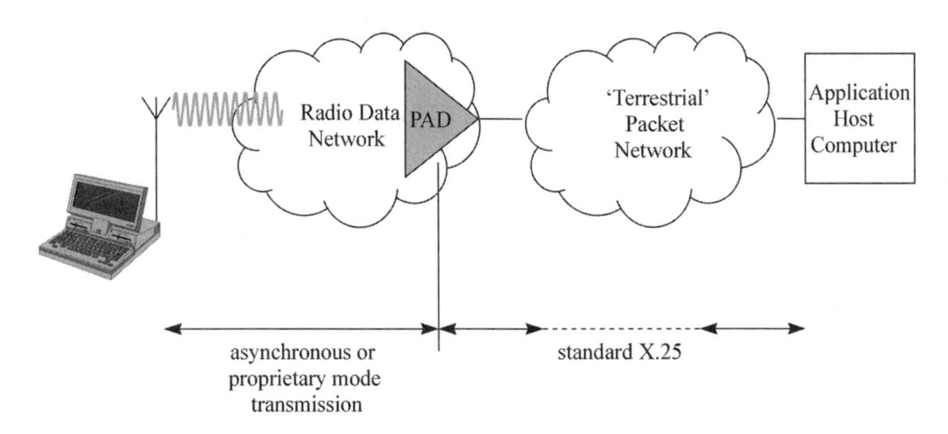

Figure 13.5 Typical arrangement of a mobile data network

Table 13.2 Bearer and teleservices supported by the various TETRA standards

	TETRA V+D (voice and data)	TETRA PDO (packet data only)
Bearer services	7.2–28.8 kbit/s circuit-switched voice or data (without error control)	
	4.8–19.2 kbit/s circuit-switched voice or data (some error control)	
	2.4–9.6 kbit/s circuit-switched voice or data (strong error control)	
	connection-oriented (CONS) point-to-point packet data (X.25)	connection-oriented (CONS) point-to-point packet data (X.25)
	connectionless (CLNS) point-to-point packet data (X.25)	connectionless (CLNS) point-to-point packet data (X.25)
	connectionless (CLNS) point-to-point or broadcast packet data in non-X.25-standard format	connectionless (CLNS) point-to-point or broadcast packet data in non-X.25-standard format
Teleservices	4.8 kbit/s speech encrypted speech	

even between different TETRA networks. The *Inter-System Interface* (*ISI*) allows for interconnection of TETRA networks operated by separate entities. The various *c-plane* and *u-plane Air Interfaces* (*AI*) are designed to conform with OSI.

Table 13.3 presents a brief technical overview of the TETRA system.

Wireless LANs

The idea of *Wireless LANs* (*WLANs*) has been around as long as *LANs* (*Local Area Networks*) themselves. Indeed, the first LAN, developed by the Xerox company based on the ALOHA-protocol, which became the basis of Ethernet, was based on a radio medium.

There are two main benefits of wireless LANs when compared with cable-based LANs:

- the ability to support mobile data terminals (for example, employees using laptop computers at various different desk locations within a given office building);
- the ability to connect new devices without the need to lay more cabling.

Two standards for wireless LANs have been developed: the IEEE 802.11 standard, and the ETSI *HIPERLAN* (*high performance LAN*) standard. They are not really intended to be used as access networks to public data networks, but are ideal as radio data access networks for campus or extended campus areas. We describe here the ETSI HIPERLAN system.

In a wireless LAN, each of the devices to be connected to the LAN is equipped with a radio transmitter and receiver suited to operate at one of the defined system radio channel frequencies. For the HIPERLAN system, five different channels are available, either in the band 5.15–5.30 GHz or in the band 17.1–17.3 GHz, but only one of the channels is used in

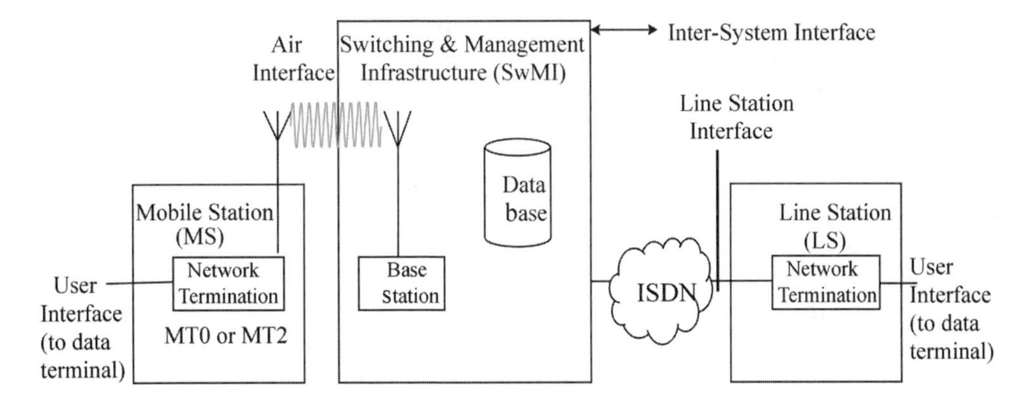

MT0, mobile termination type 0 provides a non-standard terminal interface

MT2, mobile termination type 2 provides a TETRA standard R-interface

Figure 13.6 Basic architecture of the TETRA system

a single LAN at a time. The radio channel has a total bit rate close to 24 Mbit/s, but the maximum user data throughput rate is around 10–20 Mbit/s, i.e. of similar capacity to a cable-based Ethernet or token ring LAN.

When a device wishes to send information, this is transmitted in a manner similar to that used in an Ethernet LAN. In other words, the information is simply transmitted to all other

Table 13.3 Technical overview of the TETRA system

Radio bands	Uplink: 380–390 MHz 410–420 MHz 450–460 MHz 870–888 MHz	Downlink: 390–400 MHz 420–430 MHz 460–470 MHz 915–933 MHz
Channel separation	25 kHz	
Transmitter power (max)		
Channel multiplexing	V+D: TDMA (time division multiple access), with *S-ALOHA* on the random access channel PDO: *S-ALOHA* with *data sense multiple access* (*DSMA*)	
Duplex modulation	*FDD* (*Frequency Division Duplex*), 10 MHz spacing	
Frame structure	V+D: 14.17 ms/slot, 510 bits per slot, 4 slots per frame PDO: 124 bit block length with *Forward Error Correction* (*FEC*). Continuous downlink transmission, burst uplink ALOHA	
Modulation	$\pi/4$ *DQPSK* (*Differential Quaternary Phase Shift Keying*)	
Connection set-up time	circuit switched connection, less than 300 ms connection oriented data, less than 2 s	
Propagation delay	V+D: less than 500 ms for connection-oriented services 3–10 seconds for connectionless services PDO: less than 100 ms for 128 byte packet	

terminals in the LAN, as soon as the radio channel is available. All devices participating in the LAN 'listen' to the radio channel at all times, but only 'pick up' and decode data relevant to themselves. The structure of the LAN is therefore very simple, as Figure 13.7 illustrates, but all devices must lie within about 50 metres of one another — because of the 1-watt maximum radio transmit power allowed.

The multiple radio frequencies (five per band) defined in the HIPERLAN standard allow multiple LANs to exist beside one another, and even overlapping one another. Without multiple frequencies, different LANs in adjacent offices might not be possible, and multiple LANs in the same office certainly not.

The 50 metre maximum diameter of the LAN could also be a major constraint in some circumstances. For this reason, the radio *MAC* (*Medium Access Control*) of HIPERLAN provides a forwarding (or relay) function. When the forwarding function is configured into the wireless LAN, a number of the stations are defined to be *forwarders*. The other stations are *non-forwarders*. The forwarding stations relay messages as necessary between themselves, broadcasting the message to any relevant *non-forwarding stations* within their sector, as shown in Figure 13.8.

We shall not discuss the detailed radio channel access methodology and medium access control techniques here, but as an overview, Figure 13.9 presents the protocol reference model. Note that the use of the standard IEEE 802.2 *Logical Link Control* (*LLC*) enables HIPERLAN to be used as a one-for-one replacement of an existing LAN. Unlike a normal LAN, however, two further sublayers are used beneath the LLC layer. In addition to a *Medium Access Control* (*MAC*) sublayer, a *Channel Access Control* (*CAC*) sublayer is also used. This is necessary to accommodate the control mechanisms necessary for the radio channel.

The LLC sublayer of OSI layer 2 provides for correct and secure delivery of information between two terminals connected to the LAN (error detection, correction, etc.). The MAC sublayer provides for the delivery of the information to the correct endpoint, providing for LAN addressing, data encryption and relaying as necessary. The *Channel Access Control* (*CAC*) sublayer codes the MAC information into a format suitable for transmission across a radio medium. The technique used is called *Non-Pre-emptive Priority Multiple Access*

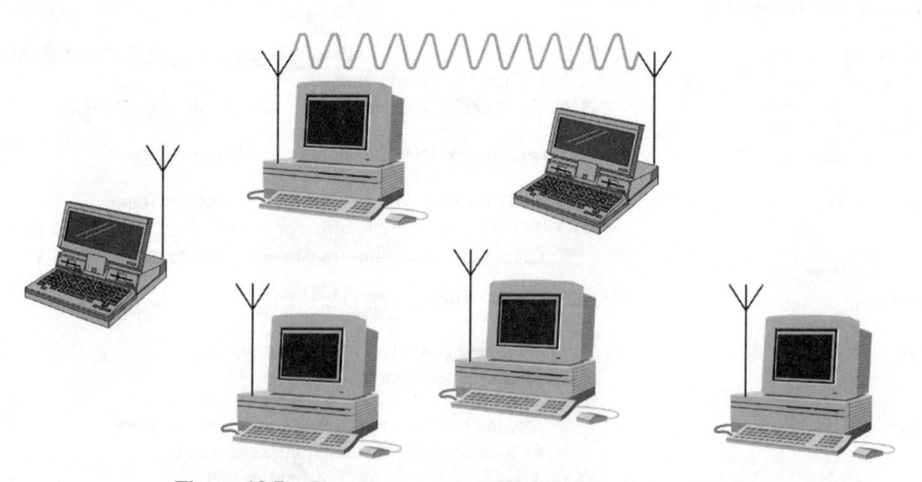

Figure 13.7 The structure of a HIPERLAN wireless LAN

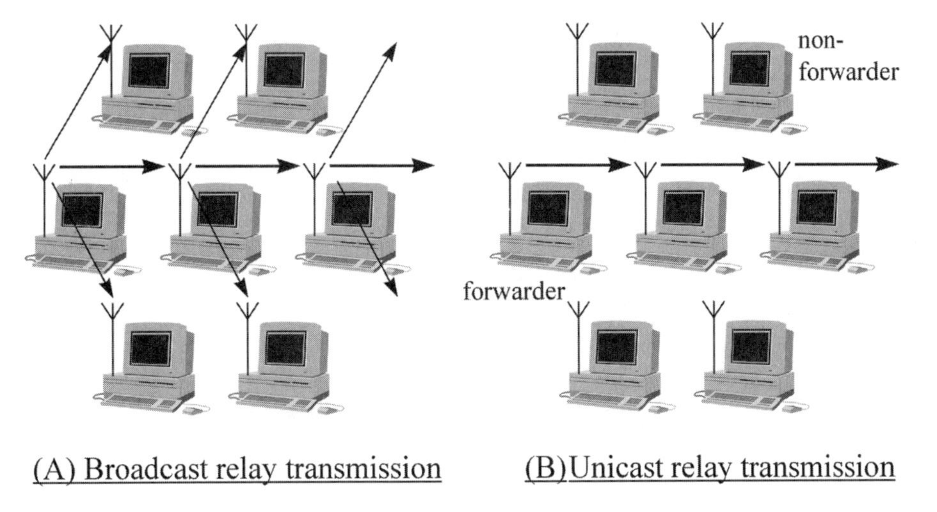

Figure 13.8 The forwarding function of HIPERLAN

(*NPMA*). NPMA breaks up the radio channel into a number of *channel access cycles*, each of which is further sub-divided into three cycle sub-phases:

- a *priority resolution phase*;
- a *contention resolution phase*;
- a *transmission phase*.

In the priority phase, any stations which do not currently claim the higest priority transmission status, are refused permission to transmit. The remaining stations compete for use of the radio channel during the next phase and any *contention* is resolved. The remaining transmission phase is then allocated to successful stations surviving both the *priority resolution* and *contention resolution* phases. This is the phase when user data is transmitted. Priorities will change from one cycle to the next to ensure all stations have an equal ability to send data.

So much for the strengths of wireless LANs. The greatest difficulty is achieving complete radio signal coverage throughout an office. *Multipath* effects, interference and propagation difficulties can lead to *blackspots* suffering very deep radio *fade* (i.e. poor transmission). For static devices, the problem of a fade caused by multipath of interference can be solved by moving the device only a small distance. For mobile terminals continuous good quality transmission may not be possible.

13.7 Third-generation Mobile Telephony

Another important development in the area of mobile data communications is the development of the third generation of mobile telephone network technology. This is to be based on wideband CDMA (W-CDMA), and is specifically designed to allow data communications (including Internet access from mobile telephone handsets). We shall have to wait to see what bit rates are possible in practice with these networks.

14

Broadband Wireless Access

As competition has increased in the world's deregulated telecommunications markets, so the network operators have sought ever more innovative ways to win market share. Fixed wireless access has allowed new operators to compete more effectively with the established ex-monopoly competition, without having to rely on them for the provision of access network services. This has helped the new operators to reduce their costs and, through better control of their own infrastructure, be more responsive to their customers' demands and service expectations. Initially, it was sufficient for operators to offer telephone and ISDN (Integrated Services Digital Network) bit rates on their access lines, but as the margins available from public telephone services have dropped due to fierce competition between providers, there has been a significant increase in demand for high speed data services, particularly for accessing the Internet and other IP (Internet Protocol)-based services. As a result, broadband wireless access has emerged. Most of the modern telecommunications operators who are applying for radio spectrum allowing them to operate Wireless Local Loop (WLL) services are planning to offer a multimedia portfolio of voice and broadband data services. We review in this chapter the different classes of equipment designed to offer 'broadband wireless access', and the considerations which should accompany the planning and operation of such a network.

14.1 The Different Philosophies for Spectrum Allocation

The type of broadband wireless access services which may effectively and economically be operated within a given country depend to a large extent upon the regulatory regime governing the allocation and use of the relevant radio spectrum.

There are two basic (but very different) manners in which regulators have chosen to allocate spectrum. In the American model (followed by the United States' regulator, the Federal Communications Commission), spectrum is auctioned off in large quantities to the highest bidder. Thus, for example, the LMDS (Local Multimedia Distribution Service) auctions run in the USA during 1999 resulted in two network operators in each region of the United States (called a *BTA*, or *Basic Trading Area*) receiving, respectively, 1150 MHz and 150 MHz of spectrum allocation — equivalent to 575 MHz and 75 MHz duplex capacity. With this amount of spectrum available, high bit rate services (perhaps even as high as 155 Mbit/s) can be offered to multiple access network customers simultaneously, no matter whether point-to-point or point-to-multipoint technology is used in the wireless access network.

By contrast, European regulators have tended instead to allocate much more limited spectrum to each network operator. Their intention in doing so has been to ensure highly efficient usage of the spectrum, more competition between operators and a reduced likelihood that the spectrum 'winner' will simply sell-on the spectrum for speculative gain. As a result, European point-to-multipoint wireless access network operators in most are having to plan their networks using typically only 14 MHz, 28 MHz, 56 MHz of spectrum (the very lucky ones — and only in some countries — might receive 112 MHz) within any given geographical region. Since the bit rate maximum bitrate which can be offered to end customers is of the order of one bit-per-second per Hertz of spectrum, thus many European operators are offering maximum bitrates around 8 Mbit/s, or 34 Mbit/s per remote customer terminal.

One thing is common to both the north American and European operators — none of them have access to the unlimited spectrum resources. No matter what the spectrum licensing regime, all the operators and their chosen equipment manufacturers are trying to squeeze maximum potential from the allocated spectrum — trying to cheat the laws of physics by offering individual customers ever higher bit rates while simultaneously increasing the number of customers, without increasing the amount of radio spectrum. In the remainder of the chapter, we consider some of the various different approaches to the challenge.

14.2 The Higher Frequency the Radio Band the Greater the Capacity!

In general, one can say that the higher the frequency band of operation of a fixed wireless access system, the higher the potential network capacity. This is because both the bit rate available per end-customer and the number of end customers can be increased. The reasons for the increase in capacity with higher frequency band of operation are two-fold:

- The historical difficulty of harnessing very high frequency radio bands has meant that there is still significant, as yet unused, spectrum in the higher frequency bands, and so more potential for very large bandwidth allocations.
- The decreasing range of radio systems with increased frequency of operation improves the potential for *frequency re-use*. Because radio base stations as a result can be placed more closely together, so a higher density of remote terminal connections becomes possible.

An approximate empirical relationship between the maximum available bit rate and the band of operation is shown in Figure 14.1, which is based on currently-deployed radio access technology.

Figure 14.1 considers cellular telephone networks to be point-to-multipoint networks operating in the 450 MHz, 900 MHz, 1800 MHz and 1900 MHz bands. Most modern cellular network technology is based upon digital transmission, at a rate around 10 kbit/s per end-user. The transmission can be used for either voice or data signal carriage. While fax transmission and basic Internet services and email might be possible with the latest types of mobile telephones, and despite the fact that the third generation of mobile network standards (called *UMTS*, or *Universal Mobile Telephone Service*) is also designed to support Internet services, cellular mobile networks of this type are not suited to the transmission of large email attachments or the downloading of bitrate-hungry Internet web pages.

The box labelled 'telephony' *Wireless Local Loop* (*WLL*) systems in Figure 14.1 represents the various systems which have appeared in recent years for various bands (and in relatively limited bandwidths) between about 1 GHz and 4 GHz (specifically 2.2 GHz, 2.6 GHz and 3.5 GHz, amongst others). These systems have mostly been designed to be alternative means of providing access lines to public telephone networks, specifically in city suburbs or remote and sparsely-populated rural areas.

The most modern telephony WLL systems are also capable of supporting $n \times 64$ kbit/s services, including *Basic Rate ISDN* (*BRI*) (at 144 kbit/s). While some of these systems also support rates even as high as 2 Mbit/s, such rates are not intended to be the 'norm' for the target end customers of such systems. At such high bit rates, too high a proportion of the operator's available spectrum (typically 14 MHz) is used by a single customer, thereby reducing the capability to 'share' spectrum between many users, and so detracting from the economic viability.

The European broadband point-to-multipoint (PMP) systems also shown in Figure 14.1 typically have bit rates per end customer up to about 8 or 16 Mbit/s. Much higher bit rates are generally excluded by the amount of spectrum made available to a single broadband wireless operator. Such systems are nonetheless ideally suited to provide an access network medium for end users of public telephone network services and Internet or IP-based services. Typically, *broadband* wireless access systems have been designed and developed for the European bands at 10 GHz, 26 GHz and 28 GHz (CEPT-frequency-raster).

End customers who generate only occasional usage of communications lines are ideally suited to share the medium of a point-to-multipoint wireless access system. (By 'occasional use' we mean that the end-customer does not make telephone calls all the time, and only now and then — typically about once every two minutes when actively surfing — calls for a web page to be downloaded to him over the Internet.) Thus, for example, a modern office local area network working at fast Ethernet speed of 100 Mbit/s (100 baseT) could be connected to a remote LAN. For short periods of demand, the two networks could swap information at 16 Mbit/s. Meanwhile, during the 'idle' periods between transmissions, other LAN connections could be using the 'spare' wireless network capacity for their communications. In a similar manner, radio spectrum allocated for carrying telephone calls can be re-allocated to different end-users in line with their actual call-usage, so that only as many radio channels are needed as the maximum number of consecutively held conversations (this is a much lower number than the total number of customers 'connected' to a wireless access system. Such re-allocation of spectrum (either for carriage of 'bursty' data or 'occasional' calling) makes ideal use of the 'sharing' capability of a point-to-multipoint system.

'Burst' data transmission or 'occasional' use is progressively more rare at higher line speeds. Most lines today operating at 34 Mbit/s, 45 Mbit/s, 52 Mbit/s or 155 Mbit/s, for example, are actually used as 'trunk' lines between switching devices. Thus, 34 Mbit/s trunks are common as trunks between routers in *IP* (*Internet Protocol*)-based data networks. 155 Mbit/s lines are typically trunks in *ATM* (*Asynchronous Transfer Mode*) networks, or are used as the *SDH* (*Synchronous Digital Hierarchy*) trunks between public telephone network switch sites. Trunk lines are not well-suited as applications for point-to-multipoint networks. This is because the telecommunications traffic on a trunk circuit is usually continuous, and thus not suited to the medium sharing principle of PMP systems. (The function of the switch which is collecting traffic onto the trunk is, after all, to make

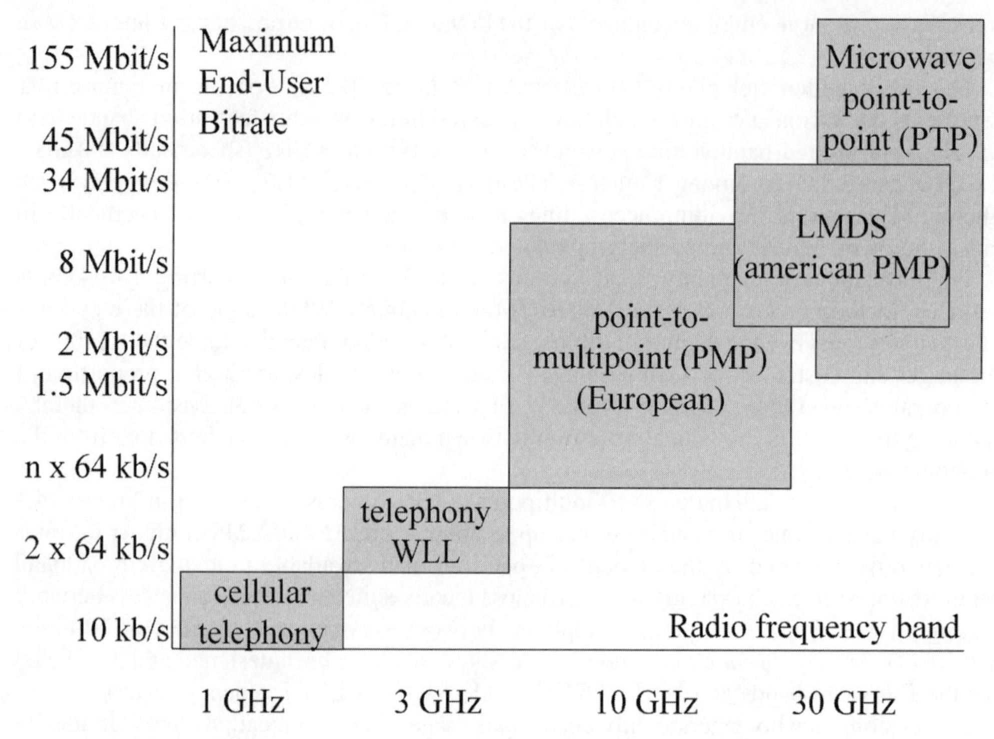

Figure 14.1 Empirical relationship between radio band and maximum supportable bit rate

maximum use of the trunk by collecting traffic from many different sources to fill it.) Trunk circuits and bit rates much higher than about 16 Mbit/s are thus the realm of point-to-point radio access network technology.

Point-to-point (*PTP*) fixed wireless access technology, as shown in Figure 14.1, can be used effectively and efficiently to carry very high speed *access lines* or trunks from public telecommunications network operators to their customers. Indeed, a number of wireless access network operators (e.g. Winstar and ART) have appeared in the USA, and made successful businesses from providing such high speed network access services using high bit rate PTP technology. Modern shorthaul PTP technology is able to support bit rates typically ranging from 2 × 2 Mbit/s (European hierarchy) or 2 × 1.5 Mbit/s (north American hierarchy) up to 155 Mbit/s.

A 155 Mbit/s, a *PTP* link might need a 56 MHz channel to be used, but the restriction of the use of the spectrum to the direct line between the two link end-points (achieved by the use of directional antennas) makes for the potential of significant *re-use* of the same spectrum for multiple other links in the same region. However, while some regulators permit the use of such high bit rate channels, others limit the maximum available channel bandwidth to around 28 MHz (typically equivalent to 34 Mbit/s, 45 Mbit/s or 52 Mbit/s).

As well as being more spectrum-efficient in the carriage of high bit rate *trunk*-type connections, point-to-point wireless technology is also typically more economic than point-to-multipoint technology for carrying such connections. (At high bit rates the relatively high cost of point-to-multipoint base stations can only be shared between relatively few remote users.)

Finally, Figure 14.1 shows point-to-multipoint equipment designed for the LMDS and MMDS (*Multichannel Multipoint Distribution Service*) bands in north America (at 28 GHz (FCC-frequency raster) and 40–42 GHz). These systems typically aim to address higher bit rates than those targetted by European PMP systems. The higher bit rates have come about due to three causes:

- the original focus of the *multipoint distribution* systems deployed in north America was for pay-TV broadcasting and video-on-demand;
- the American telecommunications market typically demands higher bit rates earlier than the European market — and end customers are already demanding 45 Mbit/s, 52 Mbit/s and even 155 Mbit/s bit rates for network access connections;
- the spectrum allocation regime of north America (auctioning huge bandwidths to the highest bidder) have allowed the operators the scope to offer higher bit rates than possible under the European spectrum regime. North American, LMDS equipment thus tends to address multiples of 1.5 Mbit/s, up to about 45 Mbit/s or 52 Mbit/s.

A number of different radio bands are in use in the north American point-to-multipoint world (called variously MMDS-, LMDS- or LMCS-). However, while the band names differ, there is coming to be commonality between the type of equipment used in each of the bands and the main applications and market targetted by the network operators (Internet access and *interactive TV*, as well as telephone/ISDN).

The operators nowadays simply apply for any spectrum they can get. The technical standards to be used in each band have largely been left by the regulator for the operators to develop and decide as they choose. So that, instead of developing equipment to pre-determined technical standards and then tendering this equipment to successful spectrum-winning operators (as in the European PMP-market), the manufacturers catering to the north American MMDS-, LMDS- and LMCS-markets have tendered for contracts to develop equipment to the individual operators specifications. As a result, the technologies differ slightly from one band to another and tend to be operator-specific (Teligent at 24 GHz, Nextlink at 28 GHz, Winstar at 39 GHz, etc.). Some operators and manufacturers, for example, choose to use *Frequency Division Duplex* (*FDD*), while others have chosen *Time Division Duplex* (*TDD* — see Chapter 13), believing TDD is better suited to carriage of Internet-based data traffic.

14.3 MMDS (Multichannel Multipoint Distribution System)

The initial intention of the MMDS bands was for terrestrial television channel distribution. As terrestrial broadcast television channels in the *VHF* (*Very High Frequency*) and *UHF* (*Ultra High Frequency*) ranges became rapidly oversubscribed during the 1980s, caused by the advent of multichannel television and pay-TV viewing, so television and cable operators started to look for alternative solutions for terrestrial television and video broadcasting. One of the earliest solutions was simply to use conventional television transmission techniques, but in the microwave radio frequencies. The *MMDS* (*Multichannel Multipoint Distribution Service*) was born. While it is used widely in some countries, other countries have moved rapidly to *Direct Broadcasting by Satellite* (*DBS*) of pay-TV channels.

The initial focus of MMDS was broadcast television, and a range of different microwave bands were allocated in the USA in the early 1990s by the FCC for *CARS* (*cable television relay service* in the bands 2 GHz, 6 GHz, 12 GHz, 18 GHz and 31 GHz) allowing in some cases very large radius broadcast 'cells' to be established—up to 25 km range. However, as demand continued to grow, and as public telecommunications operators recognised the potential for using an adaptation of the technology for providing two-way and interactive multimedia communications, telephony and Internet access, an allocation of much higher frequency and shorter range radio bands became necessary. So came about *LMDS* (*Local Multipoint Distribution Service*) at 28 GHz and 40 GHz.

Other related technologies are *MDS* (*Multipoint Distribution Service*) and *IFTS* (*Instructional fixed Television Service*).

14.4 LMDS (Local Multipoint Distribution Service)

The emergence of *LMDS* (*Local Multipoint Distribution Service*) was a consequence of the heavy demand and rapid exhaustion of the MMDS system radio bands in the United States. The central idea, multichannel distribution of wideband signals, is common to both systems, but the range of LMDS is lower. As a result, it is only suited to 'local' distribution. This has two major benefits—first, the higher frequency has meant more bandwidth is available, and secondly, the shorter range allows more frequent frequency re-use. So, the overall capacity of the network and the bit rate possible per user is somewhat higher.

The spectrum allocated by the United States Federal Communications Commission (FCC) for LMDS is in the range between 27.5 GHz and 31.3 GHz (Figure 14.2). In common usage the band is often referred to as the '28 GHz band', but this should not be confused with the CEPT 28 GHz band (which is between 27.5 and 29.5 GHz, and has a different channel and duplexing raster, as we discussed in Chapter 2).

From the start, LMDS was viewed by the United States telecommunications regulator (the FCC) as a technology which would enable new operators to offer major competitor to local exchange and cable television services. Due to the large value of this market and the huge interest amongst the operators to get hold of spectrum, the FCC elected to auction the spectrum.

The LMDS bands were first auctioned in February and March 1998, but re-auctioned in April and May 1999. A separate auction took place in each of 493 regions, termed *Basic Trading Areas* (*BTAs*). As a result of the auction, within each of the BTAs, a total of 1300 MHz of spectrum was allocated to the highest bidders (an *LMDS Block A* licence allocation of 1150 MHz and a separate *LMDS Block B* licence allocation of 150 MHz).

The band structures for Bands A and B are shown in Figure 14.2. The structure, as can be seen, is somewhat complicated (Block A is in the ranges 27.5–28.35 GHz, 29.1–29.25 GHz and 31.075–31.225 GHz; while Block B has the two chunks of spectrum 31.000–31.075 GHz and 31.225–31.300 GHz). The exact duplex channel raster and upstream/downstream channel allocations are not restricted, so there is scope for *asymmetric* use of the bandwidth (say, giving more spectrum over to the *downstream* direction for broadcasting or downloading applications. This is seen as a major benefit by some of the operators and equipment manufacturers alike, and time division duplexing (as we discussed in Chapter 13) will be deployed by some for this reason. On the other hand, the

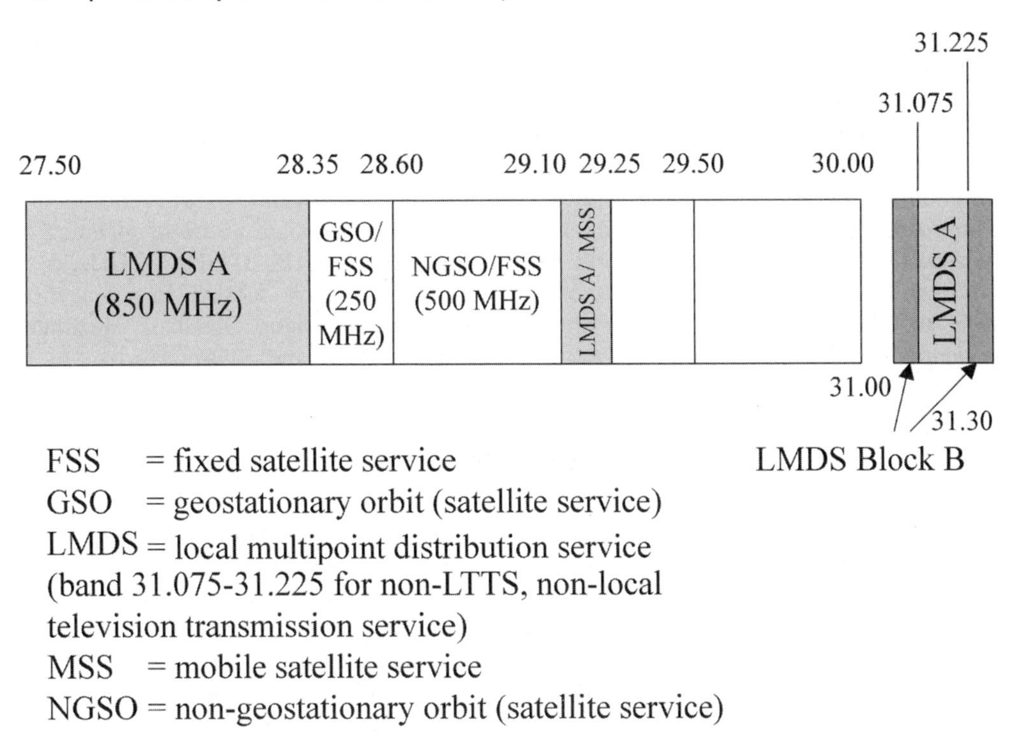

Figure 14.2 The FCC 28 GHz and 31 GHz band allocations and LMDS (Local Multipoint Distribution Service)

somewhat fragmented allocations complicate the job of equipment and network design. This has slowed the deployment of systems on the ground.

The Block B allocation would allow a 'classical' frequency division duplex use of the 150 MHz as 75 MHz duplex (equal upstream and downstream bit rate allocations), but the band was viewed by some operators as having too restricted a bandwidth to be economically viable. The Block A allocation has much larger overall bandwidth, but in fragmented pieces distributed throughout the band between 27.5 GHz and 31.2 GHz. A further problem is the co-primary status of LMDS with the *nongeostationary mobile satellite service* (*NGSO-MSS*) in the band 29.1–29.25 GHz (in effect, the double allocation of the band to two separate uses). This may mean that transmission in this particular sub-band by LMDS operators will be prohibited.

LMDS licences have been issued for a ten year term. Licence holders must demonstrate 'substantial service to their service area' in order not to lose their licences. This requirement is intended to encourage real network deployment and to prevent potential speculation with the frequencies by auction bidders who perhaps had no real intention of building a network.

One of the major owners of spectrum in the 28 GHz LMDS band in the US was a company called Nextlink. They are expected to start major network deployment during the year 2000.

The great interest shown in the LMDS band and the recognition of the potential for wireless broadband alternatives to traditional telecommunications *local loop* technologies has spilled over into a number of other spectrum bands, which had previously been allocated to operators, in some cases maybe initially with an intended focus on point-to-point communications. While these bands are not, strictly-speaking, 'LMDS bands', manufacturers of broadband wireless equipment are tending to adapt their technology for us in any of the bands. As a result the term 'LMDS equipment' is increasingly coming to be synonomous with 'broadband wireless access' or 'broadband point-to-multipoint' equipment. In this wider LMDS market, a range of frequencies are in use or planning to be used by different operators in north America (24 GHz, 28 GHz, 31 GHz, 39 GHz, over 40 GHz). Even some European operators use the colloquial term 'LMDS' in conjunction with the European and ETSI (European Telecommunications Standards Institute) standards-based point-to-multipoint bands at 3.5 GHz, 26 GHz and 40–42 GHz.

14.5 LMCS (Local Multipoint Communications System)

LMCS is thought of by some people as being the 'Canadian version of LMDS'. It is true that LMCS, like LMDS, is also a terrestrial microwave wireless system designed for pay-TV, Internet access, videoconferencing and other multimedia applications, but it is unfair to call LMCS the Canadian version of LMDS, since LMCS largely preceded LMDS in the US. As a result of a discussion paper issued in the Canada gazette in December 1994, the process of allocating bandwidth for LMCS commenced. Following an auction during 1996, a number of LMCS operators emerged. The largest of these were WIC Spectrum Inc. and Maxlink Communications Inc. Other initial licence owners were Cellular Vision Canada Ltd, Digital Vision Communications and Regional Vision Inc. The licences require compliance with the ITU radio regulations applying to the band Band 25.25–27.5 GHz.

14.6 The DAVIC (Digital Audio-Visual Council) Protocol

Spurred by the auctions of LMCS spectrum in Canada during 1996 and by the US LMDS frequency auctions, the *Digital Audio-Visual Council* (*DAVIC*) was established to ''favour the success of emerging digital audio-visual applications and services, by the timely availability of internationally agreed specifications of open interfaces and protocols that maximise interoperability across countries and applications/services. Membership is open to any interested organisation who declares himself individually and collectively committed to open competition in the development of digital audio-visual products, technology and services''.

DAVIC has issued a number of specifications (DAVIC 1.0, 1.1, 1.2, 1.3, 1.4 and 1.5). The initial version, DAVIC 1.0, is the most widely deployed version today, and is designed to support the following applications by means of MMDS or LMDS systems:

- TV distribution;
- video-on-demand;
- Basic teleshopping.

The subsequent versions of DAVIC specifications are designed to add the following further functionality:

Table 14.1 Structure and classification of the DAVIC 1.0 specification

	Part number	Document title
Group 1		DAVIC 1.0 tools
	7	High-layer and mid-layer protocols
	8	Lower-layer protocols and physical interfaces
	9	Information representation
	10	Security
	11	Usage information protocols
Group 2		DAVIC subsystems
	3	Service Provider System architecture and interfaces
	4	Delivery System architecture and interfaces
	5	Service Consumer System architecture and interfaces
Group 3		System-wide issues
	1	Description of DAVIC System functions
	2	System reference models and scenarios
	12	Dynamics, reference points and interfaces

- PSTN/ISDN and Internet Access (DAVIC 1.1);
- 3D graphics, virtual reality, copyright protection and 'watermarking' (DAVIC 1.2);
- multiplayer games and mobility (DAVIC 1.3, published September 1997);
- management architecture and protocols (DAVIC 1.4, published June 1998);
- DAVIC 1.5, published April 1999.

Like other modern protocols, the DAVIC protocol suites are designed as a series of *layers*, *functions* and *subsystems*. Table 14.1 classifies the various parts of the DAVIC 1.0 specification accordingly.

Potential protocols which might compete with the DAVIC protocol for supremacy in broadband wireless access networks are the protocols being developed by the IEEE's (Institute of Electrical and Electronic Engineers) influential 802-committee. Specifically, the 802.11 and 802.16 standards may overtake DAVIC. Another potential competing protocol standard is that being developed for broadband wireline local loop technologies (e.g. for *ADSL, Asymmetric Digital Subscriber Line*).

14.7 MVDS (Multipoint Video Distribution System)

MVDS is similar to MMDS and LMDS, and is also a microwave-based wireless distribution system designed for pay-TV, video broadcasting and interactive (Internet-like) services. The technical standards designed for MVDS are termed the *DVB* (*Digital Video Broadcasting*) standards and were prepared by ETSI (European Telecommunications Standards Institute). One of the most important standards is EN 300 748.

In principle, the standards allow the MVDS system to be operated in any of the bands between 2.5 GHz and 40 GHz, whereby the standards are segregated into the DVB-MC suite for frequencies below 10 GHz (designed for use on cable-TV networks) and the DVB-MS suite designed for terrestrial microwave (MVDS) or satellite distribution. Typically European countries seem likely to harmonise on the 40 GHz band (40.5 GHz to 42.5 GHz) for MVDS. Of the total 2 GHz bandwidth, 100 MHz is used for the *upstream* or *return* channel (this channel is a 'low speed' control channel— enabling the user to determine what high bit rate information should be 'downloaded'). Most of the available spectrum is used in the downstream direction for broadcasting videos, Internet or other interactive information.

14.8 Point-to-Multipoint

In the European arena, as well as other regions influenced by ETSI standards (Australasia, Africa, the middle East and some parts of South America and Asia) *point-to-multipoint* or *broadband point-to-multipoint* technology is more common than the US-equivalent technology LMDS, though the term LMDS is sometimes used as a slang term to cover both types of technology and radio standards.

The European market is influenced by the relatively restricted bandwidth available to each network operator. Initially, bandwidth was made available in several countries in various specific bands between about 1 GHz and 4 GHz specifically for *Wireless Local Loop* (*WLL*) access networks supporting telephone and ISDN services. However, the success of the various service offerings has varied greatly from one country to another. In those countries where WLL systems were deployed as the 'only available means' of providing public telephone service, the networks have tended to be successful and economic.

Much initial effort was placed on providing telephone network coverage for rural and remote areas. Subsequently, many of the operators have also recognised the better business potential of suburban and city areas in which a higher density and a greater number of customers can be served by a smaller number of base stations and a lower investment. Some of the most successful networks have been in Eastern Europe, Ceylon and the Philippines. These are countries where the number of telephones per head of population had previously been extremely low. The cable infrastructure was very poor.

In the more developed nations (for example, in western Europe), there has also been a move on the part of the telecommunications regulators to licence the use of similar WLL bands to new network operators, in order to enable these operators to compete more effectively with the established ex-monopoly public telephone companies. In these countries, though, WLL restricted to simple telephone service or *Basic Rate ISDN* (*BRI*) look to be doomed to failure.

The likely failure of simple telephone and ISDN-based WLL in western Europe lies in the relatively expensive economics of a wireless-based solution in comparison to the 'wireline' alternative. The problem is that microwave radio part of each subscriber terminal inflicts one or even several thousand US dollars cost on each subscriber line. When this cost is to be borne by a single residential customer with a single telephone line, it is rarely economic (except for very heavy telephone callers). The cost of the subscriber terminal increases only very slightly even if much higher aggregate bit rates, of say, 8 Mbit/s, 16 Mbit/s or even higher are offered to end customers, so that the business revenue potential can be increased greatly without increasing costs. As a result, there has been a huge renewed interest amongst wireless network operators in western Europe for offering *broadband* network access (comprising multimedia, voice, ISDN, data, Internet and interactive services), particularly targetting an attractive offer for small and medium-sized business customers, and for regional offices of multinational corporations. Broadband point-to-multipoint has been born!

14.9 Broadband Point-to-Multipoint Equipment

The 'Muliple Point-to-Point' Approach

The various equipment manufacturers have different philosophies with regard to network architecture and radio modulation techniques. The more conservative manufacturers have tended to offer systems which represent a one-for-one 'replacement of wirelines'. The approach is in effect 'multiple-point-to-point' (Figure 14.3), with user channel bit rates of 64 kbit/s or multiples thereof ($n \times 64$ kbit/s), up to the 'classical' T1 and E1 rates of

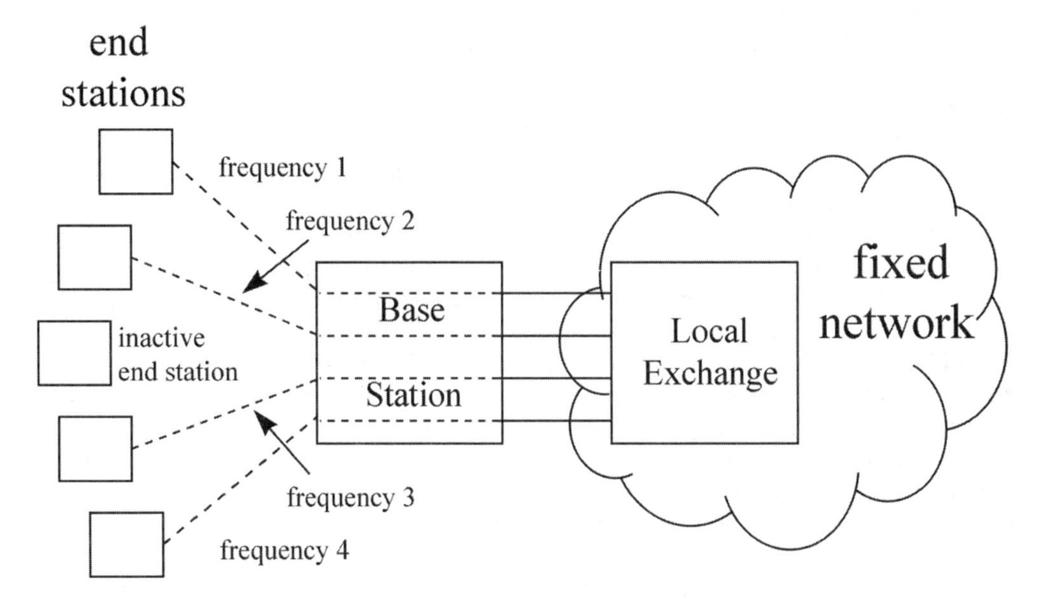

Figure 14.3 'Classical' approach to broadband wireless—multiple point-to-point connections as one-to-one replacements for wirelines

1.5 Mbit/s (North America and Japan) and 2 Mbit/s (Europe, Africa, South America, parts of Asia).

The 'wireline replacement' approach aligns with the 'classical' approach of public telecommunications operators to plan separately their 'transmission' or 'transport' networks (based on 'transparent' transmission channels, i.e. application and protocol-independent). This type of approach was adopted by most of the manufacturers who entered the WLL market early. The approach is well-suited to a radio world using frequency division duplexing, and to the strict $n \times 64$ kbit/s bit rates of the classical public telephone network operators. From a radio-modulation perspective, the approach lends itself to the adoption of either FDMA (Frequency Division Multiple Access) or TDMA (Time Division Multiple Access), as we discussed in Chapter 5.

Bosch Telecom, one of the world's major manufacturers of telecommunications transmission equipment and a very strong radio manufacturer, elected a 'multiple-point-to-point' and 'wireline replacement' network architecture (as described above) for its DMS point-to-multipoint system. (This was also initially marketed by Ericsson under the name 'Airline'.) The system is based upon FDMA modulation, with a granularity of $n \times 64$ kbit/s duplex channels, typically up to 2 Mbit/s or multiples thereof (e.g. 4×2 Mbit/s) per customer.

As an experienced radio manufacturer, Bosch (now part of Marconi) was able to 're-use' technology previously developed for satellite systems, and to refine some of the advantages of using FDMA technology (over TDMA or CDMA):

- The possibility of using very sharp spectrum masks.
- The possibility to adjust the modulation used on individual connections (i.e. radio channels), thus enabling higher modulation to be used where possible to increase the bits/s per Hertz yielded from the spectrum when lack of signal interference allow this.

The generally sharper spectrum masks of FDMA (Figure 14.4a) as compared with either TDMA or CDMA (e.g. Figure 14.4b) may mean that it is possible to use the *adjacent channels* (i.e. the next neighbouring radio frequency) in the same geographical region or *sector*. This obviously has advantages for the efficient *re-use* of radio spectrum (as we discussed in Chapter 9).

Actually, the extent to which the very sharp spectrum masks are achieved in practice depends upon the accuracy of the oscillator (the channel *centre frequency* must be precise and stable over long periods of time otherwise there will be an overlap anyway, Figure 14.4c). Nonetheless, if there is an overlap, a 'dynamic' FDMA system could use a 'dynamic interference algorithm' and simply decide not to use that portion of the spectrum which is subject to interference (Figure 14.4d).

As Figure 14.5 shows (and as we learned in chapter 5), FDMA works by dividing up the radio spectrum into a number of separate channels, each of which is used on a 'point-to-point' basis (hence 'multiple-point-to-point'). The technique allows us to allocate the radio spectrum to each user 'on-demand' (i.e. only when he actually has a need for it). The multiple-point-to-point nature of the system also allows us to adjust the modulation we use on the channel to achieve maximum bits/sec per Hertz.

Because the individual sub-channels are allocated separately to each connection as each call is *set-up*, the system can elect to use only radio channels which are free from interference. This is the manner by which the 'mask truncation' of Figure 14.4c can be carried out, and altered over time as propagation conditions change.

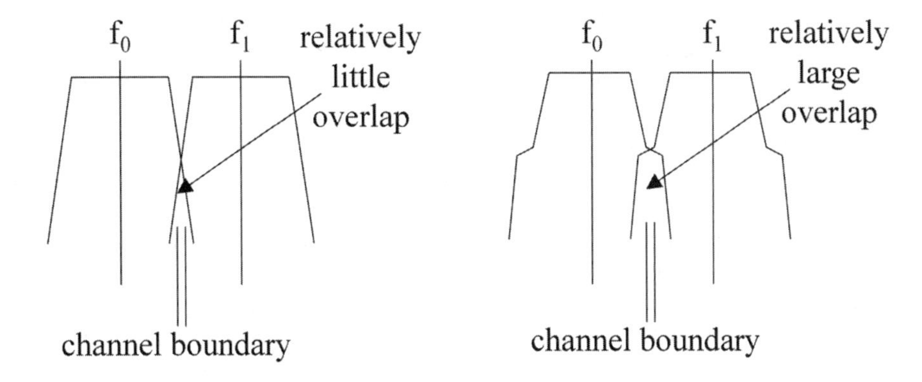

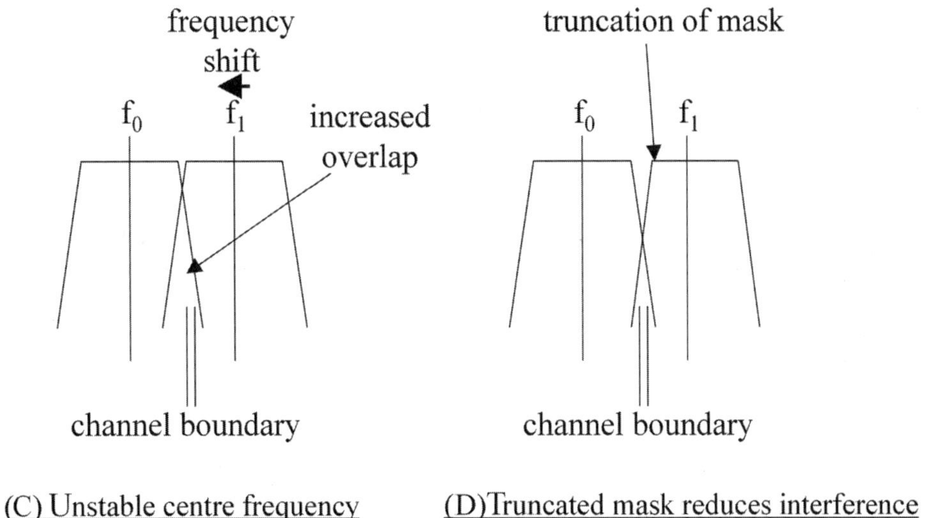

Figure 14.4 The spectrum mask of an FDMA system

The drawback of changing the modulation scheme used on each individual channel from one call to the next is that the interference sensitivity of the system varies from one call to the next. As a consequence, the radio network planning is much more complicated. To ease this problem, the equipment manufacturers may have also implemented various forms of 'interference algorithms'. This increases the complexity and cost of the system, but is not always reliable.

The disadvantage of the multiple-point-to-point approach lies in the inefficient network architecture which results. While a *concentration function* for voice services (V5.2 interface) as we discussed in Figure 1.5 of Chapter 1 (and in Chapter 11) can be realised by adding a basic switching functionality to the base station of Figure 14.3, the efficient carriage of modern data or Internet traffic is not so easy. A better approach for efficient carriage of data traffic has been sought by manufacturers using an ATM- or IP-approach, as we discuss in the next section.

FDMA and/or a multiple-point-to-point architecture is thus best suited to direct 'wireline replacement' for 'transparent carriage' of raw and 'symmetric' traffic (i.e. duplex communications — those with equal traffic in upstream and downstream directions).

The Bosch approach (indeed the DMS system) was also the initial basis of the *Spectrapoint* venture now owned by Cisco and Motorola. The business was initially set up by Texas Instruments, but taken over by Bosch (with a large injection of Bosch technology) and then sold on to Cisco and Motorola.

While most of the broadband wireless manufacturers who elected to develop based on TDMA technology have chosen an ATM- or IP-approach, as we shall discuss next, there are also a number of systems which are either FDMA or are, at least to some extent, Multiple-Point-to-Point (MPP) in nature. These include the initial version of Nortel's *Reunion* system, Floware's *Walkair* system (TDMA, but largely MPP in nature), Alcatel's 9800 and P-COMs point-to-multipoint systems. These systems variously try to exploit the capability of MPP systems for supporting very high modulation schemes, and thus for achieving very high end-customer bitrates. We discuss this approach in the last section of the chapter.

The ATM and IP Transport Approach — for Efficient Carriage of Multimedia Services

Manufacturers who entered the market for point-to-multipoint or LMDS broadband wireless access equipment within the last few years have tended to focus specifically upon the needs for high bitrate data carriage and for *multimedia* access network solutions. These are intended to allow end-customers to be offered a complete portfolio of voice, data, Internet and interactive services in a manner which is very efficient in its use of the scarce radio spectrum.

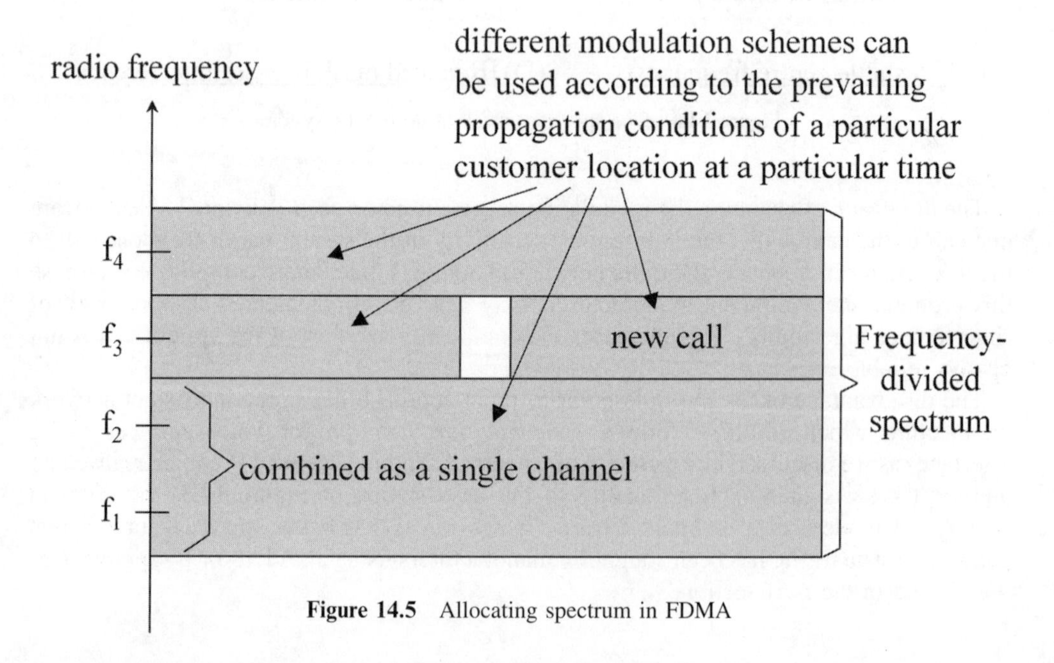

Figure 14.5 Allocating spectrum in FDMA

Generally, the approach has been to adopt TDMA and a packet-based (or *asynchronous*) transport means, such as that made possible by ATM (Asynchronous Transfer Mode) or IP (Internet Protocol). The approach is particularly effective for:

- Carriage of multimedia voice and 'bursty' data traffic.
- Carriage of *asymmetric* communications (ones in which the actual transmission is mostly only in one-direction at a time. An Internet web-page download does not require *upstream* capacity during the download, and during telephone conversation, most of the time only one person speaks at a time).
- *Overbooking* of the radio capacity, allowing for highly efficient usage of the radio spectrum, once *user information bits/s per Hertz* are considered instead of simply *bits/s per Hertz* (the distinction is that in the second case, many of the carried bits are in fact wasted, since the customer is 'silent' on the line, but nonetheless occupying the spectrum.
- Increasing the number of customers who may be connected simultaneously to the broadband wireless access network for data services within the same radio spectrum. (This is achieved by giving each a so-called *virtual connection*, which only occupies real radio bandwidth when there is actual *user information* to be transmitted).

The basics of the ATM/TDMA or IP/TDMA approach we discussed in detail in Section 13.3 in Chapter 13. As we explained in conjunction with Figures 13.1 and 13.2, one of the earliest manufacturers in this field was Netro Corporation. Their *CellMAC* protocol (air-interface protocol) which they developed specially for the function of *statistical multiplexing* has subsequently been submitted to the IEEE 802.16 committee as the basis of an industry standard.

More recently, a large number of other companies have started to develop similar products, based either on ATM/TDMA or IP/TDMA. These include Alcatel 9900

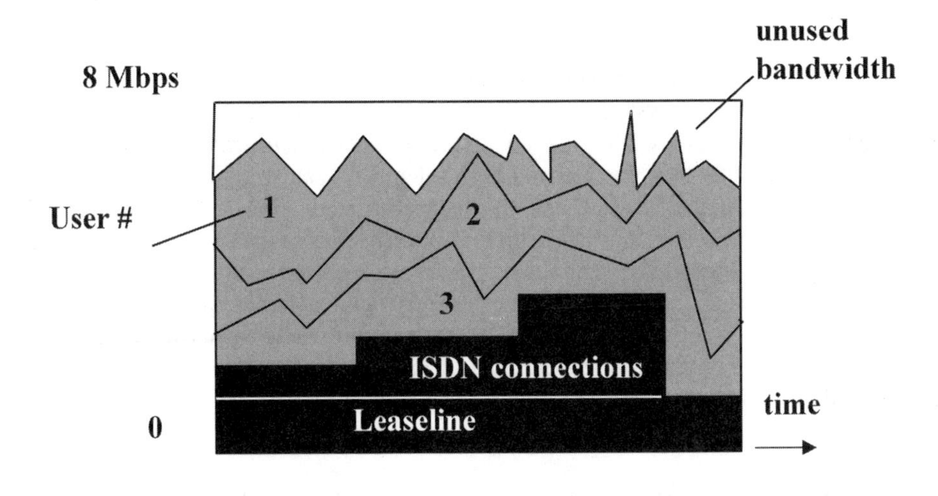

Figure 14.6 The use of ATM/TDMA in Netro's AirStar to allow statistical multiplexing and most efficient radio spectrum usage

(Evolium), Nortel Reunion, Newbridge/Stanford Telecom, Adaptive Broadband and Ensemble. In addition, the Lucent OnDemand system and the Siemens SRA MP system are based on Netro's AirStar.

Figure 14.6 illustrates how the ATM/TDMA approach is used by Netro (in their AirStar system) to allow a 7 MHz radio channel (supporting 8 Mbit/s) of capacity within a given base station sector to be shared between different customers for simultaneous and efficient provision of different types of services.

Figure 14.6 illustrates three types of users sharing a single 8 Mbit/s sector radio channel.

The first customer (shown at the bottom) has a 2 Mbit/s *Constant Bit Rate* (*CBR*) line used either as a leaseline or as a permanent connection of his PBX to the ISDN telephone network using *PRI* (*Primary Rate Interface*). Alternatively, this might be a number of customers, all with *n**64 kbit/s leaseline-type connections.

The second type of customer is using ISDN service. For these customers, individual 64 kbit/s channels are being switched in and out of use on a 'bandwidth-on-demand' basis as individual calls are made and cleared.

The remaining capacity (the *Available Bit Rate*, *ABR*), which varies between about 4 and 6 Mbit/s over time, is shared between a number of data and Internet users (numbered 1, 2 and 3 in the diagram). Each of these customers occasionally sends data at a maximum rate of around 3 Mbit/s, but the 'statistical sum' of the bit rate is always less than the available bit rate (so that we can squeeze three times 3 Mbit/s into about 5 Mbit/s!).

Amongst other manufactures, Ensemble is trying to take the TDMA and packet-based air interface approach one step further, by also encorporating time division multiplexing. We also discussed TDD in Chapter 13. By using TDD, Ensemble's aim is not only to be able to statistically multiplex upstream and downstream traffic on separate frequency division duplex (see Chapter 13), but also to be able to re-allocate the radio spectrum dynamically. This might enable, for example, all the radio spectrum to be used for a short period either exclusively for upstream or exclusively for downstream traffic carriage, as customer demand required.

Wideband CDMA (W-CDMA)

Before we leave the subject of potential implementation techniques for manufacturers of broadband wireless access systems, we should also mention wideband *CDMA* (*Code Division Multiple Access*). CDMA is an effective *spread spectrum* technique which provides for very reliable transmission and high tolerance against fading caused by signal interference. It will be the basis of the third generation of GSM, also called *UMTS* (*Universal Mobile Telephone Service*), which is aiming to bring data and Internet services to mobile telephone users. The technique, however, has not been widely offered for broadband services, since its effective use depends upon a large *spreading factor* (as discussed in Chapter 5). Carrying a high bit rate within radio spectrum of relatively restricted bandwidth (e.g. attempting a carriage of 32 Mbit/s in 28 MHz) does not allow for a large spreading factor!

14.10 BRAN (Broadband Radio Access Networks)

ETSI has so far been relatively inactive in development of broadband wireless access standards (it has been occupied mainly with the 'raw' radio aspects of standardisation —

spectrum masks, etc., and not so active in developing detailed air interfaces and protocols). However, this approach has changed recently by the launch of a new standardisation project titled *Broadband Access Networks* (*BRAN*). The goal is that BRANs shall be highly efficient radio access networks not only for voice and ISDN, but also for emerging high speed data and multimedia services. The terms of reference of the project demand that both circuit- and packet-oriented transport protocols (including both ATM and IP shall be supported).

The two main objectives of the new project group (established in 1997) are:

- To produce specifications for a new type of high quality public radio access networks to be caller *HIPERACCESS*.
- To produce and further develop specifications and standards for high quality private (and public) radio access networks based on the existing HIPERLAN (see Chapter 13) and the new HIPERLAN/2 standards.

HIPERACCESS is a relatively long-range system intended for high speed (>36 Mbit/s) point-to-multipoint access networks for connecting residential and small business users to a wide range of public telecommunications networks, including UMTS, ATM and IP-based networks. Spectrum allocations for such networks are a current subject of study within CEPT (European Conference for Posts and Telecommunications).

HIPERLAN/2 is intended to be a further development of HIPERLAN (as we discussed in Chapter 13), designed to be a short-range complementary access mechanism for public UMTS networks as well as providing for private (e.g. campus-based) wireless LAN systems. It is intended to operate in the 5 GHz range, and offer high speed (36 Mbit/s) access to UMTS-, ATM and IP-based networks.

In addition, a system called *HIPERLINK* is intended to provide for high speed interconnections of HIPERLANs and HIPERACCESS networks in the 17 GHz range.

The specifications being developed will define a standardised physical layer as well as a Data Link Control (DLC) layer. These should be based as far as possible, and be coherent with, previous work of the ATM Forum, the IEEE wireless LAN committees 802.11, 802.16 and 802 N-WEST, as well as the Internet Engineering Task Force (IETF). The initial outputs from the BRAN project are expected in late 1999 (HIPERLAN/2) and early 2000 (HIPERACCESS), but it will take a little while yet before equipment meeting the standards and in stable operation is available commercially from manufacturers.

14.11 The Future—Squeezing More Capacity from Limited Spectrum

What will be the future of broadband wireless? Undoubtedly, ever-more customers will continue to demand ever-higher bit rates. So how will these be achieved, given that the radio spectrum resources are not unlimited? By one or a combination of three different measures. We have discussed each of them several times during this book:

- By moving into new, previously unused high frequencies (there is, for example, a move within Europe to start licensing the bands from 40 GHz to 60 GHz). Specifically, 54 GHz and 56 GHz are in question as potential *very shorthaul* point-to-point wireless access links.

- By employing higher modulation to allow more bit/s to be yielded per Hertz of spectrum. The need to avoid interference my additionally require use of more selective antennas.
- By using more *selective* antenna technology. This could either be by using narrower sector angles (i.e. increasing the sectorisation of a base station, as we discussed in Chapter 9, by using point-to-point antennas or a multiple-point-to-point architecture, as we discussed earlier in this chapter, or (most high-tech of all) by using 'steerable' or 'active antenna' technology.

15

Radio System Installation

The finding and preparation of appropriate relay station sites for point-to-point radio systems, and the selection of base station sites for point-to-multipoint systems, is one of the most important tasks in setting up a wireless network. The manpower involved in the site search and the costs involved in site preparation are also major contributors to the overall costs of establishing the network. In this chapter, we discuss some of the lessons learned through major deployments in the past. In particular, we shall discuss the process of finding a site, making necessary building works, safety precautions and radio frequency applications, constructing the mast, calculating wind loadings, installing peripheral equipment, aligning antennas and bringing equipment into service.

15.1 Site Acquisition, Survey and Preparation

The large scale and rapid deployment of mobile telephone networks throughout the world during the 1990s has created a number of companies specialised in finding and preparing base station sites and installing radio equipment and antennas. This will make it easier for WLL (Wireless Local Loop) network operators to find experienced personnel and project managers capable of coordinating a rapid nationwide deployment of base stations. However, the negative impacts of the earlier deployments are the high rent expectations of landlords for rooftop space, and the increased community resistance to antenna 'jungles' and 'electro-smog'.

Mobile telephone operators and the WLL operators who will follow in their footsteps have typically 'outsourced' the complete job of acquiring base station sites and preparing them for equipment. The whole job, defined in terms of how many base station sites are involved, is usually opened to tender, with the expectation to agree fixed fees for given pre-defined activities:

- Site search and acquisition (including the location of possible buildings and negotiation with landlords and local authorities for necessary permissions).
- Site survey and planning (the checking of line-of-sight availability to the other end of proposed PTP (point-to-point) links or line-of-sight coverage of proposed PMP (point-to-multipoint) base stations. The determination of exact locations and mountings for indoor and outdoor equipment and cabling. It may be necessary with special equipment to determine what height of antenna mast should be provided).

- Site preparation (the provision of a separate electricity supply, power back-up, cabling, equipment racks and/or outdoor housing, air conditioning, antenna steelwork or mast, etc. Specific permissions from the landlord, building archtiect and local authorities may need to be obtained).
- The installation of antennas and equipment.
- The alignment of antennas, the commissioning and bringing into service of equipment.
- Testing, acceptance and documentation.

Typically, the site search is triggered by the internal network planning group within the network operator's organisation. This group has the job of overseeing the overall rollout plan for base stations. At the start of a network deployment, the priority of the network planner is usually to determine the priority with which different towns or other geographical areas should be provided with radio coverage. Later on (i.e. once 'full coverage' has been achieved), the main role of network planning shifts to ensuring adequate capacity is available in all areas. At this stage, a higher density of base stations may need to be achieved to 'fill-in' the 'blackspots' of poor coverage, or the 'hotspots' where there is inadequate network or radio capacity to meet demand.

In defining a search area for the site acquisition team, the network planning department needs to consider the 'ideal' location of the new base station taking into account the following factors:

- The desired coverage area.
- The location of the adjacent base stations.
- The needs of the freqeuncy re-use plan (the planner needs to ensure a degree of coverage 'overlap' between adjacent base stations, but needs to ensure that the interference is held within acceptable limits).

The site requirements at this stage are best defined in terms of:

- A geographical area within which the base station will ideally be located.
- The intended coverage area of the base station.
- The number of 'sectors' to be provided (an *omnidirectional* antenna has to have good line-of-sight view without obstacles from a single position on the roof, or from a high mast, but when a number of sectors are provided, obstacles on the base station rooftop can be overcome by mounting a number of shorter antenna mounting poles at each corner of the roof).
- The ideal height of the base station (the height is a main determinator of the range of the base station). Mobile telephone network base station antennas, for example, are optimally 10–20 m above the ground. Too much higher and they interfere too much with adjacent stations).
- The type of base station equipment (this governs the amount of 'indoor' space necessary for equipment, the so-called *footprint*, and the type of steelwork mountings necessary for 'outdoor' equipment).
- The location of any other adjacent base station to which the new base station is to be connected using point-to-point equipment.

Nowadays, it is increasingly common for network planning groups to use sophisticated computer planning tools and detailed digital map information, including geographical topograhy, land-use, population density, and even information about detailed building locations and heights. Such tools may even help to identify exact buildings (or exact geographical locations for potential masts) as ideal sites for further investigation and 'acquisition'.

The site acquisition activity usually then proceeds in two phases. First, a number of potential sites are identified and reviewed. Initial negotiations are entered into with landlords and other agencies. These activities result in the preparation of a shortlist of optional sites, which are forwarded to network planning for final decision, based not only on the 'ideal' network needs previously requested, but also upon the relative economics of each optional site. An initial site survey of the radio potential of the site (including photographs of the coverage panorama, as we discuss in the next section) may also be used as a basis for the decision. A final decision will then lead to detailed site survey and preparation.

The rental rates which an operator can expect to pay for rooftop space for his antennas and a small footprint for his indoor equipment can be more than the equivalent square metre rental price for top office accommodation. In the meantime, the space on the roof has come to be more valuable than the space under it! The typical range lies between about \$500 and \$5000 per month.

The acquisition of sites can be contracted either to one of the number of companies now specialised in the whole general contractorship and project management of major radio network deployments, or alternatively, there are also a number of real estate brokers who have also specialised in the radio telecommunications field.

15.2 Technical Site Survey (PMP Base Station or PTP Relay Station)

Once the exact base station (for PMP, point-to-multipoint) or relay station (for PTP, point-to-point) site has been acquired, then comes the time for the detailed technical site survey). This is typically carried out by a single site planning specialist, although he may work closely with a colleague when performing link Line-Of-Sight (LOS) checks as we discussed in Chapter 8.

It is usual to document the site survey on a standard pro-forma report, adding photographs and a map of the local area to provide complete information for final detailed network planning. The planning will resolve the exact radio frequencies which should be used in individual PMP sectors or on PTP *backhaul* links connecting the site to the *backbone* network.

Standardising the procedure allows the process to be scaled. A typical pro-forma will require the following questions to be answered:

- *Name and address of site* surveyed.
- *Contact name* and telephone number for access to the site.
- *Date of site survey* and planner's name.
- *Coordinates of the site* (obtained from building plans or from a global positioning system satellite receiver carried by the planner).

- *Elevation of the site* above mean sea level and height of the proposed antenna mounting location above the ground.
- *Proposed number of sectors* (for PMP coverage) or links (in the case of PTP) and a map or sketch showing these sectors and links and their compass orientation.
- *Line-of-sight (LOS)-check* results (see also Chapter 8) — a description of the coverage potential of a PMP base station or the unobstructed line-of-sight link potential of a PTP link.
- *Proposed antenna mountings* (exact location according to sketch of rooftop), size, height, stiffness, diameter, etc. (the height may need to be determined using a helium balloon as explained in Chapter 8).
- *Other equipment on the roof* (existing masts, machinery, air conditioning or heavy machinery which might either obstruct the view or cause electromagnetic disturbance to the radio equipment. The detailed frequencies of other radio equipment is important to note).
- *Access to the roof* (determines what climbing aids, cranes, etc. may be needed at the time of equipment delivery and installation).
- *Special considerations* applying to the site (whether there are special conditions about security access to the building on certain days or at certain times-of-day, whether there are constraints on radio operation or transmitter output power arising from a nearby petrol station, airport, gas refinery plant or whatever).
- *Location of 'indoor' equipment* (whether a room is to be provided or whether a weatherproof outdoor shelter is to be provided. If a shelter is to be provided what size, and in which exact location).
- *Peripheral equipment*. If equipment racks are required, what size and type. If air conditioning and/or heating is required, what rating? Is forced air cooling required (fans instead of full air conditioning)?
- *Availability of power*. A description of what power is available, or what is to be made available. Whether Direct Current (DC) power is to be made available and whether Uninterruped Power Supply (UPS) and battery back-up is to be provided.
- *Proposed indoor-to-outdoor cabling*. A description of the type of cable to be used, a sketch of the cable run and a description of the earthing and lightning protection measures to be provided (we return to this subject later in the chapter).
- *Network interface*. The interface and handover-point, including plug or socket type for connecting the radio equipment to the network backhaul link or backhaul line (e.g. G.652 glass fibre at 155 Mbit/s (STM-1) with ST or SC-connector or 2 Mbit/s G.703 leaseline with RJ-45 connector, etc.). Is a clock source needed for network *synchronisation* or back-up? (as we discussed in Chapter 10)?
- *Network management*. Is a separate telephone line required at the site for the use of technicians visiting the site or for connecting the network management centre to the site? Should this be achieved by means of a PTP *wayside channel* (Chapter 12)?
- *External alarms*. Should door entry alarm systems, temperature sensors and other alarms be extended by the radio equipment over telemetry wayside channels (e.g. by means of a point-to-point backhaul link, as we discussed in Chapter 12) to the network management centre?

Some of these questions may be answered by the network planning department rather than the site surveyor. Together, the answers make up a complete engineering plan for the preparation of the site, though permission may need to be sought from a range of different people before the engineering work may commence.

15.3 Necessary Permits

Prior to the installation of the antenna and outdoor radio unit equipment, it may be necessary additionally to gain the approval of:

- The building landlord.
- The building architect or the building statics engineer (can the building hold the weight and windload of the proposed mast, and does it fit in with the architect's 'vision'. Often it is a requirement that antennas should not be visible from the ground).
- The local authority building regulations and planning officer (in some cities there are restrictions on the size and number of antennas which may be installed on buildings in given areas: some authorities limit the number of antennas which may be installed on a single roof and the maximum height of the antenna mast. Some even decree that the outdoor equipment must be painted to match the building or surroundings).
- The local authority safety officer.
- The local radio communications office or regulatory administration office.
- Tall masts (and masts on buildings above a given height) may need to be topped with aircraft navigation lights.

15.4 Radio Permits and Frequency Applications

Once the site is ready for installation, an application needs to be lodged for the allocation of radio channels for each of the PMP sectors and PTP links. The application may be to an internal application to the radio planning department (this is usual in the case that a 'block' allocation of spectrum has already been made to the operator by the national radio agency in the given geographic region — this is the most common way of allocating spectrum to PMP operators, as we discussed in Chapter 9).

Alternatively, it may be necessary to apply to the national radio agency for a radio channel for each individual link. This is the normal manner of allocation of spectrum for point-to-point radio systems. The national radio agency thus assumes the overall coordination of frequency re-use planning for PTP. The application may take anything from a few days up to about four weeks, depending upon whether the applicant is known to the radio agency, and depending upon the availability of resources and level of demand.

Equipment Approvals (otherwise 'equipment homologation')

It is normal that only officially tested and approved radio equipment may be operated. Two types of testing and verification are usually required to be carried out before equipment can be approved. These verfication tests (by independent certified test houses) seek to confirm:

- the conformance of the equipment to radio equipment specifications, and;
- the conformance of the equipment to electromagnetic compatibility (EMC) and safety standards.

In the European Union, the radio approvals specifications are based upon the standards produced by ETSI (European Standards Telecommunications Institute). Each country bases its approval requirements upon the relevant ETSI standards, defining where necessary any national extensions of the standards or specific operating requirements (e.g. requiring the 'extreme temperature' range of operation defined by ETSI). The manufacturer of the equipment needs to have his equipment *type-approved*. This is achieved by submitting specimen equipment for test by one of the certified independent test laboratories. A report is produced by the test laboratory which details a number of conformance tests as defined by the relevant ETSI standard. The report can then be submitted to the approvals agency of any of the EU countries as the basis of an application for a formal *approval certificate* and equipment *approval number*.

In addition to the approval to relevant radio standards, equipment imported to and operated in the EU (European Union) countries must bear the CE-Mark to confirm its testing against and conformity with relevant safety and electromagnetic compatibility standards. The CE-mark is self-certified by the manufacturer, having convinced himself (by means of independent tests as appropriate) that all relevant standards are fulfilled.

In most other countries (including the USA), a similar equipment approvals and homologation regime applies, but other standards apply. In the USA, the relevant standards used as the basis for homologation are those defined by FCC (Federal Communications Commission) and IEEE (Institute of Electrical and Electronics Engineers).

Human Exposure to Microwave Radiation

Safety standards worldwide define the maximum safe exposure to microwave radiation. The FCC, for example, defines the exposure in its bulletin 65 edition 97–01. As a result, it may be necessary to make precautions at the site of an intended radio transmitter to ensure that the maximum exposure is not exceeded. In Germany, for example, the VDE (Verband der Deutschen Elekroindustrie, the association of the German electrical industry) defines a number of zones around a transmitting antenna. Within the first zone (the nearest to the antenna or the *minimum exclusion distance*), uncontrolled exposure to the level of radiation is dangerous to humans. It is therefore necessary to fence-off this area (near high power transmitters) so that no-one can unknowingly expose himself to excessive radiation.

Before being allowed to install a radio transmitter with an *EIRP* (*Equivalent Isotropic Radiated Power*—the addition of the dB output of the transmitter and the antenna gain in dBi) or an electric field strength greater than a certain threshold, the planner of a radio transmitter site is usually forced to apply for permission, either from local authorities or from the national radio planning and coordination agency.

One of the key FCC recommended limits for 'uncontrolled' or 'general population' exposure (OET Bulletin 65, Edition 97-01) is a power density of 1 mW per square centimetre for 30 minutes. A similar limit applies in most European countries. The power density can be predicted from the following equation (according to FCC):

Power Density (Signal Exposure In Worst Case)

$$S_{surface} = \frac{PG}{4\pi R^2}$$

where S = power density in $\mathrm{mW/cm^2}$

P = power input to antenna in mW

G = gain of antenna (integral not dB value)

R = distance to the centre of radiation in cm.

Alternatively, the German regulator (RegTP) uses the formula for power density on the surface of the antenna:

$$S_{surface} = 4p_t/A$$

From either of the equations above, we could work out the range from the antenna at which the maximum allowed exposure is exceeded. This would be equivalent to the first security zone mentioned above. Within this range it should not be possible for people unknowingly to receive excessive radiation dosage. Let us consider some typical values of PTP and PMP systems to calculate the exposure according to FCC:

PTP system at 7 GHz
p_t = 500 mW (27 dBm)
g_t = 52,267 (47 dBi, 4 metre diameter)
λ = 4.3 cm
A = 126,000 cm^2 Power density (on face of antenna radome) S=10 μW/cm^2

PTP system at 38 GHz
p_t = 32 mW (15 dBm)
g_t = 7950 (39 dBi, 30 cm diameter)
λ = 0.8 cm
A = 700 cm^2 Power density (on face of antenna radome) S=0.1 mW/cm^2

Unintentional exposure is not a major problem in this band!

PMP system at 26 GHz
p_t (base station and subscriber terminal, max) = 100 mW
g_t (base station) = 32 (15 dBi)
g_t (subscriber terminal) = 630 (28 dBi)
λ = 1.2 cm
A (base station) = 27 cm^2
A (subscriber terminal) = 170 cm^2 Power density (on face of base station antenna) = 2.01 mW, Power density (on face of subscriber antenna) = 1.00 mW.

Protection against unintended exposure may be needed for the first few centimetres in front of the base station antenna (where only a single antenna is mounted at the base station). However, where more than one transmitting antenna is installed, the calculation of the power density needs to be conducted considering the combined output power of all the transmitters.

15.5 The Positioning, Design and Installation of Steelwork or a Mast for Antenna Mounting

Microwave radio equipment needs to be mounted on sturdy masts or rigid pole mountings. Only rigid mountings can ensure the proper alignment of antennas (variation within less than 1°) even during adverse weather and strong winds. They should also be mounted in positions where obstacles are not likely to block the line-of-sight. Such obstacles can include window cleaning gantries and cranes on large office blocks. Furthermore, they should be so oriented that they do not 'look' directly into other transmitting antennas in the vicinity (on the same roof or on a neighbouring building). High power antennas, immaterial of whether they are operating in the same band, can be a cause of interference (caused by 'stray' harmonic frequencies emitted by such transmitters).

Antennas should be mounted preferably at the edge of buildings (Figure 15.1a) to avoid possible reflections from the roof itself or diffraction around the edge of the building (Figure 15.1b). Large metal roof areas can be especially problematic.

The radio transmitter and receiver is sometimes remote from the antenna, as we discussed in Chapter 6. In this case, the location of the radio equipment, as well as the antenna needs to be away from possible sources of electromagnetic interference. For example, radio receiver equipment should not be placed near the strong electric and magnetic fields caused by lift or elevator motors which are often resident on office building rooftops. The fans of large airconditioning plants (which are often found on roofs) can also be causes of disturbance.

There are four basic alternatives for the antenna mounting. These are illustrated in Figure 15.2. The first alternative is to mount the antenna directly on a wall (Figure 15.2a). The advantage of direct wall-mounting is the low cost of additional materials required for the mount and the fact that a standard mounting is used each time, so eliminating the need for special steelwork preparation. The disadvantage is that wall-mounted antennas are relatively difficult to access. This complicates the installation and alignment of the antenna as well as its subsequent maintenance. The saved costs of materials are sometimes also outweighed by the extra labour costs associated with the usual safety stipulations that a worker on a ladder must always be accompanied.

Another simple form of construction is a short steel pipe (e.g. 108 mm diameter and 1.5 m in length) mounted on the roof of the building (near the roof edge as in Figure 15.2b). It is usually relatively simple to install such steelwork. In the case of a flat roof, this leads

(A) the ideal antenna mounting in near the rooftop edge

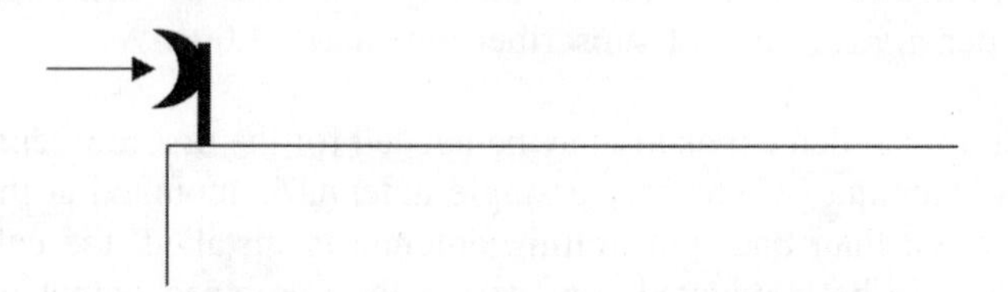

(B) mounting too far back can result in harmful reflections

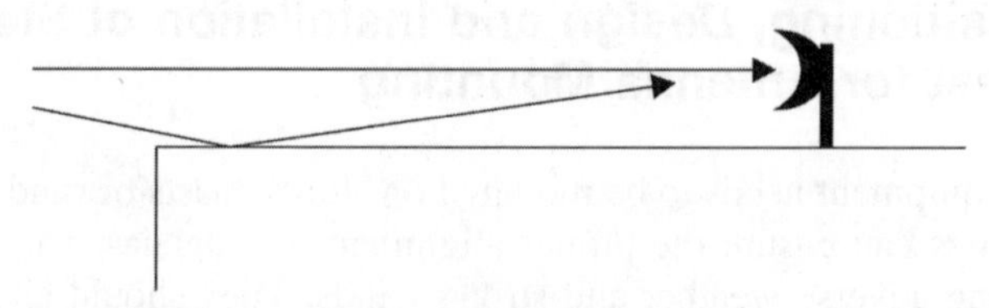

Figure 15.1 Careful positioning of antennas is important

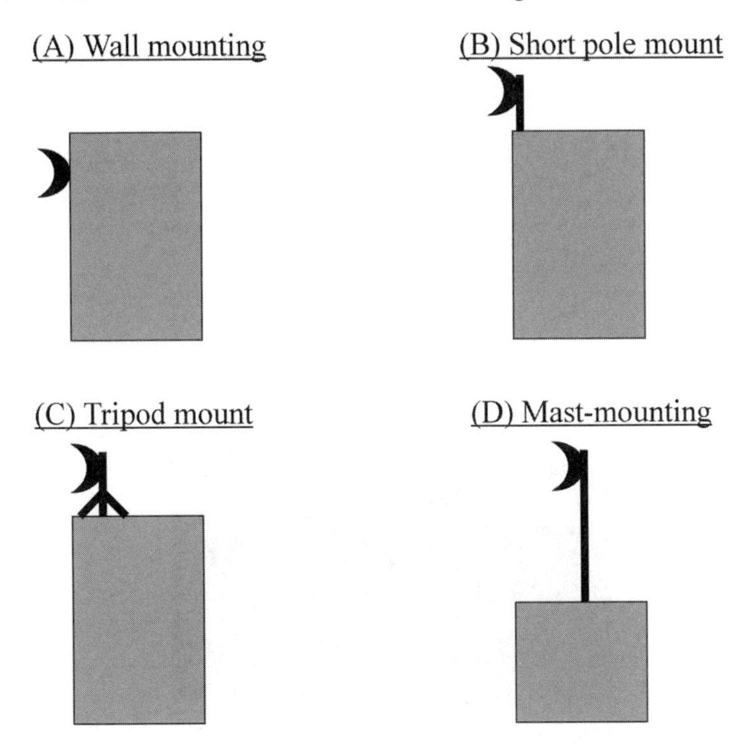

Figure 15.2 Alternative antenna mounting types

to easy access to the antenna for both installation and maintenance, but complications of access arise when the pole is mounted on or through a slanting roof surface. Before mounting the steelwork, it may be necessary to calculate the weight and other loadings which the mast will inflict upon the building structure at its anchorage points.

A further mounting method, and one used frequently by the mobile telephone network operators, is the use of a 'tripod' or 'mobile antenna mount'. These mounting devices are intended to be able to simply placed upon a flat roof, sometimes even without being secured to the roof. The stayed or tripod construction usually ensures the rigidity of the mast. It is held firmly in place either by securing by bolts, or simply by heavy concrete slabs or a base filled in-situ with water, liquid concrete or some other ballast (Figure 15.3).

The most expensive form of antenna mount (which is usually only economic at base station sites and not at customer sites) is the provision of an antenna mast (Figure 15.2d). These come in many shapes and sizes. For tall masts (e.g. upwards of 20 m) it is common to use concrete constructions. Meanwhile, for rooftop 'extension' masts it is more usual to use steel construction. I recommend using modular tubular sections, tapered from a larger diameter at the bottom to progressively smaller diameter. It is normal to provide rungs on either side of the mast for climbing it. The antennas themselves are normally mounted on 'outrigger' poles of a narrower and more suitable diameter. The outrigger prevents the equipment from being trodden on and damaged when personnel are climbing the mast. In addition, the fact that the equipment is somewhat displaced from the mast makes for easier and safer working (e.g. arms around the main pole, reaching to the equipment). Such a construction is shown in Figure 15.4.

Figure 15.3 Typical tripod antenna mounting

Antenna masts are nearly always manufactured to meet the specific requirements of a given base station site. The mast has to be designed to be the given height required and to provide for the necessary number of mounting positions for all the proposed antennas. However, in addition, the mast must be designed by a *statics engineer* (one who calculates

Figure 15.4 Typical radio mast showing platform, climbing ladder, outrigger with mounted point-to-point radio terminal and cellular radio sector antennas

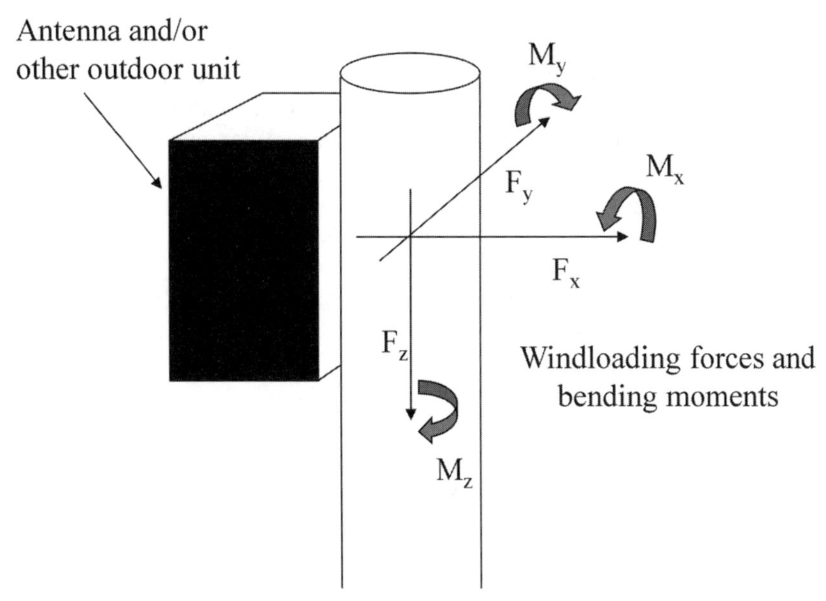

Figure 15.5 Wind load forces on an antenna mounting from a single antenna or 'outdoor' unit

the strength of structures) to withstand the weight and windloading forces which are likely to be encountered by the mast. The statics of the building on which the mast is to be mounted also need to be checked — to make sure that the proposed mast anchorage positions are strong enough to withstand the additional load.

Wind Loading

Strong winds which are incident upon large antennas will 'load' the steelwork of the antenna mounting with considerable torsional and bending forces (Figure 15.5). So that the antenna and the mounting do not either fall down or even lose their alignment, it is important for the steelwork and its anchorage to be designed with adequate strength and rigidity. The calculation of the total maximum design load on the steelwork needs to consider all the antennas which might foreseeably be mounted on the pole. This will require calculation and summation of the individual forces resulting from each individual antenna, using the formulae and tables provided by their respective manufacturers (as discussed in Chapter 6). The calculation should be made at wind speeds corresponding to local building regulations. A typical maximum design *operational wind speed* is 140 km/h (alignment rigidity within limit), with a typical survival wind speed (mast still standing) of 200 km/h.

Cabling Installation from Indoor-to-Outdoor Unit

When installing coaxial cabling from the indoor unit to the outdoor unit, it is important to:

- ensure that the cable run is well-chosen, so that the signals carried on the cable are not subjected to unnecessary and irreversible interference, and to;
- ensure that the cable is not damaged during installation.

A cable run which runs for a long parallel section beside heavy current electrical supply cables (e.g. those running up the cable shaft to an elevator motor on the roof) is almost certain to lead to problems. Similarly, indoor-to-outdoor cables which carry *Intermediate Frequency* (*IF*) signals of a frequency close to nearby installed radio transmitter equipment should be laid as far away from the disturbing radio transmitter as possible. (For example, an IF around 150 MHz is likely to suffer interference from regional trunk radio systems operating at 2 m wavelength). Such systems are frequently used for municipal services such as ambulances. Sharing the same base station site as such systems thus requires careful planning!

Should the coaxial cabling be damaged during installation, then the performance of the entire radio system may be significantly degraded. Typically, causes of damage are:

- hammer blows intended to drive in cable clips;
- 'kinks' in the cable caused during 'pulling-in' the cable;
- bending around corners tighter than the minimum specified bending radius;
- inner cable conductor damage caused when fitting coaxial cable connectors (the inner conductor is often used as the 'pin' of the connector, but can become damaged, particularly if the plug and socket are frequently 'unplugged' and re-connected.

Such problems can only be resolved by re-installation using new cable.

Precautions Against Lightning

It is important when installing outdoor equipment and the associated steelwork and cabling to make the four following provisions for protection against damage by lightning:

- a *lightning rod* should be provided at the uppermost part of the antenna mounting pole or mast;
- the steelwork mounting and the outdoor equipment should be *earthed* (i.e. *grounded*);
- any outdoor radio receiver should be designed to withstand the high voltage effects of nearby lightning strikes without damage;
- the screen of the indoor-to-outdoor cabling should be grounded regularly along its exterior path and *lightning arrestors* should be installed to prevent any lightning strike being transmitted to the indoor equipment.

The *lightning rod* is nothing more than a pointed metal extension of the antenna mounting pole. It draws the lightning strike to the pole itself (and its grounded connection) rather than allowing the lightning to directly strike the antenna or the outdoor radio equipment. The protected volume is conical in shape as illustrated in Figure 15.6.

All outdoor steelwork should be earthed (i.e. grounded) using standard lightning conductor materials. Where existing lightning conductors are available at the roof location, the earth may be provided by connection of the steelwork to the existing conductors, provided you check first that the materials are still in good (not rusted through or damaged). The screen of the indoor-to-outdoor coaxial cable should also be earthed at regular intervals along its length outside (usually every 10 metres or so). It is also normal to earth the cable after each change of direction from vertical to horizontal and directly before

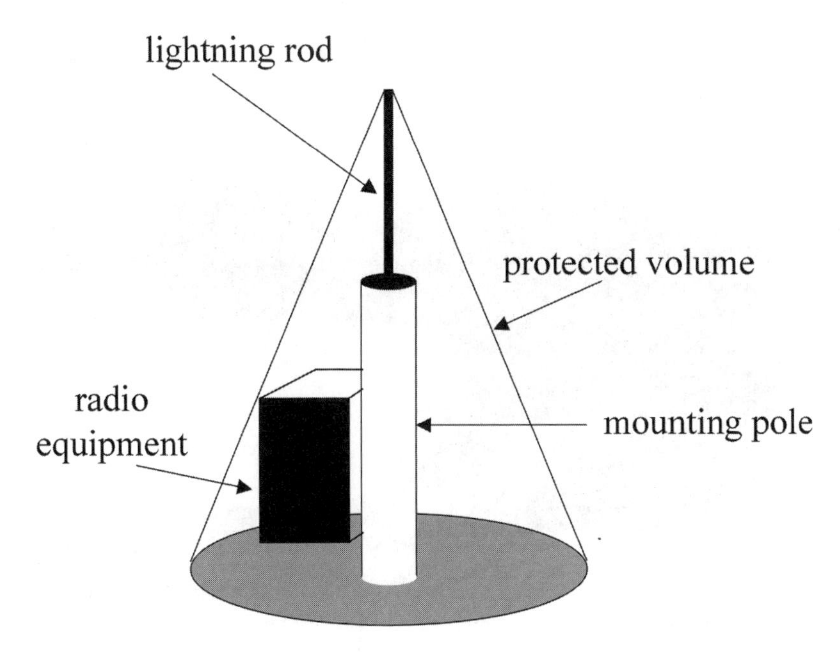

Figure 15.6 The function of a lightning rod

entry to the building (Figure 15.7). Some cable manufacturers provide special *grounding kits* for this purpose.

The design of the outdoor equipment needs to be such as to withstand nearby lightning strikes or strikes through the mounting or casing. A direct hit (despite the previously described lightning precautions) is nonetheless likely to require replacement of the equipment.

Lightning arrestors should be installed in the indoor-to-outdoor cable (Figure 15.8) to prevent a lightning strike on the cable being conveyed to the indoor equipment. Usually, two locations are sufficient — at the building entry point and either near the outdoor or the indoor equipment. Most arrestors contain a *gas discharge tube* (or equivalent) which is able to withstand and buffer the high voltage and current surge associated with a lightning strike. The tube can probably withstand nearby lightning strikes, but is likely to require replacement after a direct strike on the cable.

General Installation Precautions

When installing and aligning radio equipment, great care should be taken, since countless number of operational problems can be avoided by high quality of installation work. Mountings and bolts should be of stainless or anodised materials so that they do not rust. Anodised bolts should not be sawn-off, since this too will lead to rust. Bolts of the correct length should be used.

Reinforced cast mountings usually have a longer life than pressed steel mountings, which over time may bend further under the force of the retaining bolts. In the case of the pressed mounting plate shown in Figure 15.9a, a long-term force of the retaining bolts, plus the effects of permanent *temperature cycling* (heating and cooling) is likely to further bend

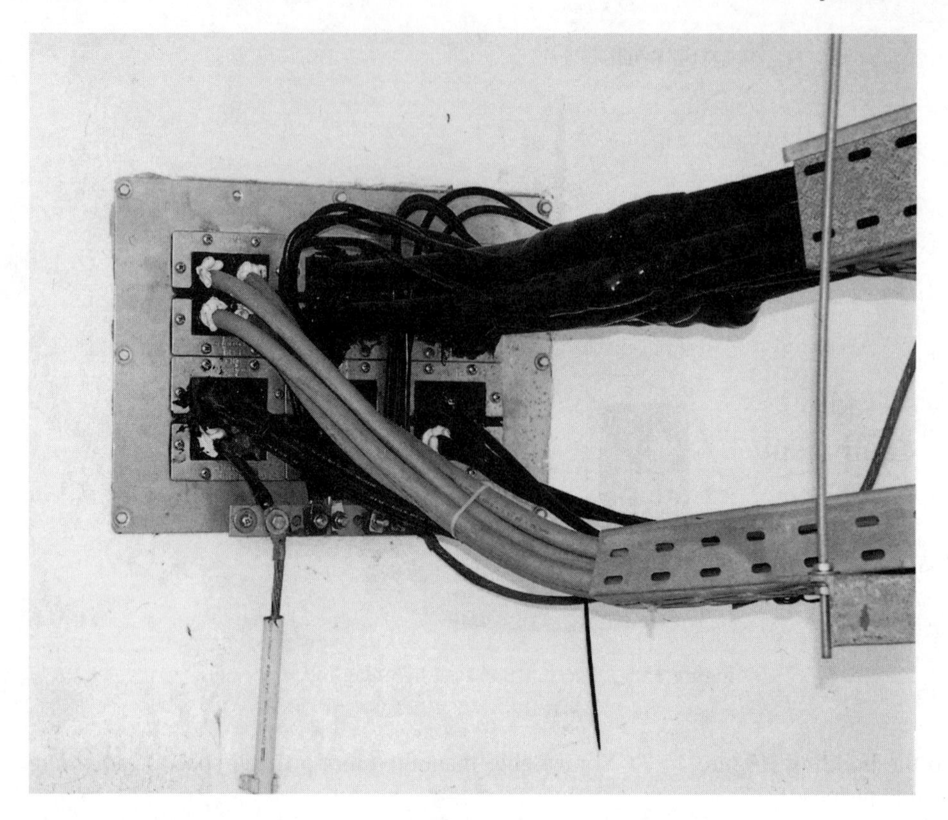

Figure 15.7 Cable ground at the building entry point

the plate around the mounting pole, causing it to become loose. This may mean the equipment or antenna will lose alignment and slip down the pole. The reinforced casting of Figure 15.9b has better long-term bending strength.

Where a choice is available, not all antenna mountings are equally easy to adjust absolutely precisely. However, precise antenna alignment is important not only for the performance of the system, but also for the frequency re-use planning, as we discussed in earlier chapters. Just a few degrees away from precisely vertical or precisely aligned might mean the loss of 3 dB of link margin or 3 dB of cross-polar discrimination (Chapter 9). Where we are relying on using both polarisations in a frequency re-use scheme, the loss of a 3 dB *installation tolerance* from a *cross-polar discrimination* of 14 dB increases the real coordination distance factor required in practical networks from 5 (14 dB) to 7 (17 dB). This obviously considerably increase the complexity of frequency re-use planning, as we discussed in Chapter 9, with the practical consequence of reducing the overall radio access network capacity.

Where cabling (e.g. indoor-to-outdoor cabling) is installed in building cable risers, building fire regulations usually demand that the riser be re-sealed with mortar between each storey of the building (i.e. flush with each floor level), to prevent rapid spread of fire throughout the building as a result of burning cable sheathing. This should be corrected after installation, even if previously the hole drilled between each floor had been left open.

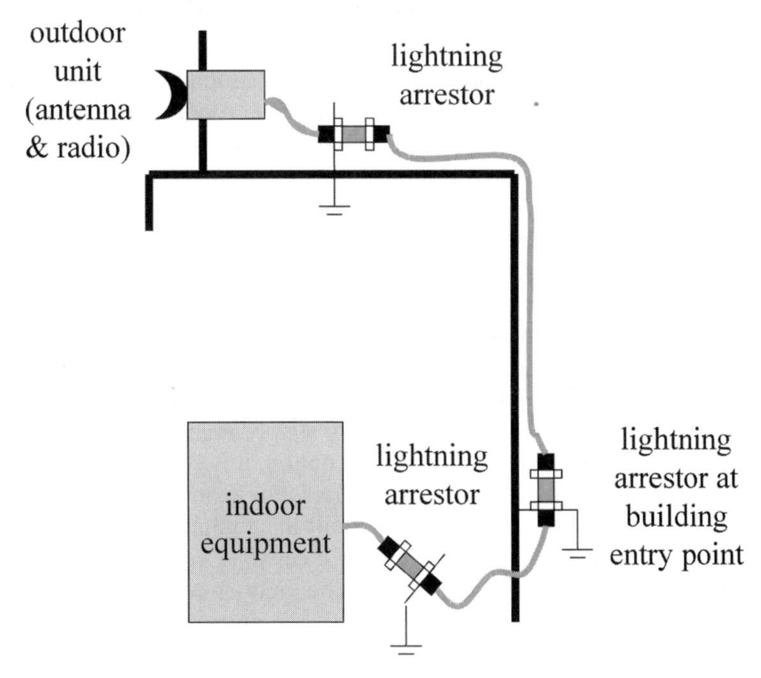

Figure 15.8 The use of lightning arrestors in the indoor-to-outdoor cable

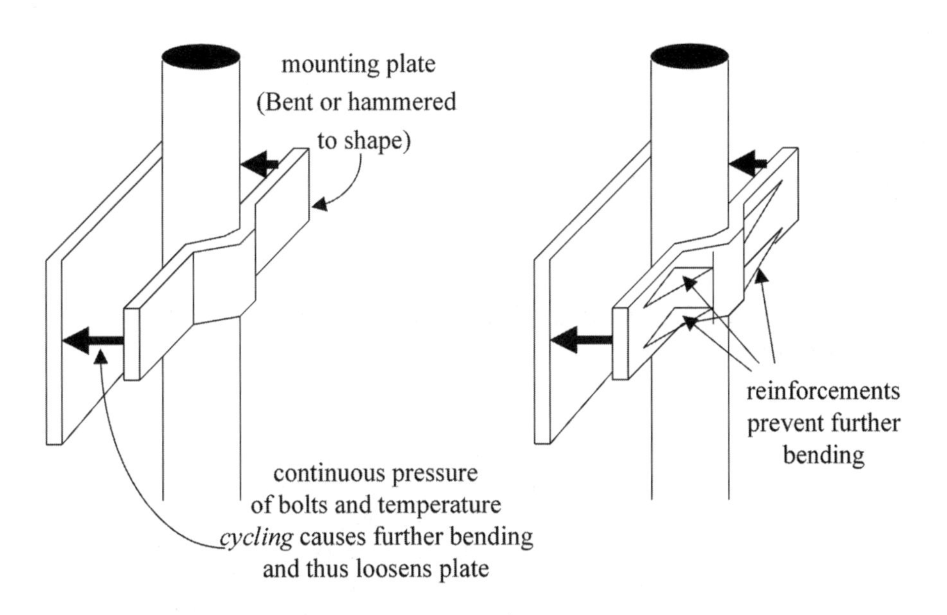

Figure 15.9 Not all mounting designs are equally good!

15.6 Peripheral Equipment and Site Preparation

Racks should for indoor equipment should be earthed (i.e. grounded). When such racks are located 'indoors', it is usual to locate them in an equipment room. This is typically a small room, used only for radio and other associated network equipment. The room should be closed-off by means of a door and free from dust. Should it not be a room which naturally is maintained between the normal operating temperature range of about 0°C to +40°C, then air conditioning may be necessary. Alternating Current (AC) power should be provided. Sometimes equipment also requires a Direct Current (DC) power source. In this case, it is also normal additionally to provide an *Uninterrupted Power Supply* (*UPS*) and batteries as a means of back-up power should the mains fail.

The provision of a separately metered mains power supply at a large number of new radio base station or relay station sites can be a time-consuming task, given that it requires close project management control of the electricity supply organisation.

In cases where the equipment design allows the option, it is sometimes advantageous to be able to install indoor equipment in outdoor shelters. For example, at a mobile telephone network base station, it is commonplace to install all the equipment (including the indoor equipment) on the roof. This avoids the need for a separate equipment room within the building, which might anyway not be available. When indoor equipment is to be installed in an outdoor shelter, it is advantageous when the equipment is compact (i.e. does not require much shelf space). In addition, since most outdoor shelters can only be opened from one side, it is also beneficial if all wiring connections (including network interfaces, power and test terminal socket) are made on the front of the equipment. In Europe, the most modern cabinets (defined by ETSI) have access from one side and are only 30 cm deep. The greatest range of outdoor cabinets is available for equipment conforming to this *form factor*.

Since most indoor equipment is only designed for an operating temperature range from about 0°C or −5°C to +40°C, it is particularly important to ensure that outdoor cabinets packed with indoor equipment are maintained within this range, even during the height of summer, with direct sunlight incident on the cabinet housing. Usually, this makes air conditioning (both cooling and heating) imperative. The cooling should be designed not to cause condensation, but since this is almost unavoidable, it should additionally be ensured that equipment ventilation holes will not accumulate such condensation and let it drip into internal electronics.

15.7 Bringing into Service and Commissioning Radio Equipment

The phase of bringing equipment into service follows the mounting of outside equipment, the alignment of the antenna(s) and the installation of inside equipment. The instructions of the equipment manufacturer need to be followed for this. Once installed, though, a standard set of commissioning tests can be conducted. For this purpose, ITU-R recommendation F.1330 defines the "performance limits for bringing-into-service of [transmission] paths implemented by digital radio-relay systems". The procedure basically defines a test in which a *BERT* (*Bit Error Rate Tester*) is used to measure the performance of the connection. The test is carried out in one or all of three phases:

Table 15.1 BIS performance objectives (BISPO)—Reference Performance Objective (RPO) (ITU-R recommendation F.1330)

PDH (Plesiochronous Digital Hierarchy)	Primary level (T1 or E1, 1.5 Mbit/s or 2 Mbit/s)	Secondary level	Tertiary level (T3 or E3, 45 Mbit/s or 34 Mbit/s)	Quarternary level (140 Mbit/s)	
SDH (Synchronous Digital Hierarchy)	1.5 to 5 Mbit/s	5 to 15 Mbit/s	15 to 55 Mbit/s	55 to 160 Mbit/s	160 to 3500 Mbit/s
Errored Seconds (ES)	max 2% of time	max 2.5% of time	max 3.75% of time	max 8% of time	Not applicable
Severely Errored Seconds (SES)	max 0.1% of time	max 0.1% of time	max 0.1% of time	max 0.1% of time	max 0.1 % of time

Note: An *errored second* is a time interval of 1 s during which one or more errors is received; a *severely errored second* is a time interval of 1 s with an error ratio exceeding 10^{-3}. (Reproduced by permission of the ITU)

- an *initial test* of 15 minutes;
- a *main testing procedure* of 24 hours;
- an *extended 7-day Bringing Into Service* (*BIS*) test.

The recommendation lays out the acceptable *Bit Error Ratio* (*BER*) during the measurement period, in particular limiting the number of *Errored Seconds* (*ES*) and *Severely Errored Seconds* (*SES*) to the maximum percentage of the time defined in Table 15.1.

Given the impracticability of having to leave BERTs at individual customer premises for 24 hours or 7 days, and then return to pick them up again, some operators may choose to limit the commissioning test to a 'quick check' (i.e. initial check) of between 15 and 60 minutes. This helps to confirm proper installation of the equipment using external test equipment at an individual customer site before the installer leaves. A useful feature of some modern radio equipment is the ability to continue monitoring the BER performance of the link, even during normal operation. When available, this feature can also be used for extended commissioning testing; but some network operators do not like to rely 100% a piece of equipment checking itself (maybe the tool is faulty as well as the equipment?). The short check with 'external' test equipment alleviates this problem.

The extended BIS test (7 days) is advisable at the main base station sites of a network, given the importance of these sites for multiple remote stations or outlying customer sites.

15.8 Project Organisation

For rapid deployment of a large number of base stations (either PTP 'hubs' or PMP base stations and the associated backhaul (PTP) connection links), the following project team structure has been used successfully by many cellular telephone network operators during the initial phase of network build-out:

- network and radio frequency planning team (plan the network locations, allocate radio frequency, organise leaseline or other wireline connections to base station sites as necessary);
- site acquisition team(s) (search for sites, negotiate with landlords and local authorities);
- project management, technical site survey and LOS-check teams (survey the site and plan the technical installation, apply for frequency and building permits);
- installation teams;
- equipment bringing-into-service teams (commission and test equipment);
- network management team (takes over operations after installation and *acceptance*).

15.9 Customer Site Installation

Where point-to-point technology is used to connect remote customer sites, there will probably remain no alternative to an installation procedure comprising an initial site survey and LOS check, followed by a second visit for installation once the frequency has been allocated by the national radio agency for the link.

On the other hand, for WLL operators planning the use of PMP technology, for which the frequency has already been allocated in advance to the relevant base station, there is the potential to try to save manpower and travelling time costs by optimising the installation procedure into a single visit, comprising:

- LOS-check (determination of best visible next base station);
- determination of site coordinates using GPS (global positioning system) receiver;
- telephone call to the network management centre to request the configuration of the new customer terminal equipment;
- installation of cabling and standard antenna mounting;
- mounting of equipment and alignment of the antenna.

The final configuration and commissioning tests can then be carried out from the network management centre after the installer has already left.

We shall consider optimised operational procedures for such installations in more detail in Chapter 16.

16

Operation and Management of Wireless Access Networks

Efficient operation, administration and management is critical to the overall economics and performance of telecommunications networks. In this chapter we consider the activities to be supported. The discussion is structured around the description and specification of a software tool designed especially to support such operational activities. We conclude the chapter with a discussion about the ideal procedures, protocols and tools making up a modern *Telecommunications Management Network* (*TMN*).

16.1 Network Services to be Supported

Wireless access networks (both point-to-point and point-to-multipoint) are rapidly appearing across the globe as a new generation of network operators strive to establish networks independent of the old monopoly telecommmunications service providers for all types of different networking services:

- leased lines;
- telephony;
- ISDN (Integrated Services Digital Network);
- Data networking, including frame relay and ATM (Asynchronous Transfer Mode);
- Internet access and Corporate Intranet networks;
- backhaul networks for cellular telephone networks (e.g. GSM-900, GSM-1800, PCS [personal communication service], etc.).

No matter which radio band the network is being operated in, no matter what the regulatory circumstances, the services being offered or the technology being used, all operators share a common basic set of problems for which comprehensive software-based planning tools are invaluable. Ideally, if choosing one of the many now appearing, an operator should choose software which can be 'customised' to the particular operator's combination of services, technology (or technolog*ies*, i.e. more than one manufacturer's radio equipment), radio band, modulation scheme and climate zone.

16.2 Basic Operations Activities

Each operator is operating according to different circumstances, offering different services using different technologies in differing radio bands, but a number of operational needs and problems are shared by all of them:

- an initial business plan and network coverage plan is needed for financiers and frequently also as part of the regulatory application for spectrum allocation (particularly in the case of point-to-multipoint spectrum allocation);
- operators need to be able to identify suitable areas for the search for acquisition of base station or relay station sites (Chapter 15);
- the analysis of the suitability of possible base station sites (coverage, interference, effect of obstacles and capacity impact on the network as a whole), and the selection of the best available site;
- the planning of the detailed realisation of the chosen base station or relay station site (Chapter 15 — this includes defining the height, design and location of the mast, the number of sectors needed and their orientation, the capacity per sector, the detailed radio frequencies to be used by each transmitter [and which do not *interfere* with neighbours], the modulation scheme to be used [Chapters 4 and 9 — if various options are available, e.g. 4-QAM, 16-QAM], antennas — the exact type orientation and polarisation to be used at a given site);
- equipment configuration planning, ordering and installation (the base station equipment needed and the exact rack layout, taking account of redundancy scheme, associated 'backbone' network planning, etc.);
- production of network coverage charts showing basic geographical coverage area (taking account of propagation and LOS limitations), interference analysis, capacity map (e.g. in total bit rate/km^2). Such charts are often used to focus sales activities on customers who could quickly be connected to the existing infrastructure;
- the radio planning, provisioning and installation of remote 'subscriber' stations (this activity includes the LOS check, the allocation of backbone network capacity and a port interface of a particular type. In addition, site survey planning, determination of site coordinates, pre-installation planning, issue of installation instructions and network configuration activities need to be supported);
- the coordination of network *provisioning* activities between different network departments corresponding to different types of network element or different sub-networks;
- network performance monitoring;
- network documentation;
- report production.

Ideally, the radio network operations support tool should be so-designed that each of the different roles involved in the operation of the wireless access network is reflected in a given 'operating window' and a tailored user interface. A corresponding password scheme for various planning and implementation roles should also reflect which data may be viewed and which may be changed by each type of user:

- business planner/marketing department/capacity planning;

- backbone network planning department/network engineering;
- provisioning/installation;
- operations;
- 'look-only'.

16.3 Detailed Requirements of Operations Software Support Tools

In the following sections, the 'ideal' requirements are set out in turn, by considering the activities undertaken and the concerns addressed by the various departments of the network operator at each stage of planning and operation of the network.

Business Planning and Regulatory Submissions

Before commencing to build the network, the network operator needs to assess his total requirement for base stations and their capacity, and to develop a phased budgetary and coverage plan.

He may also require to make formal applications (in countries where 'new' operation of PMP systems is planned) to national regulators for radio spectrum licences. This application is likely to require some sort of map of the coverage planned, probably in phases to meet stipulated minimum coverage requirements by certain key dates. (In the case of Germany, for example, when the PMP spectrum in the 2 GHz, 3.5 GHz or 26 GHz bands was opened for tender, this had to be applied for on a regional basis. The 'region' being applied for had to be defined in the application by the exact geographical coordinates of an 18-point polygon. As we have already discussed in Chapter 9, it was also a condition of the application to the German regulator that the operator guarantee to limit the signal strength (the *power flux density*) at a point 15 km beyond the defined polygon has to be less than a regulatorily defined value. A software tool might usefully help to define the region to be applied for and the power flux density in surrounding areas.)

It is useful during the business planning phase to be able to perform a number of What-if? calculations, and observe the effect on the network coverage and the business case. Thus, the ability of a tool to allow a given number of base stations to be positioned by trial-and-error to judge overall coverage would be valuable. Alternatively, an automatic function to calculate the number of base stations required to cover a given percentage of the national population or all cities of a population greater than a given threshold value would be even more useful. Ideally, the tool output is concrete enough for the network operator to use it directly as his formal business and capacity plan, needing no other database.

Although one might think that business planning and a regulatory submission is only a one-off activity, it may be repeated as often as annually. Companies adjust their investment plans and equipment forecasts based on current business. New spectrum may need to be applied for, maybe in newly available bands to meet demand in densely populated areas or to provide new coverage in previously unsupplied areas.

Reliable data, upon which the planning tool should rely would be national geographical and population density data. Initially it may suffice to use 'crude' national geographical data, saving the investment in a 'digital map' until the radio spectrum and network operating licences have been obtained.

Base Station Site Acquisition

Once the operator has gained his spectrum and completed his capital financing plan, the process of building the network commences. The operator usually defines a series of search areas (e.g. circles) for the location of possible base stations, and the defined area becomes the focus of the site acquisition activity. Site acquisition staff or subcontractors then try to locate suitable tall office blocks or prominent locations as optional base station locations.

In determining the ideal base station site search area, a 'What-if' base station coverage calculation needs to be undertaken:

- The range of the base station in the given climate zone with the available sector antenna should be adequate to reach the desired coverage area(s). Ideally, the desired coverage area can be marked using a mouse on a map (typically, a city or metropolitan area), and the tool can locate the more central area, from which (with a calculated number of base stations) the entire area can be covered.
- The topography, geography, large buildings and other major obstacles should be considered (databases containing such information can be purchased. Such data is, however, very expensive. By 'making do' with less accurate data, e.g. geographical data of plus/minus 5 m accuracy rather than plus/minus 1 m accuracy, considerable cost savings can be made).
- The location of new planned base stations relative to other existing or planned base stations should make sense. A minimum distance (the coordination distance, as we saw in Chapter 9) should be maintained to avoid radio interference.
- The 'backhaul' connection of the base station using fibre network POPs or point-to-point microwave radio links may place a constraint on maximum separation.

The ideal output of this planning effort is a map with an indicated search area and a target number of base station sites to be found, together with ideal separation distance.

Base Station Suitability Assessment

Once a number of potential base station sites have been identified by site acquisition staff, radio planning staff will wish to analyse the likely coverage of adding single or different combinations of multiple base stations to the existing network. The aim will be to find the ideal combination of new base stations which should be built to optimise coverage, capacity and radio frequency planning.

By this stage, the sites will have been visited, so that exact coordinates are known for each of the optional sites, and some assessment will have been made about nearby obstacles (location, size, etc.), possible sources of reflections from nearby buidlings, causes of multipath, etc. (as we discussed in Chapter 15).

One consideration may be the relative costs of each of the optional sites. A major factor governing cost will be the height of the mast and the complexity of civil works.

An ideal software support tool would allow the mast height of the potential base station to be varied by 'trial-and-error' and the coverage area to be plotted, accordingly. The coverage area is affected by the type of antenna and by the propagation characteristics of the radio band. In addition, the near-to-mast 'shadow' area needs to be taken into consideration (e.g. with a 7° vertical aperture, an antenna at 60 metres height will have a 'shadow' (the area on the ground before the antenna is 'visible' of about 500 metres).

The tool should also help to assess negative effects of reflections, multipath, Fresnel Zone obstruction, disturbance from neighbouring base stations or systems operated locally by other network operators, etc.

Detailed Base Station Radio Planning

Having agreed the site acquisition contract, the new base station needs to be designed and the installation project managed. The first job is to determine the radio coverage. The number and orientation of the individual sectors needs to be determined as well as the bitrate or capacity in each sector.

A frequency re-use planning and interference analysis is required to determine which radio channel frequencies and polarisations should be used in each sector. The output of this part of the process should be a radio configuration sheet detailing horizontal and vertical orientation of each sector, capacity per sector, radio modulation scheme and polarisation to be used. Detailed antenna patterns will need to be stored in the computer tool to support the frequency planning process.

In addition, it may be necessary to perform the radio frequency and link planning for a point-to-point microwave radio link for hooking the base station or 'hub' relay station up to the backbone network.

Base Station Equipment Configuration Planning

It is valuable for an installation and project planning tool to be able to convert the base station radio planning output into shelf layout diagrams and cabling schedules for indoor equipment and coaxial cable runs. This exercise should calculate the capacity of any existing base station equipment and be able to determine what further equipment is needed, taking due account of redundancy and other site-specific requirements.

Particular consideration should also be given to the trunk connection of the base station to the backbone network. Here a certain amount of topology planning may be necessary. At the least, the capacity of the trunk or trunking arrangement (multiple trunks to multiple services) should be checked. In addition, it may be necessary to assign relevant paths through a backbone SDH or ATM transport network. Thus SDH path identifiers or ATM *VCI/VPI* (*Virtual Channel Identifier/Virtual Path Identifier*) values may need to be agreed and assigned in conjunction with the backbone SDH or ATM network planning group.

It would be additionally beneficial to have software modules available within the planning tool for the configuration of other types of equipment at the base station or subscriber terminal site (e.g. mobile telephone network *Base Station Controller* (*BSC*) or *Base Station Transceiver* (*BTS*) *station* equipment, access network multiplexors, crossconnects or other equipment).

Network Coverage Charts

Once the base station is in operation, the network operator has a live network and can start to consider adding remote subscribers. To guide his sales force, it is valuable to have a rough idea of which areas can be covered so that the main sales effort can be focused on customers who can be connected quickly. The coverage analysis needs to take into account the propagation, range and interference issues as well as the 'shadows' caused by mountains, hills or other obstacles.

It is also valuable to be able to show plots of total network capacity (e.g. bit rate per km^2 (so that the network is not over-subscribed and over-loaded) Ideally, given the known number of subscriber units already installed, it should also be possible to show estimated outstanding spare capacity. This may have to be an estimate if the traffic records and patterns of the actual traffic cannot be analysed.

Subscriber Station Planning and Installation

For the radio planning of subscriber stations, it should be possible first to plot the address location of the customer's site onto a map. A 'find' feature mapping the postcode to the approximate location is valuable to determine the approximate link distance on a street map. Of course, it should also be possible to directly use geographical or map coordinates directly gained from the customer, building plans or from a GPS (Global Positioning System) receiver, but these may not be available prior to a site visit. The initial goal is to check that it makes sense to send field staff out to undertake an LOS check or installation attempt.

It should be possible to request on the streetmap the nearest base station location and/or all base stations to which there is both likely LOS visibility and sufficient range. It should also be possible to annotate the map with the range (e.g. 99.99% range, 99.97%, 99.9%, etc.) of the base station for the particular chosen type of subscriber antenna.

It is additionally helpful to view the planned link from a number of different 3D angles to get an idea of the likely obstructions. Being able to print the view from the remote station towards the base station helps the installer more quickly locate the appropriate base station, and thus more quickly align the antenna during installation.

It should also be possible to use digital map information to plot the cross section of the path, checking both the direct line of site and the various Fresnel zones (as we discussed in detail in Chapter 8). It should be possible to print a mini 'LOS report' showing the various diagrams and maps pertaining to the planned link (and possible BS options). This report could then be used by the external field staff as the basis of the final LOS check conducted from the customer site in the field. The service technician thus knows in advance which buildings, terrain or other obstacles might be obstructing the Fresnel zone 1 (Figure 16.1), and can assess their impact during the site survey.

During the site visit, a telephone call with the planning department could enable the field technician to report final exact coordinates of the installation location (using a GPS receiver), and ask for any further analysis of unexpected LOS problems (e.g. likely reflection point, interference, etc.). He can also report the exact network interfaces which need to be provided and the confirm the total bit rate/capacity need. Then, while the technician gets on with the job of installing the cable, power supply, antenna mount and other equipment, the planner can be arranging the 'provisioning' of the link (i.e. its configuration in network management).

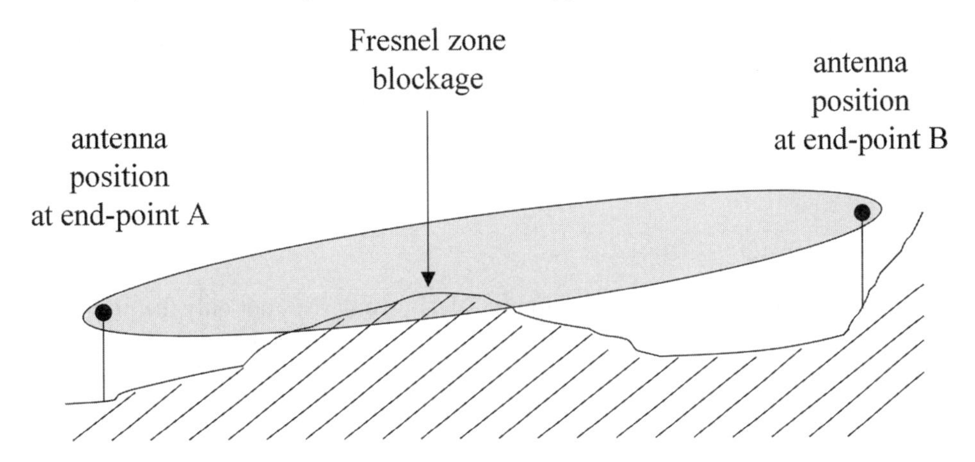

Figure 16.1 Link cross-section showing possible obstructions of the Fresnel zone

Should the link length differ from that on the original plans (e.g. due to different coordinates), then the planner can inform the field technician of the expected RSL (Received Signal Level) voltage for aligning the antenna. Otherwise, the technician takes the expected value from his LOS plan documents.

The technician commences installation, marking up his documentation with the exact location, including a description (e.g. indoor unit installed in 3rd floor equipment room, etc.). The antenna location could be marked directly on a map and entered into the computer database on his return (e.g. Figure 16.2).

Once the installer has completed the installation, measured the correct RSL voltage for antenna alignment and completed his installation report (by marking up the LOS plan

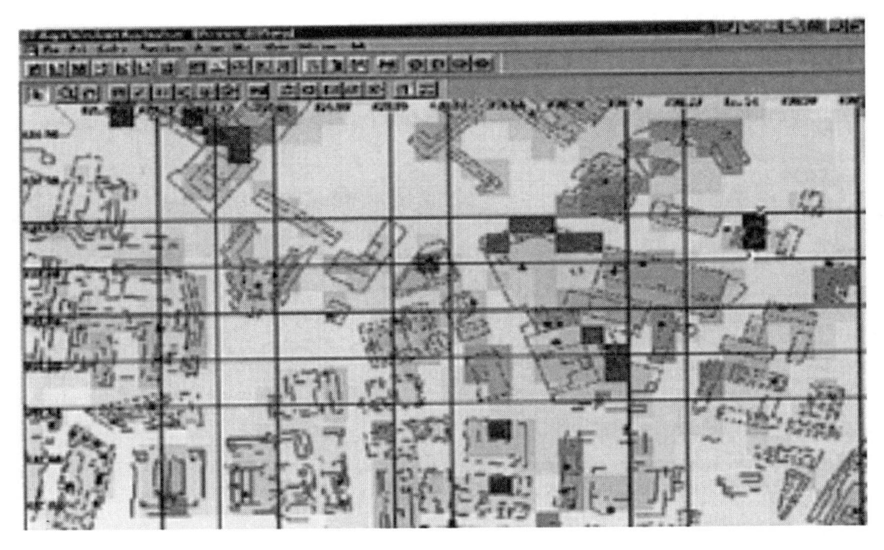

Figure 16.2 Detailed building map, superimposed with geographical contours to aid orientation of the planner. This map can be marked up by the installer

documents with exact coordinates, antenna position, orientation and elevation, cable lengths, indoor unit location, contact name and telephone number on site, results of any installation tests conducted (if any), etc.), he may ask for an 'acceptance signature' of the customer and leave. Remaining 'provisioning' activities can then be undertaken from the network management centre.

Network Provisioning

The provisioning of connections for customers ideally comprises not only the provisioning of the link within the radio base station equipment, but also in the network 'backhaul' and switching equipment (Figure 16.2 illustrates a typical network configuration of base station equipment (access network) 'backhaul' lines connecting this site to the main network 'backbone' and switching equipment.

As shown in Figure 16.3, the complete access network connection usually connects the end customer equipment to a public telephone or data switch (e.g. ATM or frame relay), or to a router giving Internet access or access to a corporate Intranet.

It can be very time consuming for the public network operator if he has to configure each subnet section (i.e. each equipment making up the customer connection) separately. For example, in the case of Figure 16.3, it might be necessary (and using separate *network management systems* for each different type of equipment) to configure separately:

- the radio *Terminal Station* (*TS*) (in particular, *activating* and *configuring* the port correctly for connection of the customer equipment);
- the radio *base station* (or *Central Station, CS*);
- the peripheral equipment at the base station (for example, a transmission network multiplexor, e.g. an SDH add drop multiplexor, or the terminal equipment of a point-to-point radio backhaul link);

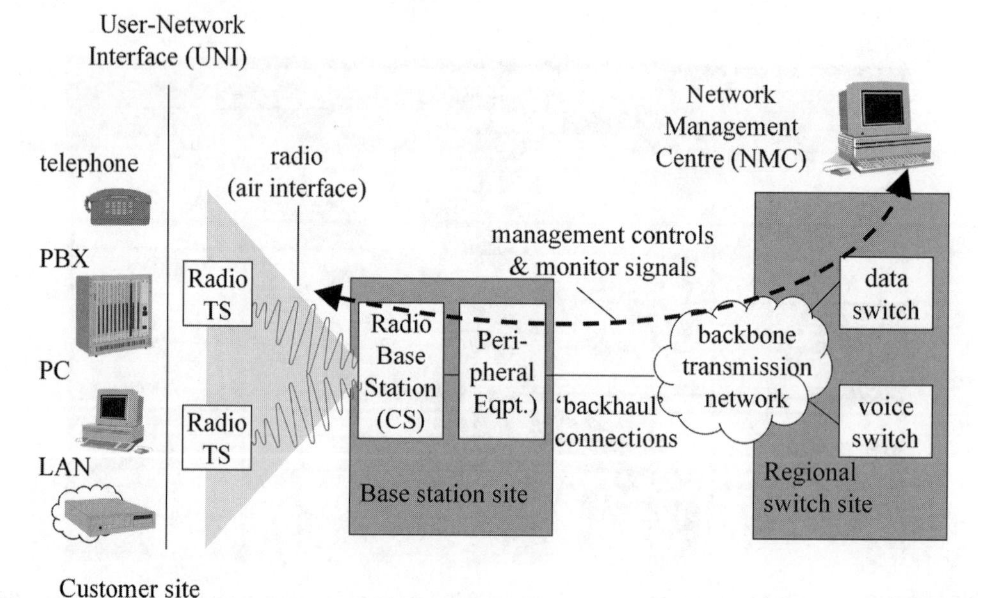

Figure 16.3 Radio Equipment integrated into the network as a whole

- the connection across the transmission backbone network;
- the port on the voice or data switch.

Since each piece of equipment is usually the responsibility of a separate operations group, this may require multiple provisioning orders. This requires careful coordination and runs the risk of making mistakes.

The ideal solution for network provisioning is to use an *umbrella network management system* capable of configuring all the equipment required for the likely types of customer connections (leaseline, ISDN PRI, ISDN BRI, Internet access, data network access, etc.). Ideally, the network management (configuration tool) system draws information directly from the network planning tool. The resultant database of exact customer connection configurations (on an end-to-end basis) is invaluable in later network troubleshooting, since the cause of alarms can be more quickly traced.

As we shall discuss later in the chapter, the TMN (Telecommunications Management Network) and CORBA (Common Object Request Broker Architecture) standards are beginning to provide a common information technology platform for management of disparate network components. However, the standards are not yet complete, and the management of all different types of network system is not today possible. For this reason, it is necessary during the network architectural planning stage to try to select components which have a basic compatibility with one another — and particularly are easily managed together. Let us explain by comparing two possible architectures incorporating point-to-multipoint radio technology into an Internet Protocol access network.

In the first of two possible architectures (Figure 16.4) we elect to use an ATM-based PMP system and an ATM backhaul and ATM backbone network. We assume that the radio terminal station converts IP signals received from the customer in Ethernet LAN format into an IP-over-ATM format. This format is then forwarded over a backhaul and backbone ATM network to a regional switch site, at which a router is located. The connection of the ATM backbone network to the router can also be by means of the ATM UNI (user-network interface).

Since all the components in Figure 16.4 are based around the ATM network architecture, they are similar to manage. It is thus more likely that an umbrella network management system can be found which is capable of managing all the various different network components. On the other hand, even if such an umbrella network management system were not available, the various human operators are likely to understand all the different

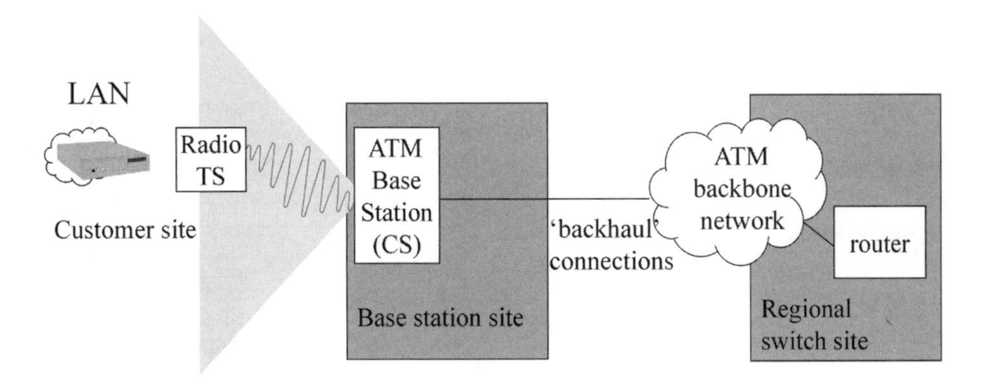

Figure 16.4 ATM-based end-to-end network architecture

technologies involved, and the terminology used for configuring each customer access line would be common to each type of equipment, so making the planning of the network and the later documentation much simpler.

By comparison, the architecture of Figure 16.5 includes a number of disparate technologies. Let us assume that the radio manufacturer uses an telephone-network-based signalling scheme on the air interface of Figure 16.5 so that radio spectrum is only allocated to a particular customer device when it is actually in use (some manufacturers call this *bandwidth-on-demand*). Thereafter, an SDH (Synchronous Digital Hierarchy) transmission network is used for backhaul and backbone transmission to the router at the regional switch site.

Because of the disparate technology types used in the access network architecture of Figure 16.5, the network is considerably more difficult to manage than the network of Figure 16.4. It is most likely that separate network management devices will be necessary for the radio equipment, for the backhaul and backbone networks, and for the router. Worse still, the terminology for the connections in each of the sub-network parts is likely to be different. Thus the router expects 'virtual connections' while the SDH backbone network can only connect 'permanent' connections. Meanwhile, the radio base station is trying to allocate bandwidth-on-demand like a telephone network. It is very unlikely that a single umbrella network management system can be found to match up the disparate terminology and technology. So the human network planner and operator has to step in with extra effort and coordination. In addition, the fact that the signal has to be 'converted' at different points along the connection has meant that more equipment is necessary (in the case of Figure 16.5, an additional SDH *Add Drop Multiplexor* (*ADM*) is required at the base station site — increasing the investment necessary at this site).

Network Operations and Network Management

The network operation of the radio part of the customer access connection (the *Access Network*, *AN*) will, most likely, be by means of the manufacturer-provided management system (a proprietary *element manager*). However, the operator will probably wish to use an 'umbrella' network management system to show him all the equipment components of the access network on a single screen.

When the operator receives a phone call reporting a fault on the customer line, he wishes to be able to call-up a graphic showing the entire connection (e.g. to the ISDN switch), and

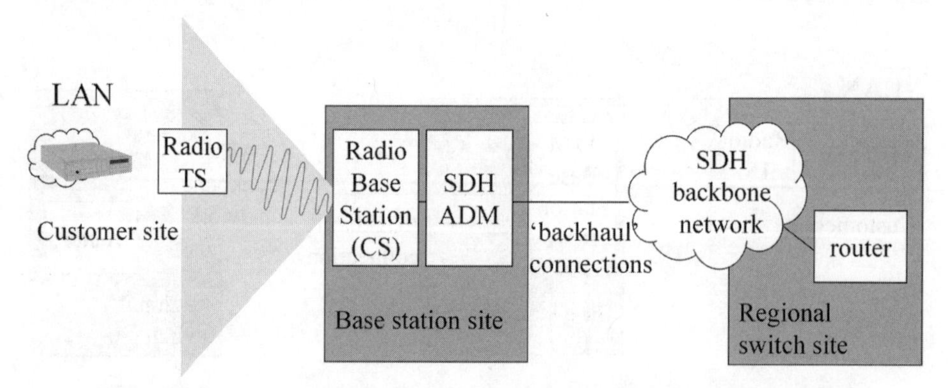

Figure 16.5 Access network architecture comprising disparate technology types

to get an overview of any current alarms. He may also wish to conduct test routines at and/or between selected equipments along the connection.

Given that the planning tool has the main database showing the connection configuration, management information and commands need to be written in a standardised form and carried by means of a suitable standardised protocol (e.g. *SNMP*, *Simple Network Management Protocol* and/or preferably, *CMIP*, *Common Management Information Protocol*) to the network management system. Correlation of the information (as we discussed in conjunction with Figures 16.4 and 16.5) is usually only possible where the different sub-networks share a common architecture, connection types and terminology.

Fault Diagnosis and Tools

For the diagnosis of faults and for the monitoring of network performance, it is useful that network management and operations staff have a number of diagnosis and test tools available. Some of the most valuable tools are:

- the ability to continuously monitor (and adjust if necessary) the actual transmitter output power;
- the ability to continuously monitor the *Receiver Signal Level* (*RSL*);
- the ability to continuously monitor the *Bit Error Ratio* (*BER*) performance of the radio (this can be achieved in a good radio design by monitoring the number of bits which had to be corrected by the *Forward Error Correction* (*FEC*) code;
- the ability to apply *loopbacks* so that external test equipment can be used to check sub-sections of a given connection. In particular, it is valuable to be able to apply at a number of different points within a link (e.g. in the high frequency (*Radio Frequency*, *RF*) part of the equipment), in the intermediate frequency part (i.e. between the outdoor unit and the indoor unit), or at the network interface level. The various loopback positions help to isolate a faulty piece of equipment or a faulty circuit board (Figure 16.6).

The various loopbacks of Figure 16.6 allow a fault to be isolated to a particular piece of equipment, despite testing from a remote location. Testing is carried out by applying a test equipment at a remote location (from the right of Figure 16.6 from a distant radio terminal). The test equipment could, for example, be a *Bit Error Rate Tester* (*BERT*) sending a test pattern onto the transmit path and checking the signal received on the receive path).

If, when a loopback is applied at position 1 of Figure 16.6, the test equipment shows the 'looped' circuit to be OK (where previously the link had been reported 'faulty'), then we are able to determine that the fault lies in the radio terminal illustrated in our figure. By changing the loopback to position 2, we are able to check whether the outdoor unit has a problem. (If the loopback gives satisfactory test results then probably the outdoor unit is also operating OK. Unsatisfactory test results at this point would suggest an outdoor unit problem.)

Having checked the outdoor unit and found it to be OK, the loopback position can steadily be moved from position 2 to position 3, and then position 4. A fault at position 3 indicates a cabling fault on the indoor-to-outdoor unit cabling. A fault at position 4

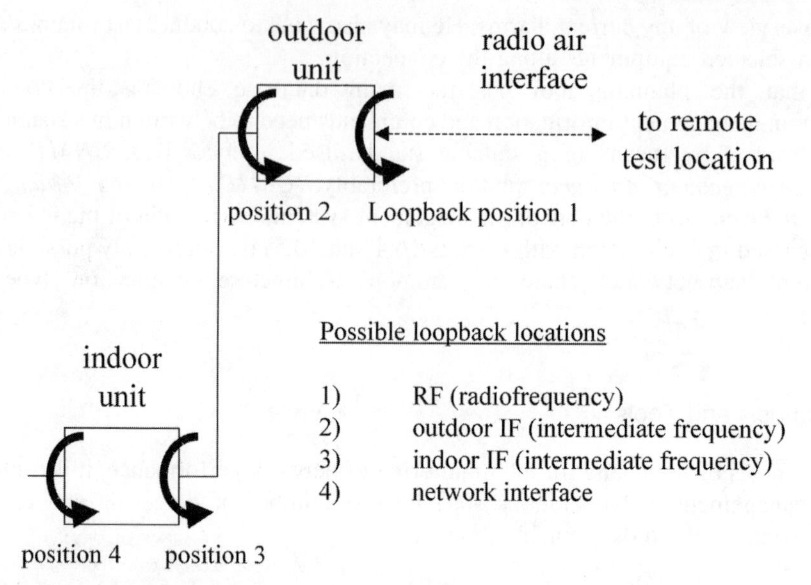

Figure 16.6 Applying loopbacks to isolate the precise equipment failure or faulty circuit board

indicates a faulty indoor unit. If, however, the loopback test checks OK at position 4, then the most likely cause of the reported fault is in the customer equipment connected to the network interface of the indoor unit.

Field Service and Logistics

The most important considerations when planning the service organisation necessary to support a given wireless access network is the number and location (home stations) of field service staff, and the number and locations of equipment spares.

The factors which are of most importance are:

- the customer expectation of the return-to-service time (*time-to-repair* or *Mean-Time-To-Repair*, *MTTR*) following the report of a fault. Typically, modern network operators commit to repair within four hours (maybe even less within important business metropols);
- the failure rate of given equipment types (typically measured as the *MTBF*, or *Mean Time Between Failure*). A *MTBF* of 10 years means that, on average, a given component will fail every ten years. Or said another way, of every ten components in service during a given year (10 years service), one of these components will fail and need to be replaced. However, despite the 10% failure rate, the number of spares held is typically 2–5% of the number in service (minimum of one spare of each component, of course). A 10% spares holding is not necessary, since we are able to return spares for repair or buy new components to replace the spare units we put into service before the next repair (the 10% of failures do not all happen at once);
- the level of hardware *redundancy* provided within the equipment. By providing redundant hardware, the customer service will be unaffected by a fault, and the service

technicians can afford to wait until the next day or a few days later before making the service call to replace the faulty hardware.

Of course, the lower the total number of different components going to make up the radio network, the fewer the number of different types of spares which must be stocked and the lower is the investment in working capital for spares. Given that the spares stock may have to be duplicated at different service 'home locations' around a given country to meet the time-to-repair response times expected by customers, only a few less components may lead to considerably lower overall stock value.

The fact that a single type of radio unit covers the entire radio band used by the network operator also has a significant impact on service logistics and the response times possible following report of a fault. Some network operators, for example, use a common staff for regional installations and for service calls. When not attending a fault, the service technician is conducting an installation. Should he get called to the fault, the radio intended for the new installation could be used instead as a replacement for the faulty one. (Or maybe, more realistically, the technician carries two units with him.)

Network Documentation

There should be a 'master' database for the access network. An export capability from the planning tool storing it can also be very valuable. Thus, for example, the export of data in a standard format to Windows95 Windows98 or WindowsNT applications allows for easy management report production using standard PC software. The database needs to be kept up-to-date by synchronisation with the network using a standard protocol such as SNMP or CMIP.

The best-designed networks allow the configuration of faulty network elements to be stored and then downloaded into the replacement hardware after a fault.

It is also valuable to be able to 'sort' the network documentation (i.e. list the network configuration information) on a per-line, per customer, per-service-type, per equipment or per-network basis.

Network Performance Monitoring

For monitoring of network operation of the radio network, it should be possible to predict interference and intermodulation, as well as signal levels along defined test routes, which may then subsequently be checked against mobile test measurement results. It should be possible also to take account of 'foreign' radio equipment (i.e. a second equipment manufacturer being used by the same operator in a neighbouring area, or the network of a second operator in the same area).

The traffic performance of the access network and the backhaul network should also be able to be predicted (e.g. telephone traffic in erlangs, Internet traffic and/or ATM network quality parameter measurements [delay, jitter, wander, Cell Delay Variation (CDV), etc.] so that these can also be compared against available traffic reports of the radio equipment and/or other network devices. It may be valuable, in addition, to be able to adjust the prediction algorithms with factors enabling values nearer to those measured in reality. The prime aim is to be able to predict the capacity of an existing network and/or the equipment volumes needed to carry a predicted future volume of traffic.

Adding Network Capacity

Network operators find it useful to have planning tools specifically designed to assist in adding capacity to the network. The best tools enable various planning, equipment design and installation project planning to be conducted, first on a 'What-if' basis, subsequently firming the chosen preferred option into the official plan.

A network capacity increase could take any of the following forms:

- adding radio units and/or antennas to the base station;
- splitting a sector by reducing the angle of the sector;
- superimposing a new sector with different radio modulation (e.g. 16-QAM instead of 4-QAM);
- changing the modulation scheme of the sector (e.g. from 4-QAM to 16-QAM) to increase the available bit rate/capacity;
- changing the channel scheme;
- adding new radio channels in opposite polarisation;
- adding equipment from another manufacturer (the operation of a second system may be desired by operators either for specific technical reasons (e.g as a 'second source' technology). A common planning platform is usually desired.

It is useful to be able to predict any configuration problems (e.g. if the base station has much more radio interface capacity than it has trunk network capacity, etc.).

It is also valuable to be able to predict radio interference problems, coverage problems (e.g. customer previously in-service is now out-of-range due to use of higher modulation scheme, or cannot operate again until his subscriber antenna is turned for the opposite polarisation, etc.). A really good tool would suggest a 'corrective action' for each arising problem.

Production of Management Reports

The staff of many network operators like to be able to generate detailed reports about network performance, capacity and configuration. For this purpose, it is common nowadays for radio equipment manufacturers and computer software tool suppliers to allow for the 'exporting' of network configuration and performance data in a standard database format (e.g. Microsoft Access, Microsoft Excel, dbase, or some other standard database format). Nonetheless, as 'standard' features of a set of planning and network management software tools, the following information is invaluable to planning and operations staff:

- individual subscriber site infomation (location and configuration, home base station, radio orientation);
- link cross-section profile of individual customer links;
- link map of individual customer links;
- base station coverage area map with range, signal level contours, interference, intermodulation, blackout areas, bit rate/capacity, etc.;
- base station site information, equipment configuration, rack layouts, backbone trunking configuration, etc.

Network Security Considerations

There are four main network security considerations when designing and operating public telecommunications networks:

- the need to ensure the *confidentiality* of the content of customers' communications;
- the prevention of unauthorised use, monitoring or re-configuration (i.e. *misuse*) of the network;
- the need to ensure the *integrity and privacy of personal data*;
- the need to allow *secret monitoring (tapping) of communications by national security agencies*.

The confidentiality of customers' communications is ensured by good network design and careful operational working practice. The network operator needs to ensure, for example, that data stored during transmission of a given communication is not retained once transmitted. In normal circumstances, technicians should not be allowed to monitor a customer's communications without the customer being aware. In practice, this may to a large extent depend upon the integrity of the operations staff only to monitor lines which have been reported faulty — and only during a period of fault diagnosis.

Where simple good practice and good will is not sufficient security for the content of a given customer's communication, special content encryption may be necessary. This we discuss later in the chapter.

Prevention of network misuse relies upon the good design of the equipment and the network as a whole. Much network misuse can be eliminated simply by being careful to make sure that all network management control devices are password-protected. There should not be any devices which can be dialled-up from anywhere, logged-into without a password, and capable of network monitoring or re-configuration. You would be surprised, however, in just how many networks this is possible. Against internal misuse (by network management staff or by infiltration of the network management centre), only good building security and good personnel screening are effective.

The integrity and privacy of personal data is legislated in most countries by data protection statutes. The practical implications for public telecommunications network operators are:

- the storage in computer-media of only the personal information necessary for operating the business (e.g. the name and address for invoicing, etc.);
- the deletion of call records (or other records linking a given communication to a particular customer) after customer invoicing and once a reasonable period in which the customer may dispute his invoice have expired.

In some countries, the national security laws also demand that public telecommunications network operators make available a capability for national security agencies to 'tap' communications (phone calls, data messages and all other types of communication over the network). It has to be possible for the agency to conduct the tapping without the knowledge of the network operator, and without alerting the awareness of the customer being tapped. Such measures are primarily focused on detecting organised crime, industrial and international espionage.

Encryption

Ecryption (sometimes called scrambling) is available for the protection of both speech and data information carried by radio systems. A cypher or electronic algorithm can be used to code the information in such a way that it appears to third parties like meaningless garbage. A combination of a known codeword (or combination of codewords) and a decoding formula are required at the receiving end to reconvert the message into something meaningful. The most sophisticated encryption devices were developed initially for military use. They continuously change the precise codewords and/or algorithms which are being used, and employ special means to detect possible disturbances and errors. One of the most secure methods was developed by the United States defence department, and is known as *DES* (*Defense Encryption Standard*).

To give maximum protection, information encryption needs to be coded as near to the source and decoded as near to the destination as possible. There is nothing to compare with speaking a language which only you and your fellow communicator understand!

In a technical sense, the earliest opportunity and best place for encryption is the caller's handset. Sometimes, either for technical or economic reasons, this point is not feasible and the encryption is first carried out deeper in a telecommunication network. Thus, for example, a whole site might be protected with only a few encryption devices on the outgoing lines, rather than equipping each PBX extension separately. Clearly, the risks are then higher.

For most commercial concerns I do not believe the security risks arising from technical interception of signals within wide area networks are that great. It is much simpler to overhear conversations on the train, read fax messages carelessly left on unattended fax machines or 'bug' someone's office than it is to intercept messages half way across a network. For maximum protection of data, the data itself should always be stored in an encrypted form, and not just encrypted at times when it is to be carried across telecommunications networks. Permanent encryption of the data renders it in a meaningless or inaccessible form for even the most determined computer hacker. Thus, for example, encrypted confidential information held on an executive's laptop computer can be prevented from falling into unwanted hands, should the laptop go missing.

Accounting

The final of the management capabilities defined by ISO (International Organization for Standardization) and which we have not yet considered is *accounting*. (There are five defined network management capabilities which network operators need to provide— captured by the acronym *FCAPS—Fault management, Configuration management, Accounting management, Performance management and Security management*).

Accounting records need to be collected by public telecommunications network operators who wish to invoice their customers based upon their usage (number of minutes or volume of data transmitted) during a given conversation or communication. It is usual to collect such accounting data records only at switching locations. Thus, as Figure 16.7 illustrates, it is not always necessary for the radio equipment or base station to collect such accounting records (or *Call Data Records*, *CDRs*).

Where the base station acts only as a access network *multiplexor* (without location switching, Figure 16.7a) no accounting records need to be collected, since all communications pass through a switch (another piece of equipment). However, where 'local switching' in the base station might mean that a communication can be switched from one end to the other without traversing any other switching equipment (Figure 16.7b),

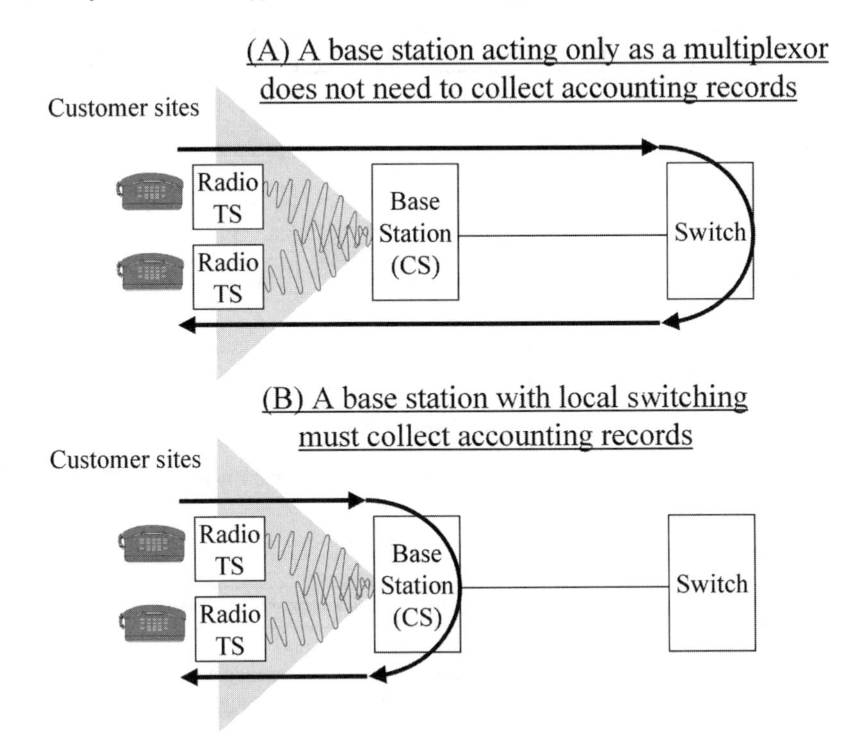

Figure 16.7 Local switching functionality at a radio base station demands the generation of accounting records — multiplexor-only functionality circumvents this responsibility

accounting records will have to be generated by the radio base station equipment for the purpose of accounting and customer usage invoicing.

Umbrella Network Management of Networks with Fixed Wireless Access — The TMN Model for Access Networks

The most advanced set of standards for 'umbrella management' of complex networks comprising different types of *Network Elements* (*NEs*) are defined in ITU-T's (International Telecommunications Union — Standardisation sector) *TMN* (*Telecommunications Management Network*) series of standards (ITU-T recommendations in the series M.3000). These incorporate a layered and structured set of functions as shown in Figure 16.8, enabling multiple types of services and customer connections to be managed, monitored and administered despite heterogeneous network technology (i.e. different *network elements*). The standards also embrace the latest in *object oriented computing*:

- CMIP (Common Management Information Protocol — a 'language' allowing network management computers to talk to one another or to network elements).
- CORBA (Common Object Request Broker Architecture — software enabling different types of 'objects' to be dealt with).
- MIB (Managed Information Base — standardised definitions of network elements and their functions and controls and monitoring parameters which may be applied to them).

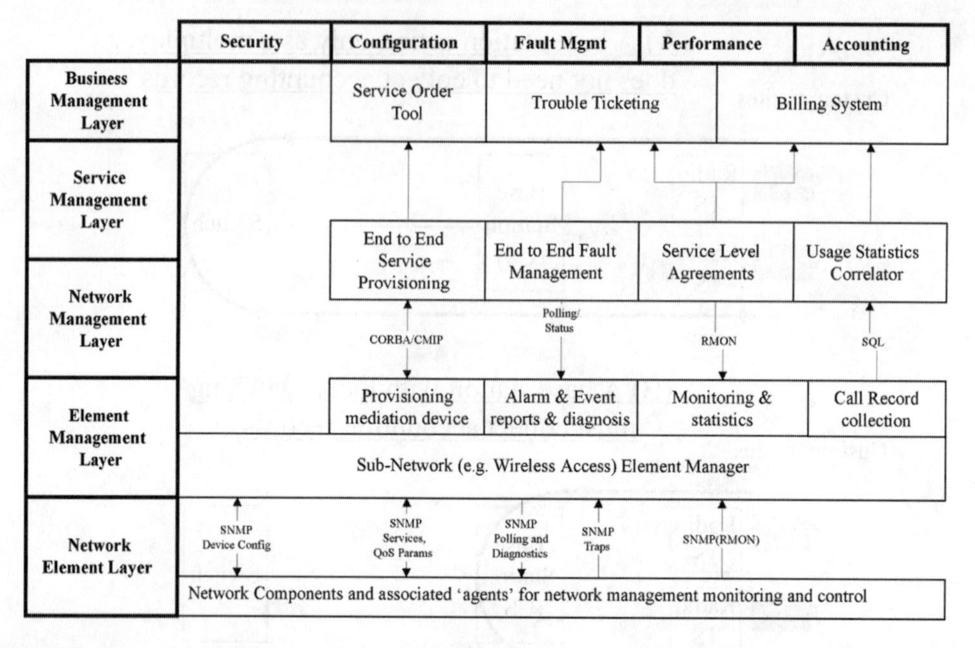

Figure 16.8 Possible structure of functions for an umbrella network management system controlling an end-to-end network comprising wireless and other network elements (equipment)

- RMON (remote monitoring) — a protocol from the IP (Internet Protocol) and Ethernet suite of protocols which allows for remote monitoring.
- SNMP (Simple Network Management Protocol — a language similar in some ways to CMIP for communicating with network elements to query their status or control their configuration).
- SQL (System Query Language — a standardised language for querying a database).

De-facto standard software products for the *mediation layers* between the *Element management Layer* (*EML*) and the *Network Management Layer* (*NML*) and *Service Management Layer* (*SML*) are beginning to appear from a number of manufacturers. These include the *Syndesis* product 'end-to-end network configurator', a product from *Micromuse* for end-to-end fault management, and various products from Hewlett Packard (Metrica) and Concord communications for end-to-end network performance management. The interfaces to these *de facto* standard software products are increasingly being incorporated into wireless equipment manufacturers' proprietary *Network Element Managers* (*NEM* — the official standardised name for what manufacturers often simply refer to as their 'network management system').

Appendix 1

Radio Bands and Channel Rasters for Fixed Wireless Systems

ITU-R Recommendations of the F-Series: Radio-Frequency Channel Arrangements

Document number	Revision (1998)	Document title	Contents	Pages
F.746	3	**Radio-frequency channel arrangements for radio-relay systems**	Basic structure and rules for channel arrangement	12
F.1242	0	Radio-frequency channel arrangements for digital radio systems operating in the range 1350 MHz to 1530 MHz	**Channel structure 1.35–1.53 GHz**	4
F.701	2	Radio-frequency channel arrangements for analogue and digital point-to-multipoint radio systems operating in frequency bands in the range 1.350 to 2.690 GHz	**Channel structure [for PMP 1.5, 1.8, 2.0, 2.2, 2.4 and 2.6 GHz]**	3
F.1098	1	Radio-frequency channel arrangements for radio-relay systems in the 1900–2300 MHz Band	**Channel structure 1.90–2.30 GHz**	4
F.1243	0	Radio-frequency channel arrangements for digital radio systems operating in the range 2290 MHz to 2670 MHz	**Channel structure 2.29–2.67 GHz**	2
Rec.283	5	Radio-frequency channel arrangements for low and medium capacity analogue or digital radio-relay systems operating in the 2 GHz Band	**Channel structure 1.70–1.90 GHz** **Channel structure 1.90–2.10 GHz** **Channel structure 2.10–2.30 GHz** **Channel structure 2.50–2.70 GHz**	3
Rec.382	7	Radio-frequency channel arrangements for radio-relay systems operating in the 2 and 4 GHz Bands	Six go and six return channels with capacities from 600 to 1800 telephone channels	4
Rec.635	4	Radio-frequency channel arrangements based on a homogeneous pattern for radio-relay systems operating in the 4 GHz Band	**Channel structure 3.40–4.20 GHz (channel bandwidth 30, 40, 60, 80 or 90 MHz)**	8
F.1099	2	Radio-frequency channel arrangements for high-capacity digital radio-relay systems in the 5 GHz (4400–5000 MHz) Band	**Channel structure 4.40–5.00 GHz (channel bandwidth 20, 40, 60 or 80 MHz)**	6
Rec.383	5	Radio-frequency channel arrangements for high capacity radio-relay systems operating in the lower 6 GHz Band	**Channel structure 5.85–6.425 GHz (channel bandwidth of 60 or 80 MHz)**	4

Rec.384	6	Radio-frequency channel arrangements for medium and high capacity analoguie or digital radio-relay systems operating in the upper 6 GHz Band	**Channel structure 6.43–7.11 GHz (80 MHz channel spacing)**	5
F.385	6	Radio-frequency channel arrangements for radio-relay systems operating in the 7 GHz Band	**Channel structure 7.425– 7.725 GHz (28 MHz channel spacing)** **Channel structure 7.435– 7.750 GHz (5, 10 or 20 MHz channel spacing)** **Channel structure 7.110– 7.750 GHz (28 MHz channel spacing)** **Channel structure 7.425– 7.900 GHz (28 MHz channel spacing)**	7
F.386	5	Radio-frequency channel arrangements for medium and high capacity analogue or digital radio-relay systems operating in the 8 GHz Band	**Channel structure 7.725– 8.275 GHz (29.65 MHz channel spacing)** **Channel structure 8.275– 8.500 GHz (14 or 28 MHz channel spacing)**	6
Rec.747	0	Radio-frequency channel arrangements for radio-relay systems operating in the 10 GHz Band	**Channel structure 10.50– 10.68 GHz**	3
F.387	7	Radio-frequency channel arrangements for radio-relay systems operating in the 11 GHz Band	**Channel structure 10.70– 11.70 GHz (40 MHz channel spacing)**	8
F.497	5	Radio-frequency channel arrangements for radio-relay systems operating in the 13 GHz frequency Band	**Channel structure 12.75– 13.25 GHz (28 MHz channel spacing)**	5
F.636	3	Radio-frequency channel arrangements for radio-relay systems operating in the 15 GHz Band	**Channel structure 14.40– 15.35 GHz (3.5, 7, 14 or 28 MHz channel spacing)**	4
F.595	5	Radio-frequency channel arrangements for radio-relay systems operating in the 18 GHz freqeuncy Band	**Channel structure 17.70– 19.70 GHz (Multiples of 3.5 GHz, 13.75 MHz 55, 110, 220 or 320 MHz channel spacing)** [Various inconsistent channel structures used in different countries and world regions]	10
F.637	2	Radio-frequency channel arrangements for radio-relay systems operating in the 23 GHz Band	**Channel structure 21.2– 23.6 GHz (Multiples of 2.5 MHz or 3.5 MHz channel spacing)**	9

(continued)

ITU-R Recommendations of the F-Series: Radio-Frequency Channel Arrangements

(continued)

Document number	Revision (1998)	Document title	Contents	Pages
F.748	2	Radio-frequency channel arrangements for radio-relay systens operating in the 25, 26 and 28 GHz	**Channel structure 24.25– 25.25 GHz [25 GHz]** **Channel structure 25.25–27.5 or 24.5–26.5 [26 GHz]** **Channel structure 27.5– 29.5 GHz (Multiples of 2.5 MHz or 3.5 MHz channel spacing)**	7
F.749	1	Radio-frequency channel arrangements for radio-relay systems operating in the 38 GHz Band	**Channel structure 37.0– 39.5 GHz [CEPT] (Multiples of 3.5 MHz channel spacing)** **Channel structure 38.6– 40.0 GHz [USA] (Multiples of 2.5 MHz channel spacing)** **Channel structure 36.0– 37.0 GHz [Russia] (Multiples of 3.5 MHz channel spacing)**	7
F.1100	0	Radio-frequency channel arrangements for radio-relay systems operating in the 55 GHz Band	**Channel structure 54.25–58.2 GHz (Multiples of 2.5 MHz or 3.5 MHz channel spacing)**	4

CEPT Frequency Allocations for Europe and Channel Raster Specifications

Band	Frequency range	Application	Applicable Channel Plan and Raster
	1880–1900 MHz	DECT (Digital European Cordless Telephony)	ERC/DEC/(94) 03
2.1 GHz	2025–2110 MHz	Fixed links	ERC Rec 13-01 Annex C
	2200–2290 MHz	Fixed links	ERC Rec 13-01 Annex C
ISM	2400–2500 MHz	ISM (Industrial, Scientific and Medical)	EU2, EU15
2.6 GHz	2520–2670 MHz	Fixed links	ERC Rec 13-01 Annex D
3.4 GHz 3.5 GHz	3400–3600 MHz	Fixed links	CEPT/ERC Rec 14-03
3.7 GHz	3600–4200 MHz	Fixed links	CEPT/ERC Rec 12-08
	5150–5250 MHz	HIPERLANs	ERC/DEC(96)03
6 GHz	5925–6425 MHz	Medium/High capacity Fixed links	CEPT/ERC Rec 14-01
7 GHz	6425–7125 MHz	Medium/High capacity Fixed links	CEPT/ERC Rec 14-02
8 GHz	7125–8500 MHz	Fixed links based on ITU-R F.385	EU2, EU27
10 GHz 10.5 GHz	10.15–10.30 GHz plus 10.50–10.65 GHz	Fixed links including point-to-multipoint	CEPT/ERC Rec 12-05
	10.45–10.50 GHz	Fixed links including point-to-multipoint	CEPT/ERC Rec 70-03
11 GHz	10.70–11.70 GHz	Fixed links	CEPT/ERC Rec 12-05
13 GHz	12.75–13.25 GHz	Fixed links based on ITU-R F.497	CEPT/ERC Rec 12-02
15 GHz	14.50–15.35 GHz	Fixed links	CEPT/ERC Rec 12-07
	17.10–17.30 GHz	HIPERLAN	EU2, CEPT/ERC Rec 70-03
18 GHz	17.70–19.70 GHz	Fixed links	CEPT/ERC Rec 12-03
23 GHz	22.00–23.60 GHz	Fixed links	ERC Rec T/R 13-02
26 GHz 25 GHz	24.50–26.50 GHz	Fixed links point-to-point and point-to-multipoint	ERC Rec T/R 13-02
28 GHz	27.50–29.50 GHz	Fixed links	ERC Rec T/R 13-02
32 GHz	31.00–33.40 GHz	High density fixed links, including point-to-multipoint	Under review at World Radio council (1999)
38 GHz	37.00–39.50 GHz	Low and medium capacity fixed links	ERC Rec T/R 12-01

(continued)

CEPT Frequency Allocations for Europe and Channel Raster Specifications

(continued)

Band	Frequency range	Application	Applicable Channel Plan and Raster
40 GHz	40.50–42.50 GHz	MVDS (multipoint video distribution systems)	ERC Rec T/R 52-01
52 GHz	51.40–52.60 GHz	High density fixed links	
56 GHz	55.78–57.00 GHz	Low power, short range, low and medium capacity fixed links	ERC Rec T/R 22-03
58 GHz	58.20–59.00 GHz	High density fixed links	
65 GHz	64.00–66.00	High density fixed links	ERC Rec T/R 22-03

Appendix 2

ETSI (European Telecommunications Standards Institute) Radio Specifications for Fixed Wireless

European Standards (EN – European Norm) and European Telecommunications Standards (ETS)

Standard designation	Subject
ETS 300 197	38 GHz Digital Radio Relay Systems (DRRS); Parameters for the transmission of digital signals and analogue video signals
ETS 300 198	23 GHz Digital Radio Relay Systems (DRRS); Parameters for the transmission of digital signals and analogue video signals
EN 300 220-01	25 MHz to 1 GHz with power levels up to 500 mW; technical characteristics and test methods
ETS 300 234	STM-1 high capacity Digital Radio Relay Systems (DRRS) operating in bands with 30 MHz channel spacing
ETS 300 328	2.4 GHz wideband transmission in the ISM (industrial, scientific and medical) band using spread-spectrum techniques
ETS 300 339	EMC (electromagnetic compatibility) and radio spectrum matters (ERM) for radio communications equipment
ETS 300 407	55 GHz Digital Radio Relay Systems (DRRS); Parameters for the transmission of digital signals and analogue video signals
ETS 300 408	58 GHz Digital Radio Relay Systems (DRRS); Parameters for the transmission of digital signals and analogue video signals (without coordinated frequency planning)
ETS 300 430	18 GHz STM-1 high capacity Digital Radio Relay Systems (DRRS) with a channel spacing of 55 MHz
ETS 300 431	24.25 to 29.50 GHz Band; Digital Fixed Point-to-Point radio link equipment

(continued)

(continued)

Standard designation	Subject
ETS 300 432	18 GHz STM-1 high capacity Digital Radio Relay Systems (DRRS) with a channel spacing of 110 MHz
I-ETS 300 440	1 GHz to 25 GHz radio equipment; technical characteristics and test methods
ETS 300 454	Wide band audio radio links; technical characteristics and test methods
ETS 300 630	1.4 GHz low capacity Point-to-Point Digital Radio Relay Systems (DRRS)
EN 300 631–1	1 GHz to 3 GHz Antennas for Point-to-Point Digital Radio Relay Systems (DRRS)
ETS 300 632	24.25 GHz to 29.50 GHz fixed radio link equipment for the transmission of analogue video signals
ETS 300 633	2.1 GHz to 2.6 GHz low and medium capacity Point-to-Point Digital Radio Relay Systems (DRRS)
ETS 300 635	STM-N transmission of the Synchronous Digital Hierarchy (SDH) by radio
ETS 300 636	1 GHz to 3 GHz Point-to-Multipoint TDMA (Time Division Multiple Access) radio systems
ETS 300 638	10 GHz to 14 GHz Point-to-Point Digital Radio Relay Systems (DRRS) with 20 MHz alternate channel spacing for transmission of digital signals and analogue video
ETS 300 639	13 GHz/15 GHz/18 GHz Sub-STM-1 Digital Radio Relay Systems (DRRS) with 28 MHz co-polar and 14 MHz cross-polar channel spacing
ETS 300 674	5.8 GHz ISM (Industrial, scientific and medical) band; Road Transport and Traffic Telematics (RTTT)
ETS 300 748	Multipoint Video Distribution System (MVDS); Digital Video Broadcasting (DVB) at 10 GHz and above
ETS 300 749	Microwave Multipoint Distribution System (MMDS); Digital Video Broadcasting (DVB) below 10 GHz
ETS 300 751	SWIFT (System for Wireless Infotainment Forwarding and Teledistribution)
ETS 300 752	TFTS (Terrestrial Flight Telecommunications System); packet mode data
ETS 300 785	SDH (Synchronous Digital Hierarchy); transmission of sub-STM-1 by radio
ETS 300 786	13 GHz/15 GHz/18 GHz Sub-STM-1 Digital Radio Relay Systems (DRRS) with 14 MHz co-polar spacing
ETS 300 826	HIPERLAN (High performance radio local area network) at 2.4 GHz; EMC (electromagnetic compability) and radio spectrum matters (ERM)
EN 300 828	TETRA (terrestrial trunked radio); EMC (electromagnetic compability) and radio spectrum matters (ERM)
ETS 300 833	1 GHz to 3 GHz Antennas for Point-to-Point Digital Radio Relay Systems (DRRS)
ETS 300 836	HIPERLAN type 1 (high performance radio local area network); conformance testing specification
EN 301 021	3 GHz to 11 GHz Point-to-Multipoint TDMA (time division multiple access) Digital Radio Relay Systems (DRRS)
EN 301 040	Lawful Interception Interface (LI) for TETRA (terrestrial trunked radio)

(continued)

(continued)

Standard designation	Subject
EN 301 055	1 GHz to 3 GHz Point-to-Multipoint DS-CDMA (direct sequence code division multiple access) Digital Radio Relay Systems (DRRS)
EN 301 091	Road Transport and Traffic Telematics (RTTT) in the band 76.0 GHz to 77.0 GHz; EMC (electromagnetic compatibility) and radio spectrum matters (ERM)
EN 301 124	3 GHz to 11 GHz Point-to-Multipoint DS-CDMA (direct sequence code division multiple access) Digital Radio Relay Systems (DRRS)
EN 301 126	Conformance testing for Digital Radio Relay Systems (DRRS)
EN 301 126-1	Conformance testing for Digital Radio Relay Systems (DRRS); Part 1 — Point-to-Point systems
EN 301 128	13 GHz/15 GHz/18 GHz PDH (plesiochronous digital hierarchy) low and medium capacity Digital Radio Relay Systems (DRRS)
EN 301 129	SDH (synchronous digital hierarchy) Digital Radio Relay Systems (DRRS); system performance monitoring parameters
EN 301 213-1	24.25 GHz to 29.50 GHz Point-to-Multipoint Digital Radio Relay Systems (DRRS)
EN 301 213-2	24.25 GHz to 29.50 GHz Point-to-Multipoint FDMA (Frequency Division Multiple Access) Digital Radio Relay Systems (DRRS)
EN 301 213-3	24.25 GHz to 29.50 GHz Point-to-Multipoint TDMA (time division multiple access) Digital Radio Relay Systems (DRRS)
EN 301 215	Point-to-Multipoint Antennas for Digital Radio Relay Systems (DRRS) in the band 11 GHz to 60 GHz

ETSI Technical Reports (ETR)

Standard designation	Subject
ETR 069	HIPERLAN (high performance radio local area network); services and facilities
ETR 133	HIPERLAN (high performance radio local area network); system definition
ETR 139	Radio in the Local Loop (RLL)
ETR 151	EMC (electromagnetic compatibility) testing of telecommunication equipment above 1 GHz
ETR 226	HIPERLAN (high performance radio local area network); architecture for time-bound services
ETR 239	SDH (Synchronous Digital Hierarchy); list of documents relevant to SDH transmission equipment
ETR 241	PDH (plesiochronous digital hierarchy); functional architecture of 2 Mbit/s-based PDH transport networks
ETR 306	Access networks for residential customers
TR 101 016	Performance prediction models for Digital Radio Relay Systems (DRRS); comparison and verification
TR 101 031	HIPERLAN (high performance radio local area network); requirements and architectures for wireless ATM access and interconnection
TR 101 035	SDH (Synchronous Digital Hierarchy) aspects regarding Digital Radio Relay Systems (DRRS)
TR 101 127	STM-1 high capacity Digital Radio Relay Systems (DRRS) in frequency bands with 30 MHz channel spacing and Co-Channel Dual-Polarised (CCDP) operation

Appendix 3

IEEE Publications and FCC Standards

IEEE Publications

Dokument designation	Subject
139—1988	Recommended Practice for Measurement of Radio Frequency Emmission from Industrial, Scientific and Medical (ISM) equipment installed on User's Premises
145—1993	Standard Definitions of Terms for Antennas
149—1979	Standard Test Procedures for Antennas
211—1997	Standard Definitions of Terms for Radio Wave Propagation
291—1991	Standard Methods for Measuring Electromagnetic Field Strength of Sinusoidal Continuous Waves, 30 Hz to 30 GHz
299—1997	Standard for Measuring the Effectiveness of Electromagnetic Shielding Enclosure
8802-11: 1999 (ISO/IEC) IEEE 802.11	Wireless LAN Medium Access Control (MAC) and Physical Layer (PHY) Specifications
IEEE 802.11a-1999	Wireless LAN High Speed Physical Layer (PHY) in the 5 GHz Band
IEEE 802.11a-1999	Wireless LAN High Speed Physical Layer (PHY) in the 2.4 GHz Band

FCC Standards

Standard designation	Subject
47 CFR www.foc.gov/oet/info/rules/	47th code of federal regulations (Federal Rules and Regulations)

Appendix 4

Waveguide Specifications

Waveguide designation	Radio Band name	Frequency range	Attenuation	Waveguide dimensions (inside, in mm)
IEC R 3 (EIA WR 2300)	400 MHz	320–490 MHz	0.001 dB/m	580 × 290
IEC R 14 (EIA WR 650)	L-Band	1.14–1.70 GHz	0.010 dB/m	165 × 83
IEC R 22 (EIA WR 430)	W-Band	1.72–2.60 GHz	0.019 dB/m	109 × 55
IEC R 32 (EIA WR 284)	S-Band	2.60–3.95 GHz	0.037 dB/m	72 × 34
IEC R 48 (EIA WR 187)	C-Band	3.95–5.85 GHz	0.070 dB/m	48 × 22
IEC R 70 (EIA WR 137)	6 GHz, 7 GHz	5.85–8.17 GHz	0.114 dB/m	35 × 16
IEC R 100 (EIA WR 90)	10 GHz, 11 GHz	8.20–12.4 GHz	0.217 dB/m	22.9 × 10.2
IEC R 140 (EIA WR 62)	Ku-Band (13 GHz, 15 GHz)	12.4–18.0 GHz	0.351 dB/m	15.8 × 7.9
IEC R 220 (EIA WR 42)	K-Band (18 GHz)	18.0–26.5 GHz	0.723 dB/m	10.7 × 4.3
IEC R 260 (EIA WR 34)	23 GHz, 26 GHz	22.0–33.0 GHz	0.868 dB/m	8.6 × 4.3
IEC R 320 (EIA WR 28)	Ka-Band (38 GHz)	26.5–40.0 GHz	1.162 dB/m	7.1 × 3.6

Appendix 5

Coaxial Cable Waveguides (RG-Nomenclature)

Listed in the table below are the values of some coaxial cables commonly used for indoor unit (IDU)-to-outdoor unit (ODU) connections. Other sources (see the Bibiography) provide more complete information. The nomenclature (RG) derives from the EIA (Electronic Industries Association).

Cable type (RG/U type)	Outside diameter (mm)	Dielectric & jacket material	Shielding	Nominal impedance (ohms)	Loss at 100 MHz in dB per 100 m	Loss at 3 GHz in dB per 100 m
6A	8.4	Polyethylene & PVC	double	75	9.5	72.2
8	10.3	Polyethylene & PVC	single	52	7.2	62.3
12A	12.1	Polyethylene & PVC/armour	single	75	7.2	62.3
34B	16.0	Polyethylene & PVC	single	75	4.6	52.5
58	5.0	Polyethylene & PVC	single	53.5	15.1	124.6
211	18.5	PTFE & Braid	single	50	3.0	24.6
212	8.4	Polyethylene & PVC	double	50	9.5	72.2
213	10.3	Polyethylene & PVC	single	50	7.2	62.3
214	10.8	Polyethylene & PVC	double	50	7.2	62.3
215	12.1	Polyethylene & PVC/armour	single	50	7.2	62.3
216	10.8	Polyethylene & PVC	double	75	7.2	62.3
217	13.8	Polyethylene & PVC	double	50	4.6	42.6

(continued)

(continued)

Cable type (RG/U type)	Outside diameter (mm)	Dielectric & jacket material	Shielding	Nominal impedance (ohms)	Loss at 100 MHz in dB per 100 m	Loss at 3 GHz in dB per 100 m
218	22.1	Polyethylene & PVC	single	50	2.7	29.5
219	24.0	Polyethylene & PVC/armour	single	50	2.7	29.5
220	28.4	Polyethylene & PVC	single	50	2.1	23.0
221	30.4	Polyethylene & PVC/armour	single	50	2.1	23.0
222	8.4	Polyethylene & PVC	double	50	42.3	285.4
223	5.5	Polyethylene & PVC	double	50	14.1	118.1
224	15.6	Polyethylene & PVC/armour	double	50	4.6	42.6
225	10.9	PTFE & Braid	double	50	6.9	45.9

Appendix 6

Forward Error Correction Codes

General

The basic principle of *Forward Error Correction* (*FEC*) is the addition of a number of *code* bits to a user signal to reduce the probability of a *bit error* arising during transmission of that user signal. For a user signal string of length k bits to be encoded we must find at least $2k$ separate and *orthogonal* codes — one distinct *codeword* for each possible bit combination in the original signal.

The *Hamming distance* (d) defines the difference in value of the individual codewords. Specifically, the Hamming distance is the minimum number of bit positions by which the individual codewords differ from one another. The larger the Hamming distance, the greater is the capability of the code to *detect* and *correct* errors in the presence of *noise* or *errors* introduced during transmission.

The number of errors that can be *corrected* is less than or equal to $(d-1)/2$. The number of errors that can be *detected* is less than or equal to d.

There are five main types of forward error correction codes:

- *block codes* (including Hamming codes);
- *cyclic codes*;
- *BCH* (*Bose–Chaudhuri–Hocquenghem*) codes;
- *Reed–Solomon* codes;
- *Convolutional codes* (including *Viterbi codes*)

We briefly discuss the principles and mechanism of each of these types of codes.

Block Codes

Block codes break up the user data stream into a series of blocks of k bits in length. To each block, a code block of $(n-k)$ bits is added to make an n-bit *codeword*. The code is described by the notation (n, k) block code.

Thus, for example, a (15, 4) code has 15 bit codewords, 11 bits of which are real coded user data, and four bits are *parity bits*. An example of a (7, 3) code was given in Chapter 6.

In mathematical notation, the coding takes place as follows:

$$\mathbf{c} = \mathbf{d}[\mathbf{G}]$$

where **c** is a vector representing the *codeword*, **d** is the vector of the user data, and [**G**] is the *generator matrix*.

The generator matrix is a matrix of dimensions k by n, made up of the *identity matrix* [**I**] and the *parity matrix* [**P**] (the matrix [**P**] identifies, for each of the parity bits, which bit positions in the user-data signal are used to calculate the parity):

$$[\mathbf{G}] = [\mathbf{IP}]$$

An example of the various matrices and vectors for a simple (7, 3) block code might be as shown below. (Actually, these are the actual matrices relevant to the example we discussed in Chapter 6, but with the bits rearranged so that the data bits appear in the first three positions of the codeword and the parity bits thereafter):

$$\mathbf{d} = (0\ 1\ 1\ 1)$$

$$\mathbf{G} = \begin{bmatrix} 1\,0\,0\,0\,0\,1\,1 \\ 0\,1\,0\,0\,1\,0\,1 \\ 0\,0\,1\,0\,1\,1\,0 \\ 0\,0\,0\,1\,1\,1\,1 \end{bmatrix}$$

(The last three columns generate the parity bits. The rows of the column indicate which bits of the data are to be used for the calculation. The first parity bit will be based on bits 1, 2 and 4.)

$$\mathbf{c} = (0\ 1\ 1\ 1\ 1\ 0\ 0)$$

The matrix multiplication is performed *modulo 2*. The effect of this multiplication is to add $n-k$ parity bits to the end of the user data signal of k bits.

At the receiving end, the user data part of the signal (as received) is again subjected to the same parity calculation. The $n-k$ parity check result generated by the receiver is represented by the matrix formula $\mathbf{d}'[\mathbf{P}]$, where $\mathbf{d}'$ is that part of the received codeword representing the user data. The result should be equal to the received parity bit code $\mathbf{c_p}$. This can be determined by electronic hardware by a modulo 2 addition (which is equivalent to a modulo 2 subtraction). The result should be 0:

$$\mathbf{d}'[\mathbf{P}] + \mathbf{c_p} = [\mathbf{0}]$$

For our example above, let us assume that the data is received with a bit error at position three of the user data. This is equivalent to the second example of Figure 6.14:

$$\textbf{Received codeword} = (\underbrace{0\ 1\ 0\ 1}_{\mathbf{d}'}\ \overbrace{1\ 0\ 0}^{\mathbf{c_p}})$$

$$\mathbf{d}' = (0\ 1\ 0\ 1)$$

$$\mathbf{P} = \begin{pmatrix} 0\,1\,1 \\ 1\,0\,1 \\ 1\,1\,0 \\ 1\,1\,1 \end{pmatrix}$$

$$\mathbf{c_p} = (1\ 0\ 0)$$

$$\mathbf{d}'[\mathbf{P}] + \mathbf{c_p} = (0\ 1\ 0) + (1\ 0\ 0) = (1\ 1\ 0)$$

If the result is not 0 (as in our example above), then the values in the resulting vector can be used to determine the exact bit position of the error(s), and thus be used to correct them. Because of the rearrangement of the position of the bits, the correction values (3, 5, 6 and 7) will not, however, have the same meaning as in Chapter 6. Instead, these values will represent errors respectively at positions (1, 2, 3 and 4).

Our efforts to explain the process using matrix notation is because this aids the design of digital electronic circuitry.

Cyclic Codes

Cyclic codes are a special type of block code, in which each of the valid codewords are a simple lateral shift of one another. Thus, if one valid code of a (10, 3) cyclic code is:

$$\mathbf{c} = (1\ 0\ 1\ 0\ 1\ 0\ 0\ 1\ 0\ 0)$$

then

$$\mathbf{c} = (0\ 1\ 0\ 1\ 0\ 0\ 1\ 0\ 0\ 1)$$

is also a valid code, as is

$$\mathbf{c} = (1\ 0\ 1\ 0\ 0\ 1\ 0\ 0\ 1\ 0)$$

etc.

Using a cyclical structure of different codes enables us to correct larger blocks of errors than those possible with non-cyclic codes. A generator matrix with particular qualities is required to generate a *cyclic code*. In particular, the generator matrix for a cyclic code always has a value '1' as its last element (i.e. in the kth row of the nth column).

If we consider our example of the last section on block codes, we had a valid (7, 3) codeword of:

$$\mathbf{c} = (0\ 1\ 1\ 1\ 1\ 0\ 0)$$

(For the data value d = 0111 the parity values were 100)

For this to be a cyclic code, all the other codes:

$$\mathbf{c} = (0\ 0\ 1\ 1\ 1\ 1\ 0) \qquad \mathbf{c} = (1\ 1\ 0\ 0\ 0\ 1\ 1)$$
$$\mathbf{c} = (0\ 0\ 0\ 1\ 1\ 1\ 1) \qquad \mathbf{c} = (1\ 1\ 1\ 0\ 0\ 0\ 1)$$
$$\mathbf{c} = (1\ 0\ 0\ 0\ 1\ 1\ 1) \qquad \mathbf{c} = (1\ 1\ 1\ 1\ 0\ 0\ 0)$$

must also be valid code words. But they are not. Another valid codeword is usually $\mathbf{c} = (0\ 0\ 0\ 0\ 0\ 0\ 0)$.

The data value (0 0 1 1), for example, yields the codeword (0 0 1 1 0 0 1), and not the allowed cyclic code (0 0 1 1 1 1 0).

The codeword $\mathbf{c} = (0\ 0\ 0\ 1\ 0\ 0\ 0)$ could also not be a valid (7, 3) cyclic codeword. Why? Because if it were, then $\mathbf{c} = (0\ 0\ 0\ 0\ 1\ 0\ 0)$ would also have to be a valid codeword. This is not possible, for a parity value of '1' is not possible for non-zero user data (the first four bits of the codeword). Not all block codes are cyclic codes!!

So, never mind the non-cyclic codes, how do we conceive the *generator matrix* of a cyclic code? By starting with the *generator polynomial* (the last row of the matrix).

For an (n, k) code we expect $n-k$ bits of user data and a k-bit parity code.

The *generator polynomial* has the form $x^{n-k} + ax^{n-k-1} + \ldots + 1$ (a value is '1' or '0') and appears as the last row of the *generator matrix* in the form (for an example (7, 4) cyclic code):

$$\begin{bmatrix} x\,x\,x\,x\,x\,x\,x \\ x\,x\,x\,x\,x\,x\,x \\ 0\,0\,1\,1\,1\,0\,1 \end{bmatrix}$$

The first $n-k-1$ elements of the row must be zeros, and the last element must be a '1'. In other words, the generator polynomial is a polynomial of order k (first element x^k). The generator polynomial in this case is $x^4+x^3+x^2+1$. This determines the properties of the code, and defines one of the valid codewords, in this case (0 0 1 1 1 0 1). The full generator matrix is:

$$\begin{bmatrix} 1\,0\,0\,1\,1\,1\,0 \\ 0\,1\,0\,0\,1\,1\,1 \\ 0\,0\,1\,1\,1\,0\,1 \end{bmatrix}$$

The valid codewords are:

$$\begin{array}{ll} \mathbf{c} = (0\,0\,0\,0\,0\,0\,0) & \mathbf{c} = (1\,0\,0\,1\,1\,1\,0) \\ \mathbf{c} = (0\,0\,1\,1\,1\,0\,1) & \mathbf{c} = (1\,0\,1\,0\,0\,1\,1) \\ \mathbf{c} = (0\,1\,0\,0\,1\,1\,1) & \mathbf{c} = (1\,1\,0\,1\,0\,0\,1) \\ \mathbf{c} = (0\,1\,1\,1\,0\,1\,0) & \mathbf{c} = (1\,1\,1\,0\,1\,0\,0) \end{array}$$

BCH Codes (Bose–Chaudhuri–Hocquenghem)

BCH-Codes are cyclic codes designed to provide for multiple error correction. The most commonly used BCH-codes are Reed–Solomon codes.

Reed–Solomon Codes

When talking about *Reed–Solomon* codes, it is common to talk about the coding of *symbols*. In many data applications, for example full *bytes* or *octets* (of 8 bits each) are used to represent a given alphanumeric *symbol*. For 8 bits there are $q = 256$ ($q = 2^8$) possible different symbols (values). Thus, we must have 256 different codewords available.

A Reed–Solomon code is a *block code* and *cyclic code* of type (n, k) as we previously discussed. The input number of data bits is $n-k$ and the codeword symbols are n bits in length. It is usual that the value of n be restricted to no more than $q-1$ (i.e. $n < q$). In our example, this means a maximum value of $n = 255$.

The code is often quoted in terms of its byte size. A typical code might be (64, 24), thus 40 bytes of data are protected with 24 bytes of code (64 bytes total). Each codeword would be $64 \times 8 = 512$ bytes long. However, of the 2^{512} combinations available, only 64^{24} combinations (i.e. $2^{6 \times 24} = 2^{144}$) combinations would actually be used. This enables up to $24/2 = 12$ symbol errors to be corrected.

Convolutional Codes

Convolutional codes work on a shift register basis. A string of consecutive bits of the input user data stream (of K bits in length, where K is the *constraint length* of the code). This string of bits are subject to a series of *modulo 2* summation actions (in effect, parity checks). The number of summation codes to which the K-bit sequence is subjected is denoted by v. The coding which results from a 1/v coder using v different code summations is called a 1/v *convolutional code*.

Figure A6.1 illustrates $\frac{1}{2}$ *convolutional coder* with a *constraint length* of $K = 3$.

The two codes being used (in alternation to produce the output bitstream) are:

$$c_1 = d_0 + d_1 + d_2$$
$$c_2 = d_0 + d_2$$

Table A6.1 illustrates the output from this coder.

Convolutional codes of *constraint length* $K = 7$ are often used, with a codeword length of 28 or 35 bits (v = 4 or 5).

The *shift register* nature of convolutional codes makes them very easy to implement in a digital electronic circuit design. In addition, the complete codeword does not have to have been received before decoding can commence. This minimises the signal delays caused during transmission.

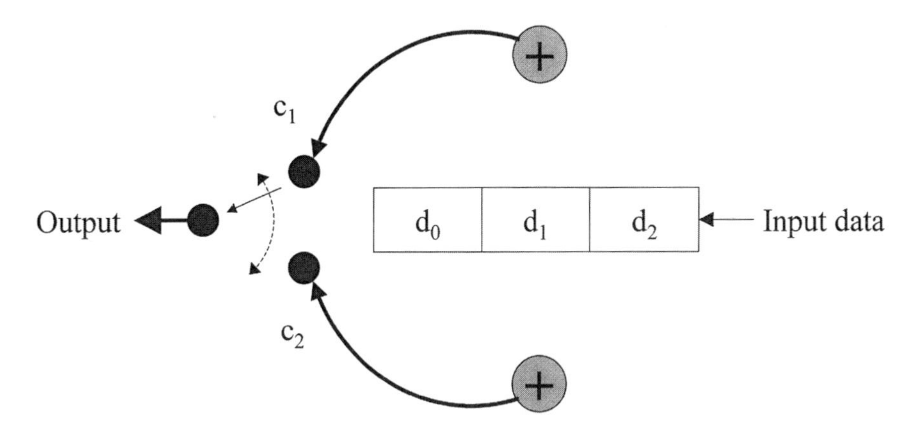

Figure A6.1 $\frac{1}{2}$ **convolutional coder with constraint length $K = 3$**

Table A6.1 Convolutional code and output

Time interval	1		2		3		4		5		6		7		8	
Input bitstream	0		1		1		0		1		0		0		1	
Bits in register ($d_0d_1d_2$)	**000**		**000**		011		110		101		010		100		001	
Output variable	c_1	c_2	c_1	c_2	c_1	c_2	c_1	c_2	c_1	c_2	c_1	c_2	c_1	c_2	c_1	c_2
Output bitstream	0	0	1	1	0	1	0	1	0	0	1	0	1	1	1	1

Viterbi Codes

Viterbi codes are a particular type of convolutional code. In particular, Viterbi determined a methodology for the correct decoding (or rather the most likely correct interpretation or *maximum likelihood* decoding) of incorrectly received codewords.

Coding Gain

One talks of the *coding gain* achieved by using *Forward Error Correction* (*FEC*). This gain is the effective improvement in the *receiver sensitivity* of the receiver (we defined this term in Chapter 6). The improvement in sensitivity results from the correction of some of the errors (resulting in an improvement in the signal quality) at a given received signal power. Typical coding gains due to FEC are around 3 to 6 dB.

Appendix 7

Wireless propagation, Frequency Re-use, Design and Operations

This appendix presents a listing of the ITU-R recommendations which are most relevant to fixed wireless access systems — in the areas of: the terminology of radiocommunication; assessment and prediction of radio propagation; the planning of frequency re-use; as well as the design and operation of radio equipment.

Background and General Interest

Document number	Revision (1998)	Document title	Contents	Pages
Rec.592	2	**Terminology used for Radio- Relay Systems**	Definition of system terminology related to radio relay systems; radio-relay system, point-to-point, point-to-multipoint, diversity, modulation etc.	3
PN.310	**9**	**Definitions of Terms relating to propagation in non-ionised media**	Definition of radio system vocabulary	4
P.1144	0	**Guide to the application of the propagation methods of radiocommunication study group 3**	Classification of propagation prediction methods and reference to relevant ITU-R recommendations	5
PN.1057	0	Probability distributions relevant to radiowave propagation modelling	Gaussian, normal, log-normal, Rayleigh, combine log-normal/Rayleigh, Nakagami-Rice, gamma, exponential, Nakagami-m, Pearson distributions described and formulae provided	14

Transmission Loss Model

Document number	Revision (1998)	Document title	Contents	Pages
P.341	4	**The concept of transmission loss for radio links**	Transmission loss concept model	6
PN.525	2	**Calculation of Free-Space attenuation**	Formula and units conversion formulae for calculation of free-space attenuation	3

Terrestrial Path Prediction Methods

Document number	Revision (1998)	Document title	Contents	Pages
P.530	7	**Propagation data and prediction methods required for the design of terrestrial Line-of-Sight systems**	Description and modelling of atten- uation effects including that due to atmospheric gases, diffraction fading, hydrometers, rain, polarization, multipath etc.	25
Rec.581	2	The concept of 'Worst Month'	Definition of the term 'worst month'	1
Rec.841	0	Conversion of annual statistics to worst-month statistics	Conversion methodology and formulae	3
Rec.678	1	Characterization of the natural variability of propagation phenomena	Expected extimation of long term, e.g. year-to-year variation of propagation	1
P.1146	0	The prediction of field strength for land mobile and terrestrial broadcasting services in the frequency range from 1 to 3 GHz	Calculations and charts for field strength based on distance (relevant for calculation of interference according to P.452)	25
P.1238	0	Propagation data and prediction models for the planning of indoor radiocommunication systems and radio local area networks in the frequency range 900 MHz to 100 GHz	Path loss model, floor penetration loss, shadow fading, delay spread, effects of polarization, obstructed paths, effects of building materials and furnishings, movement of objects.	10

Radiometeorology — Radio Disturbance Effects caused by the Weather

Document number	Revision (1998)	Document title	Contents	Pages
Rec.838	0	**Specific attenuation model for rain for use in prediction methods**	Formula for calculation the attenuation caused by rain	2
PN.837	1	**Characteristics of precipitation for propagation modelling**	Characterisation of the world in 'climate zones' A to Q [map] and definition of rainfall rate intensities for these regions	4
P.676	3	**Attenuation by atmospheric gases**	Gaseous attenuation at various radio frequencies (up to 1000 GHz)	17
P.453	6	The Radio Refractive index: its formula and refractivity data	Formula for calculation of refractivity, n; world maps for monthly variations (input values for other calculations)	9
P.835	2	Reference standard atmospheres	Formulae for: temperature at height, water vapour density, pressure	5
P.836	1	Water vapour: surface density and total columnar content	Water vapour density at ground level; world maps for monthly variations (input values for other calculations)	16
P.840	2	Attenuation due to clouds and fog	Formulae and world maps for monthly variation (frequencies from 10 GHz up to 200 GHz — recommendation designed mainly for satellite system planning where freezing atmospheres are present)	7

Ground and Obstacle Effects

Document number	Revision (1998)	Document title	Contents	Pages
P.526	5	**Propagation by Diffraction**	*Fresnel zones*, diffraction and numerical calculations over spherical surface and over obstacles and irregular terrain (for calculating losses caused by obstacles in the radio path, or partial obstructions)	15
P.527	3	Electrical characteristics of the surface of the earth	Permittivity, relative permittivity, conducivity, penetration and spread of waves, sea water and sea ice, soil, temperature and seasonal variation	5
P.832	1	World Atlas of ground conductivities	Maps	48
PN.833	1	Attenuation in vegetation	Calculation of 'specific attenuation in vegetation' (data for frequencies up to about 3 GHz)	
P.1058	1	Digital topographic databases for propagation studies	Specification of standard terms and standard methodology and format for presentation of data (coordinates, terrain, ground coverage, urban area, building heights etc.)	7

Radio Noise

Document number	Revision (1998)	Document title	Contents	Pages
P.372	6	Radio Noise	Terms related to radio noise, explanation of the sources of noise, formulae for noise, noise brightness (including global maps), noise related to frequency, atmospheric gas noise, noise variability.	73

Frequency Sharing, Interference Signal Prediction and Coordination

Document number	Revision (1998)	Document title	Contents	Pages
P.452	8	**Prediction procedure for the evaluation of microwave interference between stations on the surface of the earth at frequencies above about 0.7 GHz**	Description and calculation of interference effects caused by scatter paths, multipath, diffraction, clutter, refraction, ducting etc.	5
P.619	1	Propagation data required for the evaluation of interference between stations in space and those on the surface of the earth	Interference effects affecting terrestrial radio systems which are operating in bands also used in satellite communications systems (relevant for frequency planning and allocation bodies)	5
P.620	3	**Propagation data required for the evaluation of coordination distances in the frequency range 0.85–60 GHz**	Methodology for radio link planning relevant to PTP radio and cell planning of PMP sytems. The methodology gives the distance (the 'coordination distance') at which the interference on the same frequency can be considered to be negligible	14

Fixed Radio-Relay Link Planning — Interference and its resolution

Document number	Revision (1998)	Document title	Contents	Pages
F1093	1	**Effects of multipath propagation on the design and operation of line-of-sight digital radio-relay systems**	Effects of multipath propagation on the quality of transmission, prediction methods and counter-measures	12
F.1094	1	**Maximum allowable error performance and availability degradations to digital radio-relay systems arising from interference from emissions and radiations from other sources**	Methods for calculating the interference effects of other radios operating in the same band or from *spurious emissions*	4
F.1241	0	Performance degradation due to interference from other services sharing the same frequency bands on a primary basis with digital radio-elay systems operating at or above the primary rate and which may form part of the international portion of a 27 500 km hypothetical reference path	Defines specific performance targets for *errored second ratio (ESR), severely errored second ratio (SESR)* and *background block error ratio (BBER).*	2
F.1095	0	**A procedure for determining coordination area between radio-relay stations of the fixed service**	Procedure for radio frequency planning bodies (in particular regulatory bodies and national radiocommunications agencies) for preventing and/or resolving interference between countries caused by using the same frequencies in neighbouring countries.	4
F.1096	0	**Methods of calculating line-of-sight interference into radio-relay systems to account for terrain scattering**	Mathematical model and procedure for calculating the interference effects caused by terrain scattering.	14
F.1097	0	Interference mitigation options to enhance compatibility between radar systems and digital radio-relay systems	Tips for selection of antennas, shielding of antennas, use of filters etc.	5
F.1190	0	Protection criteria for digital radio-relay systems to ensure compatibility with radar systems in the radiodetermination service	Design criteria for radio systems	5

Frequency re-use, sharing and planning

Document number	Revision (1998)	Document title	Contents	Pages
Rec. 758	1	**Considerations in the development of criteria for sharing between the terrestrial fixed service and other services**	Guidelines for frequency planning organisations. Definition of frequency sharing system parameters	26
F.699	4	**Reference radiation patterns for line-of-sight radio-relay system antennas for use in coordination studies and interference assessment in the frequency range from 1 to about 40 GHz**	Reference radiation pattern for modelling antenna performance to be used in the absence of information about a particular antenna	6
F.1245	0	**Mathematical model of average radiation patterns for line-of-sight point-to-point radio-relay system antennas for use in certain coordination studies and interference assessment in the frequency range from 1 to about 40 GHz**	Reference radiation pattern for modelling PTP (point-to-point) antenna performance to be used in the absence of information about a particular antenna	
F.1336	0	**Reference radiation patterns of omnidirectional and other antennas in point-to-multipoint systems for use in sharing studies**	Reference radiation pattern for modelling PMP (point-to-multipoint) antenna performance to be used in the absence of information about a particular antenna	12
Rec.759	0	Use of frequencies in the band 500 to 3000 MHz for radio-relay systems	Frequency planning guidelines	4
F.1107	0	Probabilistic analysis for calculating interference into the fixed service from satellites occupying the geostationary orbit	Geometric considerations, interference calculations and computer simulation	19
F.1108	2	Determination of the criteria to protect fixed service receivers from the emissions of space stations operating in non-geostationary orbits in shared frequency bands	Simulation methods	31
F.1338	0	Threshold levels to determine the need to coordinate between particular systems in the broadcasting-satellite service (sound) in the geostationary-satellite orbit for space-to-earth transmissions and the fixed service in the band 1452–1492 MHz	Recommended limits of radio signal strength in W/m^2 as relevant to frequency sharing	3

(continued)

(continued)

Document number	Revision (1998)	Document title	Contents	Pages
F.760	1	Protection of terrestrial line-of-sight radio-relay systems against interference from the broadcasting-satellite service in the bands near 20 GHz	Recommended limits of radio signal strength in W/m^2 as relevant to frequency sharing	3
F.1334	0	Protection criteria for systems in the fixed service sharing the same frequency bands in the 1 to 3 GHz range with the land mobile service	Considerations in planning for frequency sharing	10
F.1246	0	Reference bandwidth of receiving stations in the fixed service to be used in coordination of frequency assignments with transmitting space stations in the mobile-satellite service in the 1–3 GHz range	Reference bandwidth and interference calculations	10
F.1335	0	Technical and operational considerations in the phased transitional approach for bands shared between the mobile-satellite service and the fixed service at 2 GHz	Planning guidelines specific to the shared usage of the nominated band	33
F.1247	0	Technical and operational characteristics of systems in the fixed service to facilitate sharing with the space research, space operation and earth exploration- satellite services operating in the bands 2025–2110 MHz and 2200–2290 MHz	Planning guidelines specific to the shared usage of the nominated band	10
F.1248	0	Limiting interference to satellites in the space science services from the emissions of trans-horizon radio-relay systems in the bands 2025–2110 MHz and 2200–2290 MHz	Planning guidelines specific to the shared usage of the nominated band	5
Rec.761		Frequency sharing between the fixed service and passive sensors in the band 18.6–18.8 GHz	Planning guidelines specific to the shared usage of the nominated band	1
F.1249	0	Maximum equivalent isotropically radiated power of transmitting stations in the fixed services operating in the frequency band 25.25–27.5 GHz shared with the inter-satellite service	Planning guidelines specific to the shared usage of the nominated band	26

Quality Objectives for Radio Paths and Digital Connections via Radio — Hypothetical Reference Models and Target Performance Values

Document number	Revision (1998)	Document title	Contents	Pages
Rec.390	4	**Definition of terms and references concerning hypothetical reference circuits and hypothetical reference digital paths for radio-relay systems**	Basic definition of the purpose and use of the hypothetical reference models	2
Rec.556	1	**Hypothetical reference digital path for radio-relay systems which may form part of an integrated services digital network with a capacity above the second hierarchical level**	Definition of the reference path where radio connection is of a bitrate of at least E1 (2 Mbit/s) or T1 (1.5 Mbit/s)	2
F.557	4	**Availability objective for radio-relay systems over a hypothetical reference circuit and a hypothetical reference digital path**	The definition of unavailability is set based upon the onset of ten consecutive *severely errored second* (*SES*) events. Defines minimum recommended availability for a 2500 km path of 99.7% of the time.	4
Rec.594	4	Error performance objectives of the hypothetical reference digital path for radio-relay systems providing connections at a bitrate below the primary rate and forming part or all of the high grade portion of an integrated services digital network	Description of 'factors to be taken into account when determining performance requirements' of digital radio-relay systems and specification of maximum error values.	3
F.634	4	Error performance objectives for real digital radio-relay links forming part of the high-grade portion of international digital connections at a bitrate below the primary rate within an integrated services digital network	Description of 'factors to be taken into account when determining performance requirements' of digital radio-relay systems and specification of maximum error values.	7
Rec.695	0	Availability objectives for real digital radio-relay links forming part of a high-grade circuit within an integrated services digital network	Definitive target values still under study	2
Rec.696	2	Error performance and availability objectives for hypothetical reference digital sections forming part or all of the medium-grade portion of an integrated services digital network connection at a bitrate below the primary rate utilizing digital radio-relay systems	Defines concrete quality performance values for ISDN networks	4

(continued)

(continued)

Document number	Revision (1998)	Document title	Contents	Pages
Rec.697	2	**Error performance and availability objectives for the local-grade portion at each end of an integrated services digital network connection at a bitrate below the primary rate utilizing digital radio-relay systems**	Defines concrete quality performance for radio systems used as ISDN WLL (wireless local loop) systems, including radio and equipment outages, i.e. including MTBF, mean-time-between-failures due to equipment problems	4
F.1092	1	Error performance objectives for constant bitrate digital path at or above the primary rate carried by digital radio-relay systems which may form part of the international portion of a 27 500 km hypothetical reference path	Performance targets for (particularly satellite) radio relay systems used as high-order connections in the international portion of an ISDN network.	3
F.1189	1	Error performance objectives for constant bitrate digital paths at or above the primary rate carried by digital radio-relay systems which may form part or all of the national portion of a 27 500 km hypothetical reference part	Performance targets for [particularly point- to-point] radio-relay systems used in national trunk and backbone networks	4
F.1330	0	**Performance limits for bringing into service of the parts of international plesiochronous digital hierarchy and synchronous digital hierarchy paths and sections implemented by digital radio-relay systems**	Defines the minimum quality standards which should be confirmed during commissioning of a radio system before accepting it into live network service.	7

Fixed Radio System Design

Document number	Revision (1998)	Document title	Contents	Pages
F.1101	0	**Characteristics of digital radio-relay systems below about 17 GHz**	Design guidelines for radio systems operating in frequency ranges below 17 GHz — description of main fading, interference and noise problems; component needs, modulation, filters, design trade-offs etc.	19
F.1102	0	**Characteristics of digital radio-relay systems above about 17 GHz**	Design guidelines for radio systems operating in frequency ranges above 17 GHz — description of main fading, interference and noise problems; component needs, modulation, filters, design trade-offs etc. Architecture and comparative economics of systems for WLL	10
F.750	3	Architectures and functional aspects of radio-relay systems for synchronous digital hierarchy (SDH)-based networks	Standardised functional specification and design architecture and associated terminology of SDH-radio system	48
F.751	2	Transmission characteristics and performance requirements of radio-relay systems for synchronous digital hierarchy-based networks	System design targets of standardised SDH radio system	5
F.752	1	Diversity techniques for radio-relay systems	Definition of diversity and protection methodologies for improving the reliability and availability of radio-relay systems	10
F.1191	1	**Bandwidths and unwanted emissions of digital radio-relay systems**	Definitions of *occupied bandwidth, necessary bandwidth, allocated frequency bands, channel separation, guardband, out-of-band emission, spurious emissions.*	10

Operation of Fixed Radio Systems

Document number	Revision (1998)	Document title	Contents	Pages
F.596	1	Interconnection of digital radio-relay systems	Defines the interface for interconnecting radio relay-systems (e.g. 'chaining' of point-to-point systems) as the G.703 or G.957 network interface between indoor units	2
F.700	2	Error performance and availability measurement algorithm for digital radio-relay links at the system bitrate interface	Defines correct measurement methods for assessment of bit error performance	2
Rec.753	0	Preferred methods and characteristics for the supervision and protection of digital radio-relay systems	Defines fault detection, performance measuring and system protection means for maintaining user data transfer during equipment failure	7

ITU-R Recommendations Covering Radio Systems for Specific Applications

Document number	Revision (1998)	Document title	Contents	Pages
Rec. 754	0	Radio-relay systems in Bands 8 and 9 [30–3000 MHz] for the provision of telephone trunk connections in rural areas	Guidelines for system design and equipment choice	6
F.1103	0	Radio-relay systems operating in bands 8 and 9 [30–3000 MHz] for the provision of subscriber telephone connections in rural areas	Guidelines for system design and equipment choice	4
F.755	1	**Point-to-multipoint systems used in the fixed service**	Guidelines for system design and equipment choice	9
Rec.756	0	**TDMA point-to-multipoint systems used as radio concentrators**	Guidelines for system design and equipment choice	12
F.1104	0	Requirements for point-to-multipoint radio systems used in the local grade portion of an ISDN connection	Guidelines for system design and equipment choice	6
F.757	1	Basic system requirements and performance objectives for fixed wireless local loop applications using cellular type mobile technologies	Guidelines for system design and equipment choice	10
F.1244	0	Radio local area networks (RLANs)	Guidelines for system design and equipment choice. (*UHF* (*ultra high frequency*) band 300–3000 MHz, SHF (*super high frequency*) 3–30 GHz and EHF (*extra high frequency*) Band 60 GHz)	26
F.1332	0	Radio-frequency signals transported through optical fibres	Description and operation of fibre-radio systems, particularly use of fibre for the indoor-unit-to-outdoor-unit connection	8

Appendix 8

World Climate Zones and their Precipitation Characteristics

Characteristics of Precipitation for Propagation Modelling (ITU-R Recommendation PN.837–1 [ITU-R Climate Zone definition] Tables and Figures reproduced by permission of the ITU)

Figs. 1 to 3 define the rain climate regions for the prediction of precipitation effects.

Table 1 is used to obtain the expected distribution of rain rate for the rain climate region.

Table 1 Rain climatic zones. Rainfall intensity exceeded (mm/h) (Reference to Figs. 1 to 3) (Reproduced by permission of the ITU)

Percentage of time (%)	A	B	C	D	E	F	G	H	J	K	L	M	N	P	Q
1.0	<0.1	0.5	0.7	2.1	0.6	1.7	3	2	8	1.5	2	4	5	12	24
0.3	0.8	2	2.8	4.5	2.4	4.5	7	4	13	4.2	7	11	15	34	49
0.1	2	3	5	8	6	8	12	10	20	12	15	22	35	65	72
0.03	5	6	9	13	12	15	20	18	28	23	33	40	65	105	96
0.01	8	12	15	19	22	28	30	32	35	42	60	63	95	145	115
0.003	14	21	26	29	41	54	45	55	45	70	105	95	140	200	142
0.001	22	32	42	42	70	78	65	83	55	100	150	120	180	250	170

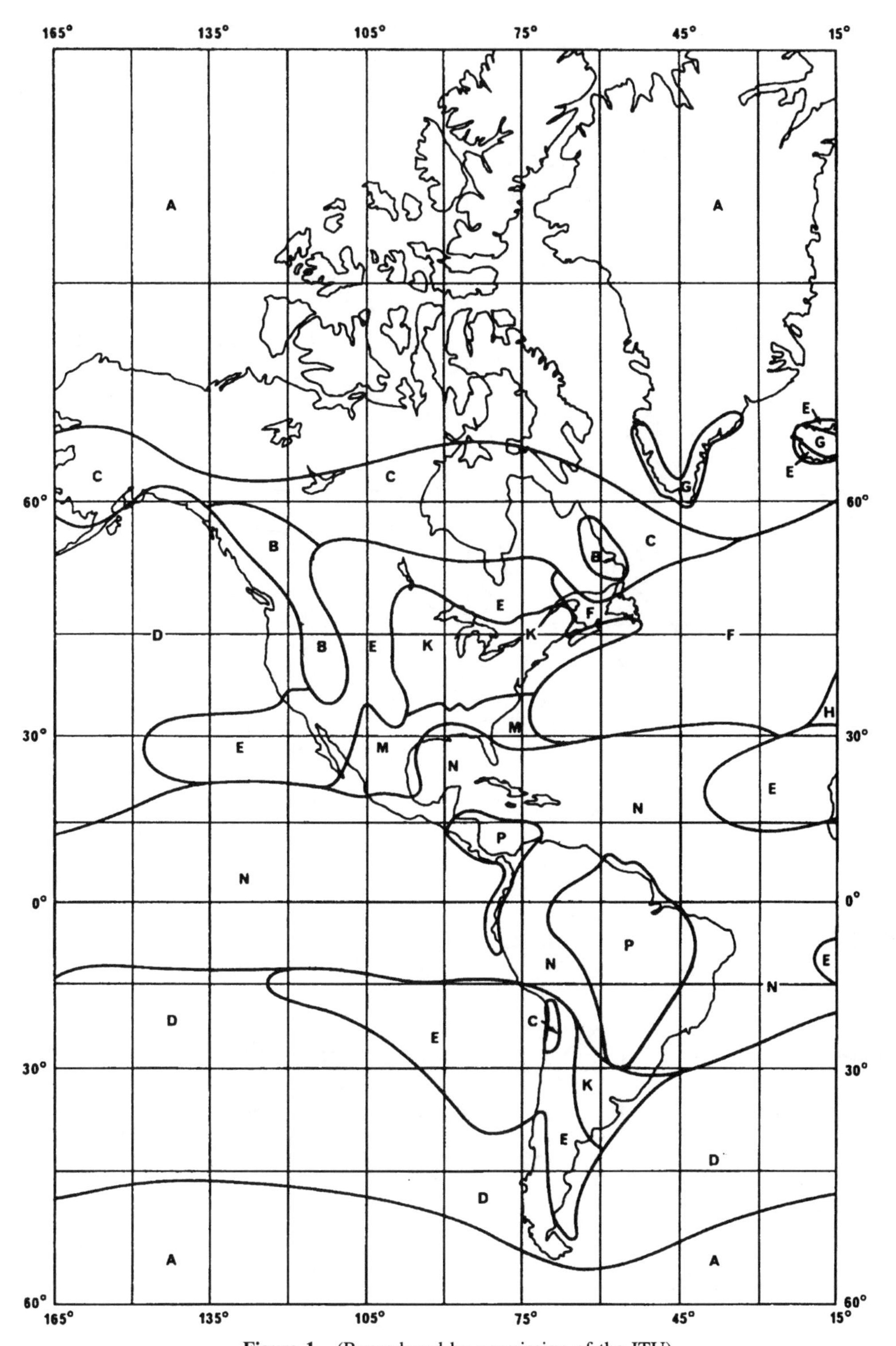

Figure 1 (Reproduced by permission of the ITU)

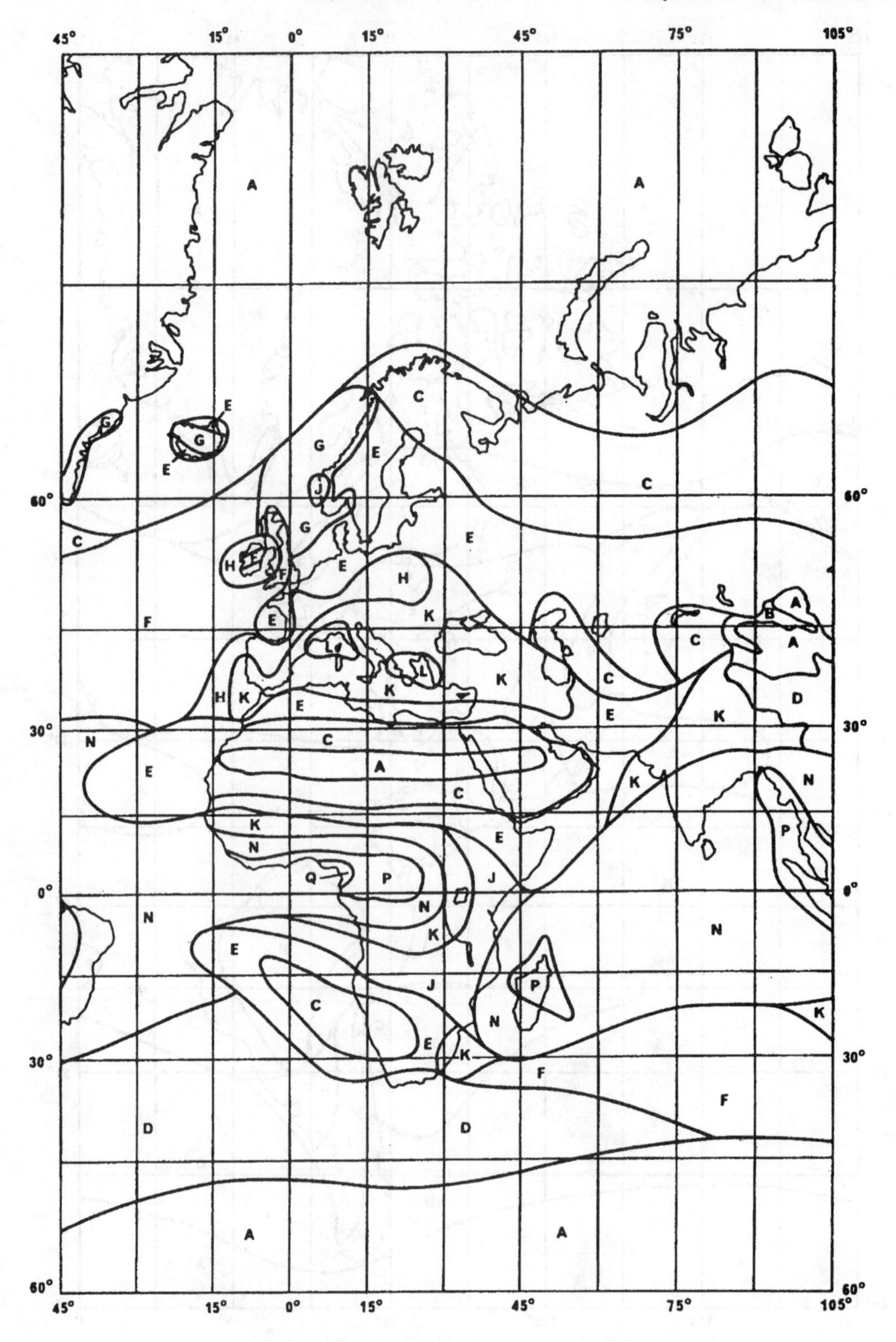

Figure 2 (Reproduced by permission of the ITU)

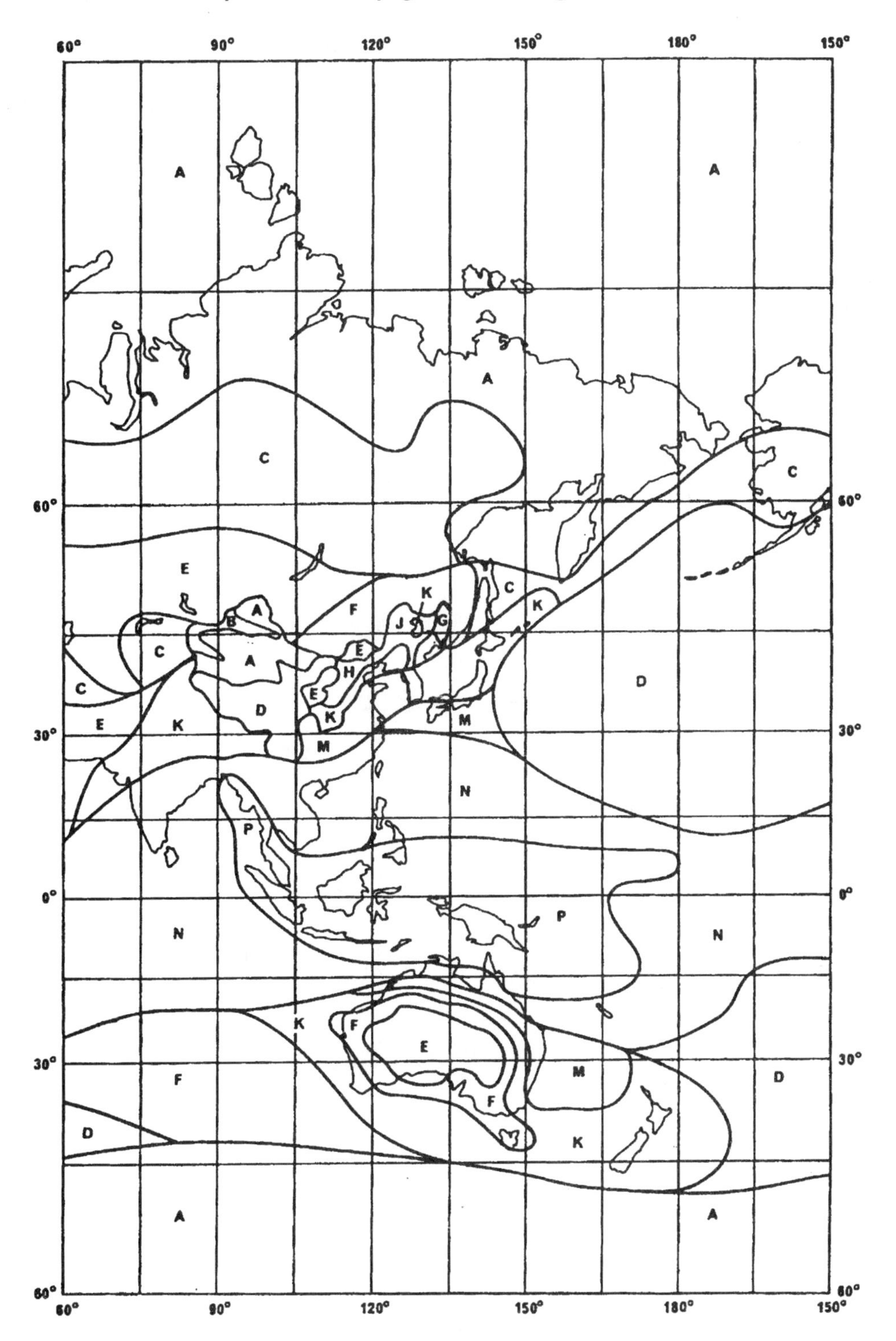

Figure 3 (Reproduced by permission of the ITU)

Appendix 9

Rainfall Attenuation Prediction Model

Specific Attenuation Model for Rain for use in Prediction Methods (ITU-R Recommendation 838)

Specific attenuation due to rain γ_R (dB/km) is obtained from the rain rate R (mm/h):

$$\gamma_R = k\,R^{\alpha} \tag{1}$$

The coefficients k and α are given in Table 1 for linear polarisations (horizontal — H, vertical — V) and horizontal paths. Values of k and α at intermediate frequencies are obtained by interpolation (logarithmic scale for frequency, logarithmic scale for k and linear scale for α).

For linear and circular polarisation and for non-horizontal paths, the values can be calculated from the values in Table 1 using the following equations:

$$k = \left[k_H + k_V + (k_H - k_V)\cos^2\theta\cos 2\tau\right]/2 \tag{2}$$

$$\alpha = \left[k_H\alpha_H + k_V\alpha_V + (k_H\alpha_H - k_V\alpha_V)\cos^2\theta\cos 2\tau\right]/2k \tag{3}$$

where θ is the path elevation angle and τ is the polarization tilt angle relative to the horizontal ($\tau = 45^\circ$ for circular polarization).

Table 1 Regression coefficients for estimating specific attenuation in equation (1) (Reproduced by permission of the ITU)

Frequency (GHz)	k_H	k_V	α_H	α_V
1	0.0000387	0.0000352	0.912	0.880
2	0.000154	0.000138	0.963	0.923
4	0.000650	0.000591	1.121	1.075
6	0.00175	0.00155	1.308	1.265
7	0.00301	0.00265	1.332	1.312
8	0.00454	0.00395	1.327	1.310
10	0.0101	0.00887	1.276	1.264
12	0.0188	0.0168	1.217	1.200
15	0.0367	0.0335	1.154	1.128
20	0.0751	0.0691	1.099	1.065
25	0.124	0.113	1.061	1.030
30	0.187	0.167	1.021	1.000
35	0.263	0.233	0.979	0.963
40	0.350	0.310	0.939	0.929
45	0.442	0.393	0.903	0.897
50	0.536	0.479	0.873	0.868
60	0.707	0.642	0.826	0.824
70	0.851	0.784	0.793	0.793
80	0.975	0.906	0.769	0.769
90	1.06	0.999	0.753	0.754
100	1.12	1.06	0.743	0.744
120	1.18	1.13	0.731	0.732
150	1.31	1.27	0.710	0.711
200	1.45	1.42	0.689	0.690
300	1.36	1.35	0.688	0.689
400	1.32	1.31	0.683	0.684

Appendix 10

Safety, Environmental and EMC (Electromagnetic Compatibility) Standards Relevant to Fixed Wireless

International Standards

EMC (Electromagnetic Compatibility) Standards

Standard designation	Subject
CISPR	International Special Committee for Radio Interference (CISPR) publication 22 — defines Class A and Class B equipment

Safety Standards

Standard designation	Subject
IEC 60950	Safety of Information technology Equipment

European and ETSI (European Telecommunications Standards Institute) standards

EMC (Electromagnetic Compatibility) Standards

Standard designation	Subject
EN 50081	Generic EMC (Electromagnetic Compatibility) emissions standard
EN 50082-1 and -2	Generic EMC (Electromagnetic Compatibility) immunity standard Part 1: residential environment Part 2: industrial environment
ETS 300 339	General Electromagnetic Compatibility (EMC) for radiocommunications equipment
ETS 300 385	Electromagnetic Compatibility (EMC) standard for digital fixed radio links with data rates at around 2 Mbit/s and above
ETS 300 386-2	Electromagnetic Compatibility (EMC) and Radio Spectrum Matters (ERM): Product Family Standard
ETS 300 683	EMC (Electromagnetic Compatibility) standard for short range radio devices (SRD) operating between 9 kHz and 25 GHz

Safety Standards

Standard designation	Subject
EN 41003	Safety standard
EN 60950	European equivalent of ISO 60950 (Safety of Information technology Equipment)

Environmental Standards

Standard designation	Subject
ETS 300 019	Environmental conditions and environmental tests for Telecommunications equipment

Appendix 11

Radio Spectrum Charges for PTP and PMP System Operation

Country	PTP Charges	PMP Bands & Charges
Australia		28/31 GHz band auctioned mid 1999 3.4 GHz to be auctioned autumn 2000
Austria		26 GHz; to be auctioned in 2000
Belgium	Fees for Category 2 networks (Faisceaux Hertziens — WLL) Per emitting station with 1 telephone channel (or its equivalent): 7728 BEF per annum Per emitting station with 2–12 telephone channels (or their equivalent): 14160 BEF per annum Per emitting station with 13–24 telephone channels (or their equivalent): 28320 BEF per annum Per emitting station with >24 telephone channels (or their equivalent): 56640 BEF p.a.	Fees for Category 2 networks (Faisceaux Hertziens — WLL) Per emitting station with 1 telephone channel (or its equivalent): 7728 BEF per annum Per emitting station with 2–12 telephone channels (or their equivalent): 14160 BEF per annum Per emitting station with 13–24 telephone channels (or their equivalent): 28320 BEF per annum Per emitting station with >24 telephone channels (or their equivalent): 56640 BEF p.a.
Finland	Wireless Local Loop Charge 9450 FIM per 25 kHz per annum *K1* K2 plus occasional examination and inspection fees and equipment import fees [K1 is a frequency band coefficient; up to 3 GHz K1 = 0.6; up to 11 GHz K1 = 0.5; up to 40 GHz K1 = 0.4] [K2 is a coverage area coefficient K2 = 1 for national coverage]	Wireless Local Loop Charge 9450 FIM per 25 kHz per annum *K1* K2 plus occasional examination and inspection fees and equipment import fees [K1 is a frequency band coefficient; up to 3 GHz K1 = 0.6; up to 11 GHz K1 = 0.5; up to 40 GHz K1 = 0.4] [K2 is a coverage area coefficient K2 = 1 for national coverage]
France		3.5 GHz; 26 GHz: to be tendered or auctioned in 2000
Germany	430 DM one-time charge plus 350 DM per annum per link plus application charges	2.6 GHz, 3.5 GHz, 26 GHz (allocations of 14 MHz to 2 × 28 MHz per operator) 17.5 million DM one-time charge (nationwide coverage, or proportional to population of the coverage area) plus 1741 DM per 'Sendefunkanlage' (transmitter) per annum
Hungary		3.5 GHz; 26 GHz to be tendered in 2000

(continued)

(continued)

Country	PTP Charges	PMP Bands & Charges
Netherlands		2.6 GHz, 3.5 GHz, 26 GHz: to be allocated during 2000 (allocations of 70 MHz to 112 MHz per operator) by auction in December 1999
Portugal		3.7 GHz; 26 GHz: allocated in October 1999
Spain		26 GHz: two licences already allocated in 1999, further licences in process of allocation (November 1999)
Switzerland		3.4 GHz; 26 GHz to be auctioned by Internet commencing February 2000 (minimum bid for 28 MHz of spectrum covering all of Switzerland 508 032 CHF per annum (8 year licence)
UK	(for Fixed Links) 230 to 2775 per link per year depending upon band, bitrate and level of congestion in the area of operation	(for Radio Fixed Access Operators) 2.0–2.1 GHz: £150 per base station per annum 2.4–2.5 GHz: £75 per base station per annum 3.4–4.5 GHz: £300 per base station per annum 10.0–11 GHz: £300 per base station per annum
USA		161 LMDS licences were sold to 40 bidders for a total of $45 million

Appendix 12

Radio Regulations Agencies

Country	Agency	Telephone number	Address
Worldwide responsibility	ITU-R (International Telecommunications Union — Radiocommunication sector)	+41 22 730 5111	www.itu.int/itu-r/index.html International Telecommunications Union, Place des Nations, CH-1211 Geneve 20, Switzerland
Argentina	CNC (Comisión Nacional de Comunicaciones)	+54 1 347 9544	www.cnc.gov.ar Perú 103 — Piso 19, 1067 Buenos Aires, Argentina
Australia	ACA (Australian Communications Authority)	+68 2 6256 5555	www.sma.gov.au Central Office, Purple Building, Benjamin Offices, PO Box 78 Belconnen ACT 2616, Australia
Austria	Oberste Post und Fernmeldebehörde (bm:wv) Bundesministerium für Wissenschaft und Verkher	+43 1 797 31-0	www.bmv.gv.at/level2tel.htm Kelsenstraße 7, A-1030 Vienna, Austria
Belgium	Frequency Management Department, IBPT (Belgian Institute of Postal and Telecommunications Services)	+32 2 226 8888	www.bipt.be Avenue de l'Astronomie, 14, Boite 21, 1210 Brussels, Belgium
Brazil	ANATEL (Agencia Nacional de Telecomunicacoes)	+55 61 312 2336	www.anatel.gov.br/default.htm SAS Quadra 06 Bloco H3o andar CEP 70.313-900 Brasilia, Brazil
Canada	CRTC (Canadian Radio-television Commission)	+1 877 249 2782 +1 819 997 0313	www.crtc.gc.ca Les Terrasses de la Chaudière, Central Building, 1 Promenade du Portage, Hull, Quebec J8X 4B1, Canada
Denmark	NTA (National Telecom Agency)	+45 35 43 03 33	www.tst.dk Holsteingade 63, DK-2100 Copenhagen, Denmark
Finland	Radioliikenne (Radio Administration) TAC [THK] (Telecommunications Administration Centre)	+358 969 661	www.thk.fi Itämerenkatu 3 A, Helsinki, Finland

(continued)

(continued)

Country	Agency	Telephone number	Address
European Region Agencies	ERO (European Radiocommunications Office)	+45 3525 0300	www.ero.dk Midtermolen 1, DK 2100 Copenhagen, Denmark
	ETSI (European Telecommunications Standards Institute)	+33 4 92 94 42 00	www.etsi.org 650 Route des Lucioles, F-06921 Sophia Antipolis Cedex, France
France	ART (Autorité de Régulation des Télécommunications)	+33 1 40 47 71 61	www.art-telecom.fr 7 Square Max Hymans, 75015 Paris, France
Germany	RegTP (Regulierungsbehörde für Telekommunikation un Post)	+49 228 14–0	www.regtp.de Heinrich-von-Stephan-Straße 1, 53175 Bonn, Germany
India	TRAI (Telecommunications Regulatory Affairs India)	+91 11 335 7815	16th floor, Jawahar Vyapar Bhawan 1, Tolstoy marg, New Delhi 110 001, India
Ireland	ODTR (Office of the Director of Telecommunications Regulation)	+353 1 804 9600	www.odtr.ie Abbey Court, Blocks D E F Irish Life Centre, Lower Abbey Street, Dublin 1, Ireland
Italy	AGC (L'Autorità per le garanzie nelle Communicazioni)	+39 06 692 0991	www.comune.napoli.it/agcom Via de'Crociferi 19, I-00187 Roma, Italy
Japan	MPT (Ministry of Posts and Telecommunications)	+81 3 3504 4200	www.mpt.go.jp/index-e.html 1-3-2 Kasumigaseki, Chiyoda-ku, Tokyo 100, Japan
Malaysia	JTM (Jabatan Telekomunikasi Malaysia)	+60 3 255 6687	www.jtm.gov.my Jalan Semantan, 50668 Kuala Lumpur, Malaysia
Mexico	CFT (Comisión Federal der Telecomunicaciones)	+52 5 261 4000	www.cft.gob.mx Bosques de Radiatas 44, Col. Bosques de las Lomas, 05210 México D.F. Mexico
Netherlands	HDTP (Radiocommunications Agency)	31 50 587 7000	www.minvenw.nl/detp/indexeng.html Telecommunications and Post Department, Ministry of Transport, Public Works and Water Management, Emmsingel 1, PO Box 450, 9700 AL Groningen, Netherlands
Norway	NPT (Norwegian Post and Telecommunications Authority)	+47 22 82 46 00	www.npt.no Revierstredet 2, PO Box 447 Sentrum, N-0104 Oslo, Norway

(continued)

(continued)

Country	Agency	Telephone number	Address
Peru	OSPITEL (Organismo Supervisor der Inversión Privada en Telecomunicaciones)	+51 1 421 4152	www.ospitel.gob.pe Av Camino Real 348, Torre El Pilar, Oficina 1302, Lima 27, Peru
Portugal	ICP (Instituto das Comunicacòes de Portugal)	+351 1 721 1000	www.icp.pt Avenida José Malhoa N 12, P-1070 Lisbon, Portugal
South Africa	SATRA (South African Telecommunications Regulatory Authority)	+27 11 321 8551	www.satra.gov.za 164 Katherine Street, Sandton 2146, Private Bag X1, Marlboro 2063, South Africa
Spain	CMT (Comisión del Mercado de las Telecomunicaciones)	+34 91 372 4242	www.cmt.es Velázquez 164, E-28002 Madrid, Spain
Sweden	PTS (National Posts & Telecommunications Agency)	+46 8 678 5505	www.pts.se PO Box 5398, S-102 49 Stockholm, Sweden
Switzerland	OFCOM [BAKOM] (Federal Office for Communications)	+41 327 55 11	www.bakom.ch Bundesamt für Kommunikation, Zukunftstrasse 44, 2501 Biel–Bienne, Switzerland
UK	RA (Radiocommunications Agency)	+44 20 7211 0211	www.radio.gov.uk Wyndham House, 189 Marsh Wall, London E14 9SX, UK
USA	FCC Wireless Telecommunications Bureau (Federal Communications Commission)	+1 888 225 5322 +1 202 418 0190	www.fcc.gov/sitemap.html 445 12th Street SW, Washington DC 20554, USA
	IEEE (Institute of Electrical and Electronics Engineers)	+1 732 981 0060	www.ieee.org IEEE Operations Center, 445 Hoes Lane, Piscataway NJ 08855–1331, USA

Glossary of Terms

Access network (AN)	A part of a public telecommunications network used to connect end-user premises to public switching centre locations
Adjacent channel	The radio channel whose centre frequency is situated next above or next below the frequency in operation on a given radio link
AGC (Automatic Gain Control)	An amplifier in a radio receiver used to ensure that the signal strength is constant at the input to the signal demodulator
AM (Amplitude Modulation)	A means of modulation — for encoding analogue signals onto a radio carrier signal
antenna	A means for transmitting or receiving radio waves (used in preference to 'aerial')
antenna diagram or antenna pattern	The radiation pattern of an antenna, showing the relative radiation intensity in each radiated direction
antenna efficiency	The ratio of the maximum effective area of an antenna to the aperture area
aperture (of antenna)	An imaginary area (on or near an antenna), based upon which, convenient assumptions can be made regarding the radio field values
array antenna	An antenna comprising a number of similar radiating elements arranged in a fixed pattern
availability	A target set for the percentage of time during which a radio system is to achieve a given Bit Error Ratio (BER) quality target for transmission
baseband	The original frequency of a user-signal to be carried by a radio system
Baud rate	The rate of change of the carrier signal modulated to carry the digital user information (also known as the *symbol rate*)
beam of an antenna	The major lobe of the antenna diagram
beamwidth	See 'half power beamwidth'
BER (Bit Error Ratio)	A measure of the quality of a digital signal
boresight	The main radiation direction of an antenna, particularly the compass bearing in the horizontal plane
carrier	The radio signal onto which the user signal or data is modulated
CDMA (Code Division Multiple Access)	A multiple access scheme for point-to-multipoint wireless systems
CEPT	European Conference for Posts and Telecommunications
channel spacing	The frequency separation between adjacent radio channels
CIR (Carrier-to-Interferor Ratio)	An important parameter used in radio planning which determines the acceptable minimum separation of radio transmitters using the same frequency

climate zone	A region of the world characterised by ITU-R as subject to a given amount of rain each year (ITU-R recommendation PN.837)
co-channel	Refers to radio systems operating on the same radio channel
coverage area	The area associated with a given service and radio frequency served by a radio base station
dBi	The gain of an antenna relative to an *isotropic* antenna
dBm	A power ratio measured relative to 1 milliWatt
dBW	A power ratio measured relative to 1 Watt
deciBel (dB)	A ratio of signal power or strength or some other measure of a ratio. The value is equal to $10 \log(x)$, where x is the ratio expressed as an absolute number value
directivity	The value of the directive gain of an antenna in a particular direction
DRRS (Digital Radio Relay System)	The term used in standards documentation to describe 'fixed link' terrestrial wireless systems
duplex	A type of communication in which transmission is possible in both directions at all times (including simultaneously)
duplex spacing	The frequency difference between radio channels used for the two directions of transmission of a single duplex radio communication
EHF	Extra High Frequency (30–300 GHz)
EIRP (Effective Isotropic Radiated Power)	The equivalent power for signal transmission which would have to be applied to an isotropic antenna in order to achieve the same radiation intensity (i.e. signal strength) at a particular point. The EIRP value is usually quoted in dB, calculated as the sum of the antenna gain in the given direction and the actual transmitted power in dB
fading	The attenuation or weakening of a radio signal caused by interference, rain or other atmospheric effects
far-field region	The region of the field of an antenna in which the field distribution is essentially independent of the distance from the antenna
FDMA (Frequency Division Multiple Access)	A multiple access scheme for point-to-multipoint wireless systems
feed (of an antenna)	The part of an antenna which couples the radio terminal in order to create the antenna *illumination* (i.e. incidence of the radio waves onto the antenna itself)
fixed link	A radio link between two fixed (as opposed to two mobile or one mobile and one fixed) endpoints
FM (Frequency Modulation)	A means of modulation — for encoding analogue signals onto a radio carrier signal
free-space propagation	The propagation of an electromagnetic wave in an homogeneous ideal dielectric medium
frequency tolerance	The maximum permissible variation of the centre frequency of a radio system allowed by the standard or specification

Fresnel zone	A defined ellipsoid area around the main *line-of-sight* path of a radio link which should be kept free of obstacles (first fresnel zone)
front-to-back ratio	The ratio of the gain of the antenna to its directivity in a direction 180° to the main radiation direction
FSK (Frequency-Shift Keying)	A means of modulation — for encoding digital signals onto a radio carrier signal
gain	Usually the maximum gain of an antenna — the ratio of the signal strength along the main *pole* of the antenna relative to the signal strength which would be achieved at the same point by an *isotropic* antenna
half duplex	A type of communication in which transmission is possible in both directions, but only in one direction at-a-time
half-power beamwidth	The angle between the two points either side of the main lobe of an antenna at which the radiation intensity is half of the maximum (i.e. 3 dB below the maximum antenna gain)
Hertz (Hz)	A measure of radio frequency equal to one cycle per second
HF	High Frequency (3–30 MHz)
horn antenna	An antenna having the shape of a horn
indoor unit (IDU)	The part of a radio terminal usually installed 'indoors' and comprising the main digital components and the modem
Intermediate Frequency (IF)	A frequency between *baseband* and *Radio-frequency* (*RF*) used to convey the signal from the *indoor unit* (*IDU*) to the *outdoor unit* (*ODU*), or vice-versa
isotropic antenna	An antenna which radiates in all directions equally (also termed *omnidirectional*)
ITU-R	International Telecommunications Union—Radiocommunications sector
LF	Low Frequency (30–300 kHz)
Line-of-sight (LOS)	The path between the radio antennas and terminals at either end of a microwave or millimetre wave radio system should be clear. There should be a 'line-of-sight'
link	A radiocommunications connection of specified characteristics between two end-points
LMCS	Local Multipoint Communications Service
LMDS	Local Multipoint Distribution Service
lobe pattern	See 'antenna diagram'
main lobe	A radiation lobe on the antenna diagram in the main (wanted) direction and corresponding to the maximum antenna gain
MMDS	Multichannel Multipoint Distribution Service
MF	Medium Frequency (300–3000 kHz)

near-field region	The region of the field of an antenna in which the radiating and reactive near-field regions predominate
noise temperature	The temperature of a resistor having an audible thermal noise power per unit bandwidth equal to that of the antenna or radio component
outdoor unit (ODU)	The part of a radio terminal installed outdoors, usually comprising the high frequency radio unit and sometimes also an integrated antenna
path loss (or transmission loss)	The loss between the input to the transmitting antenna and the output from the receiving antenna of a radio link
PMP (point-to-multipoint)	A radio system providing for shared communications between many remote *Terminal Stations* (*TS*) and a single *Central Station* (*CS*) or *Base Station* (*BS*)
polarisation (of an antenna)	The polarisation of the wave radiated by an antenna (usually either vertical or horizontal or Co-Channel Dual Polarised (CCDP)
PSK (Phase-Shift Keying)	A means of modulation — for encoding digital signals onto a radio carrier signal
PTP (point-to-point)	A radio system providing for a communications link between two fixed end-points
QAM (Quadrature Amplitude Modulation)	A means of modulation — for encoding digital signals onto a radio carrier signal
Radiation Pattern Envelope (RPE)	The allowed radiation pattern of an antenna as defined in a technical standard or specification (i.e. limits on the antenna diagram)
range	The maximum distance over which a radio system of a given operating frequency can be expected to work reliably, taking account of the availability and BER targets as well as the local climate
SHF	Super High Frequency (3–30 GHz)
side lobe	A radiation lobe on the antenna diagram in an unwanted direction
simplex	A type of communication in which transmission is possible only in one direction
spectrum mask	The mask defining the allowed frequency content of a radio signal
spurious emission	Unintended radiations of a radio outside the allowed spectrum mask
TDMA (Time Division Multiple Access)	A multiple access scheme for point-to-multipoint wireless systems
total loss	The ratio (usually expressed in deciBels) between the radio signal output of the transmitter and the signal power input to the receiver, taking account of all intermediate losses
UHF	Ultra High Frequency (300–3000 MHz)
ULF	Ultra Low Frequency (300–3000 Hz)
VHF	Very High Frequency (30–300 MHz)
VLF	Very Low Frequency (3–30 kHz)
waveguide	Tubular metal used for connecting antennas to radio units for carrying high frequency radio signals
WLL	Wireless Local Loop

Bibliography

Books and Publications

CE Marking for Telecommunications: A Handbook to the Telecommunications Directive. IEEE Press.

Reference Manual for Telecommunications Engineering. Roger Freeman, Wiley-Interscience.

ITU-R recommendations (published by International Telecommunications Union, Geneva)

Internet and World Wide Web Sources

Andrew Corporation (manufacturer of antennas and cables) Planning software available on web page www.andrew.com

Conference of European Posts and Telecommunications (CEPT)/European Radiocommunications Office (ERO) www.ero.dk

Electronic Industries Alliance (EIA) www.eia.org

European Telecommunications Standards www.etsi.frwww.etsi.org

International Electrotechnical Commission (IEC) (technical standards) www.iec.ch

International Telecommunications Union (ITU) (for ITU-R and ITU-R recommendations and ITU Handbook) www.itu.int

RFS (manufacturer of antennas and cables) www.rfs.de

Times Microwave Corporation (manufacturer of cables) Handbook available from webpage www.timesmicrowave.com

USA, FCC rules and regulations, spectrum allocations: www.fcc.gov www.fcc.gov/sitemap.html

USA, IEEE standards www.ieee.org http://standards.ieee.org/catalog/index.html

Index